INTRODUCTION TO COMPUTATIONAL MOLECULAR BIOLOGY

INTRODUCTION TO COMPUTATIONAL MOLECULAR BIOLOGY

JOÃO SETUBAL and JOÃO MEIDANIS

University of Campinas, Brazil

PWS PUBLISHING COMPANY

I(T)P

An International Thomson Publishing Company

BOSTON • ALBANY • BONN • CINCINNATI • DETROIT • LONDON
MELBOURNE • MEXICO CITY • NEW YORK • PACIFIC GROVE • PARIS
SAN FRANCISCO • SINGAPORE • TOKYO • TORONTO

PWS PUBLISHING COMPANY
20 Park Plaza, Boston, MA 02116–4324

Library of Congress Cataloging-in-Publication Data

Setubal, João Carlos.
Introduction to computational molecular biology / João Carlos Setubal, João Meidanis.
p. cm
Includes bibliographical references (p. 277) and index.
ISBN 0-534-95262-3
1. Molecular biology--Mathematics. I. Meidanis, João. II. Title.
QH506.S49 1997 96-44240
574.8'8'0151--dc20 CIP

Sponsoring Editor: *David Dietz*
Editorial Assistant: *Susan Garland*
Marketing Manager: *Nathan Wilbur*
Production Editor: *Andrea Goldman*
Manufacturing Buyer: *Andrew Christensen*
Composition: *SuperScript Typography*
Prepress: *Pure Imaging*
Cover Printer: *Coral Graphics*
Text Printer/Binder: *R. R. Donnelley & Sons Company/Crawfordsville*
Interior Designer: *Monique A. Calello*
Cover Designer: *Andrea Goldman*
Cover Art: *"Digital 1/0 Double Helix" by Steven Hunt. Used by permission of the artist.*

Printed and bound in the United States of America
97 98 99 00 — 10 9 8 7 6 5 4 3 2 1

For more information, contact:
PWS Publishing Company
20 Park Plaza
Boston, MA 02116

International Thomson Publishing Europe
Berkshire House I68-I73
High Holborn
London WC1V 7AA
England

Thomas Nelson Australia
102 Dodds Street
South Melbourne, 3205
Victoria, Australia

Nelson Canada
1120 Birchmont Road
Scarborough, Ontario
Canada M1K 5G4

International Thomson Editores
Campos Eliseos 385, Piso 7
Col. Polanco
11560 Mexico D.F., Mexico

International Thomson Publishing GmbH
Königswinterer Strasse 418
53227 Bonn, Germany

International Thomson Publishing Asia
221 Henderson Road
#05-10 Henderson Building
Singapore 0315

International Thomson Publishing Japan
Hirakawacho Kyowa Building, 31
2-2-1 Hirakawacho
Chiyoda-ku, Tokyo 102
Japan

Contents

PREFACE

Biology easily has 500 years of exciting problems to work on.
— Donald E. Knuth

Ever since the structure of DNA was unraveled in 1953, molecular biology has witnessed tremendous advances. With the increase in our ability to manipulate biomolecular sequences, a huge amount of data has been and is being generated. The need to process the information that is pouring from laboratories all over the world, so that it can be of use to further scientific advance, has created entirely new problems that are interdisciplinary in nature. Scientists from the biological sciences are the creators and ultimate users of this data. However, due to sheer size and complexity, between creation and use the help of many other disciplines is required, in particular those from the mathematical and computing sciences. This need has created a new field, which goes by the general name of *computational molecular biology*.

In a very broad sense computational molecular biology consists of the development and use of mathematical and computer science techniques to help solve problems in molecular biology. A few examples will illustrate. Databases are needed to store all the information that is being generated. Several international sequence databases already exist, but scientists have recognized the need for new database models, given the specific requirements of molecular biology. For example, these databases should be able to record changes in our understanding of molecular sequences as we study them; current models are not suitable for this purpose. The understanding of molecular sequences in turn requires new sophisticated techniques of pattern recognition, which are being developed by researchers in artificial intelligence. Complex statistical issues have arisen in connection with database searches, and this has required the creation of new and specific tools.

There is one class of problems, however, for which what is most needed is *efficient algorithms*. An algorithm, simply stated, is a step-by-step procedure that tries to solve a certain well-defined problem in a limited time bound. To be efficient, an algorithm should not take "too long" to solve a problem, even a large one. The classic example of a problem in molecular biology solvable by an algorithm is sequence comparison: Given two sequences representing biomolecules, we want to know how similar they are. This is a problem that must be solved thousands of times every day, so it is desirable that a very efficient algorithm should be employed.

The purpose of this book is to present a representative sample of computational problems in molecular biology and some of the efficient algorithms that have been proposed to solve them. Some of these problems are well understood, and several of their algorithms have been known for many years. Other problems seem more difficult, and no satisfactory algorithmic approach has been developed so far. In these cases we have concentrated in explaining some of the mathematical models that can be used as a foundation in the development of future algorithms.

The reader should be aware that an algorithm for a problem in molecular biology is a curious beast. It tries to serve two masters: the molecular biologist, who wants the algorithm to be *relevant,* that is, to solve a problem with all the errors and uncertainties with which it appears in practice; and the computer scientist, who is interested in proving that the algorithm efficiently solves a well-defined problem, and who is usually ready to sacrifice relevance for provability (or efficiency). We have tried to strike a balance between these often conflicting demands, but more often than not we have taken the computer scientists' side. After all, that is what the authors are. Nevertheless we hope that this book will serve as a stimulus for both molecular biologists and computer scientists.

This book is an introduction. This means that one of our guiding principles was to present algorithms that we considered *simple,* whenever possible. For certain problems that we describe, more efficient and generally more sophisticated algorithms exist; pointers to some of these algorithms are usually given in the bibliographic notes at the end of each chapter. Despite our general aim, a few of the algorithms or models we present cannot be considered simple. This usually reflects the inherent complexity of the corresponding topic. We have tried to point out the more difficult parts by using the star symbol ($\star$) in the corresponding headings or by simply spelling out this caveat in the text. The introductory nature of the text also means that, for some of the topics, our coverage is intended to be a starting point for those new to them. It is probable, and in some cases a fact, that whole books could be devoted to such topics.

The primary audience we have in mind for this book is students from the mathematical and computing sciences. We assume no prior knowledge of molecular biology beyond the high school level, and we provide a chapter that briefly explains the basic concepts used in the book. Readers not familiar with molecular biology are urged however to go beyond what is given there and expand their knowledge by looking at some of the books referred to at the end of Chapter 1.

We hope that this book will also be useful in some measure to students from the biological sciences. We do assume that the reader has had some training in college-level discrete mathematics and algorithms. With the purpose of helping the reader unfamiliar with these subjects, we have provided a chapter that briefly covers all the basic concepts used in the text.

Computational molecular biology is expanding fast. Better algorithms are constantly being designed, and new subfields are emerging even as we write this. Within the constraints mentioned above, we did our best to cover what we considered a wide range of topics, and we believe that most of the material presented is of lasting value. To the reader wishing to pursue further studies, we have provided pointers to several sources of information, especially in the bibliographic notes of the last chapter (and including WWW sites of interest). These notes, however, are not meant to be exhaustive. In addition, please note that we cannot guarantee that the World Wide Web Universal Resource Locators given in the text will remain valid. We have tested these addresses, but due to the dynamic nature of the Web, they could change in the future.

BOOK OVERVIEW

Chapter 1 presents fundamental concepts from molecular biology. We describe the basic structure and function of proteins and nucleic acids, the mechanisms of molecular genetics, the most important laboratory techniques for studying the genome of organisms, and an overview of existing sequence databases.

Chapter 2 describes strings and graphs, two of the most important mathematical objects used in the book. A brief exposition of general concepts of algorithms and their analysis is also given, covering definitions from the theory of NP-completeness.

The following chapters are based on specific problems in molecular biology. Chapter 3 deals with *sequence comparison.* The basic two-sequence problem is studied and the classic dynamic programming algorithm is given. We then study extensions of this algorithm, which are used to deal with more general cases of the problem. A section is devoted to the multiple-sequence comparison problem. Other sections deal with programs used in database searches, and with some other miscellaneous issues.

Chapter 4 covers the *fragment assembly problem.* This problem arises when a DNA sequence is broken into small fragments, which must then be assembled to reconstitute the original molecule. This is a technique widely used in large-scale sequencing projects, such as the Human Genome Project. We show how various complications make this problem quite hard to solve. We then present some models for simplified versions of the problem. Later sections deal with algorithms and heuristics based on these models.

Chapter 5 covers the *physical mapping problem.* This can be considered as fragment assembly on a larger scale. Fragments are much longer, and for this reason assembly techniques are completely different. The aim is to obtain the location of some markers along the original DNA molecule. A brief survey of techniques and models is given. We then describe an algorithm for the consecutive ones problem; this abstract problem plays an important role in physical mapping. The chapter finishes with sections devoted to algorithmic approximations and heuristics for one version of physical mapping.

Proteins and nucleic acids also evolve through the ages, and an important tool in understanding how this evolution has taken place is the *phylogenetic tree.* These trees also help shed light in the understanding of protein function. Chapter 6 describes some of the mathematical problems related to phylogenetic tree reconstruction and the simple algorithms that have been developed for certain special cases.

An important new field of study that has recently emerged in computational biology is *genome rearrangements.* It has been discovered that some organisms are genetically different, not so much at the sequence level, but in the order in which large similar chunks of their DNA appear in their respective genomes. Interesting mathematical models have been developed to study such differences, and Chapter 7 is devoted to them.

The understanding of the biological function of molecules is actually at the heart of most problems in computational biology. Because molecules fold in three dimensions and because their function depends on the way they fold, a primary concern of scientists in the past several decades has been the discovery of their three-dimensional structure, in particular for RNA and proteins. This has given rise to methods that try to predict a molecule's structure based on its primary sequence. In Chapter 8 we describe dynamic programming algorithms for RNA structure prediction, give an overview of the difficulties of protein structure prediction, and present one important recent development in the field called protein threading, which attempts to align a a protein sequence with a known structure.

Chapter 9 ends the book presenting a description of the exciting new field of DNA computing. We present there the basic experiment that showed how we can use DNA molecules to solve one hard algorithmic problem, and a theoretical extension that applies to another hard problem.

A word about general conventions. As already mentioned, sections whose headings are followed by a star symbol ($\star$) contain material considered by the authors to be more difficult. In the case of concept definitions, we have used the convention that terms used throughout the book are in **boldface** when they are first defined. Other terms appear in *italics* in their definition. Many of our algorithms are presented first through English sentences and then in pseudo code format (pseudo code conventions are described in Section 2.3). In some cases the pseudo code provides a level of detail that should help readers interested in actual implementation.

Summaries are provided for the longer chapters.

EXERCISES

Exercises appear at the end of every chapter. Exercises marked with one star ($\star$) are hard, but feasible in less than a day. They may require knowledge of computer science techniques not presented in the book. Those marked with two stars ($\star\star$) are problems that were once research problems but have since been solved, and their solutions can be found in the literature (we usually cite in the bibliographic notes the research paper that solves the exercise). Finally exercises marked with a diamond ($\diamond$) are research problems that have not been solved as far as the authors know.

At the end of the book we provide answers or hints to selected exercises.

ERRORS

Despite the authors' best efforts, this book no doubt contains errors. If you find any, or have any suggestions for improvement, we will be glad to hear from you. Please send error reports or any other comments to us at bio@dcc.unicamp.br, or at

J. Meidanis / J. C. Setubal
Instituto de Computação, C. P. 6176
UNICAMP
Campinas, SP 13083-970
Brazil

(The authors can be reached individually by e-mail at meidanis@dcc.unicamp.br and at setubal@dcc.unicamp.br.) We thank in advance all readers interested in helping us make this a better book. As errors become known they will be reported in the following WWW site:

http://www.dcc.unicamp.br/~bio/ICMB.html

ACKNOWLEDGMENTS

This book is a successor to another, much shorter one on the same subject, written by the authors in Portuguese and published in 1994 in Brazil. That first book was made possible thanks to a Brazilian computer science meeting known as "Escola de Computação," held every two years. We believe that without such a meeting we would not be writing this preface, so we are thankful to have had that opportunity.

The present book started its life thanks to Mike Sugarman, Bonnie Berger, and Tom Leighton. We got a lot of encouragement from them, and also some helpful hints. Bonnie in particular was very kind in giving us copies of her course notes at an early stage.

We have been fortunate to have had financial grants from FAPESP and CNPq (Brazilian Research Agencies); they helped us in several ways. Grants from FAPESP were awarded within the "Laboratory for Algorithms and Combinatorics" project and provided computer equipment. Grants from CNPq were awarded in the form of individual fellowships and within the PROTEM program through the PROCOMB and TCPAC projects, which provided funding for research visits.

We are grateful to our students who helped us proofread early drafts. Special thanks are due to Nalvo Franco de Almeida Jr. and Maria Emília Machado Telles Walter. Nalvo, in addition, made many figures and provided several helpful comments.

We had many helpful discussions with our colleague Jorge Stolfi, who also provided crucial assistance in typesetting matters. Fernando Reinach and Gilson Paulo Manfio helped us with Chapter 1. We discussed book goals and general issues with Jim Orlin. Martin Farach and Sampath Kannan, as well as several anonymous reviewers, also made many suggestions, some of which were incorporated into the text. Our colleagues at the Institute of Computing at UNICAMP provided encouragement and a stimulating work environment.

The following people were very kind in sending us research papers: Farid Alizadeh, Alberto Caprara, Martin Farach, David Greenberg, Dan Gusfield, Sridar Hannenhalli, Wen-Lian Hsu, Xiaoqiu Huang, Tao Jiang, John Kececioglu, Lukas Knecht, Rick Lathrop, Gene Myers, Alejandro Schäffer, Ron Shamir, Martin Vingron (who also sent lecture notes), Todd Wareham, and Tandy Warnow. Some of our sections were heavily based on some of these papers.

Many thanks are also due to Erik Brisson, Eileen Sullivan, Bruce Dale, Carlos Eduardo Ferreira, and Thomas Roos, who helped in various ways.

J. C. S. wishes to thank his wife Silvia (a.k.a. Teca) and his children Claudia, Tomás, and Caio, for providing the support without which this book could not have been written.

This book was typeset by the authors using Leslie Lamport's LaTeX 2_ε system, which works on top of Don Knuth's TeX system. These are truly marvelous tools.

The quotation of Don Knuth at the beginning of this preface is from an interview given to Computer Literacy Bookshops, Inc., on December 7, 1993.

João Carlos Setubal
João Meidanis

CHAPTER

1

BASIC CONCEPTS OF MOLECULAR BIOLOGY

In this chapter we present basic concepts of molecular biology. Our aim is to provide readers with enough information so that they can comfortably follow the biological background of this book as well as the literature on computational molecular biology in general. Readers who have been trained in the exact sciences should know from the outset that in molecular biology nothing is 100% valid. To every rule there is an exception. We have tried to point out some of the most notable exceptions to general rules, but in other cases we have omitted such mention, so as not to transform this chapter into a molecular biology textbook.

1.1 LIFE

In nature we find both living and nonliving things. Living things can move, reproduce, grow, eat, and so on — they have an *active* participation in their environment, as opposed to nonliving things. Yet research in the past centuries reveals that both kinds of matter are composed by the same atoms and conform to the same physical and chemical rules. What is the difference then? For a long time in human history, people thought that some sort of extra matter bestowed upon living beings their active characteristics — that they were "animated" by such a thing. But nothing of the kind has ever been found. Instead, our current understanding is that living beings act the way they do due to a complex array of chemical reactions that occur inside them. These reactions never cease. It is often the case that the products of one reaction are being constantly consumed by another reaction, keeping the system going. A living organism is also constantly exchanging matter and energy with its surroundings. In contrast, anything that is in equilibrium with its surrounding can generally be considered dead. (Some notable exceptions are vegetative forms, like seeds, and viruses, which may be completely inactive for long periods of time, and are not dead.)

Modern science has shown that life started some 3.5 billions of years ago, shortly (in geological terms) after the Earth itself was formed. The first life forms were very simple, but over billions of years a continuously acting process called evolution made them evolve and diversify, so that today we find very complex organisms as well as very simple ones.

Both complex and simple organisms have a similar molecular chemistry, or *biochemistry*. The main actors in the chemistry of life are molecules called **proteins** and **nucleic acids**. Roughly speaking, proteins are responsible for what a living being is and does in a physical sense. (The distinguished scientist Russell Doolittle once wrote that "we *are* our proteins.") Nucleic acids, on the other hand, encode the information necessary to produce proteins and are responsible for passing along this "recipe" to subsequent generations.

Molecular biology research is basically devoted to the understanding of the structure and function of proteins and nucleic acids. These molecules are therefore the fundamental objects of this book, and we now proceed to give a basic and brief description of the current state of knowledge regarding them.

1.2 PROTEINS

Most substances in our bodies are proteins, of which there are many different kinds. *Structural proteins* act as tissue building blocks, whereas other proteins known as *enzymes* act as catalysts of chemical reactions. A catalyst is a substance that speeds up a chemical reaction. Many biochemical reactions, if left unattended, would take too long to complete or not happen at all and would therefore not be useful to life. An enzyme can speed up this process by orders of magnitude, thereby making life possible. Enzymes are very specific — usually a given enzyme can help only one kind of biochemical reaction. Considering the large number of reactions that must occur to sustain life, we need a lot of enzymes. Other examples of protein function are oxygen transport and antibody defense. But what exactly are proteins? How are they made? And how do they perform their functions? This section tries briefly to answer these questions.

A protein is a chain of simpler molecules called **amino acids**. Examples of amino acids can be seen in Figure 1.1. Every amino acid has one central carbon atom, which is known as the alpha carbon, or C_α. To the C_α atom are attached a hydrogen atom, an amino group (NH_2), a carboxy group (COOH), and a *side chain*. It is the side chain that distinguishes one amino acid from another. Side chains can be as simple as one hydrogen atom (the case of amino acid glycine) or as complicated as two carbon rings (the case of tryptophan). In nature we find 20 different amino acids, which are listed in Table 1.1. These 20 are the most common in proteins; exceptionally a few nonstandard amino acids might be present.

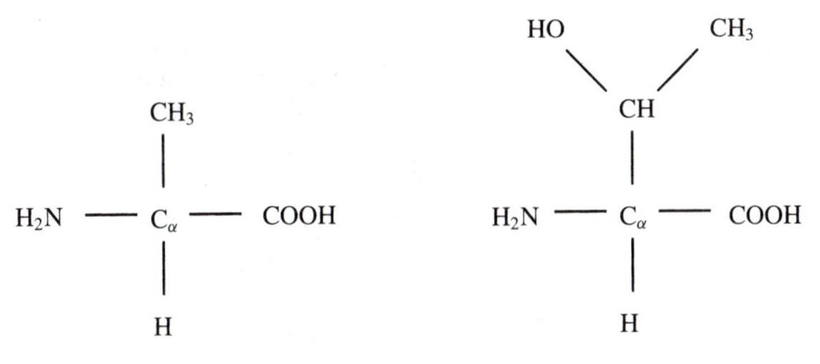

FIGURE 1.1

Examples of amino acids: alanine (left) and threonine.

In a protein, amino acids are joined by *peptide bonds*. For this reason, proteins are polypeptidic chains. In a peptide bond, the carbon atom belonging to the carboxy group of amino acid A_i bonds to the nitrogen atom of amino acid A_{i+1}'s amino group. In such a bond, a water molecule is liberated, because the oxygen and hydrogen of the carboxy group joins the one hydrogen from the amino group. Hence, what we really find inside a polypeptide chain is a **residue** of the original amino acid. Thus we generally speak of a protein having 100 residues, rather than 100 amino acids. Typical proteins contain about 300 residues, but there are proteins with as few as 100 or with as many as 5,000 residues.

TABLE 1.1

The twenty amino acids commonly found in proteins.

	One-letter code	*Three-letter code*	*Name*
1	A	Ala	Alanine
2	C	Cys	Cysteine
3	D	Asp	Aspartic Acid
4	E	Glu	Glutamic Acid
5	F	Phe	Phenylalanine
6	G	Gly	Glycine
7	H	His	Histidine
8	I	Ile	Isoleucine
9	K	Lys	Lysine
10	L	Leu	Leucine
11	M	Met	Methionine
12	N	Asn	Asparagine
13	P	Pro	Proline
14	Q	Gln	Glutamine
15	R	Arg	Arginine
16	S	Ser	Serine
17	T	Thr	Threonine
18	V	Val	Valine
19	W	Trp	Tryptophan
20	Y	Tyr	Tyrosine

The peptide bond makes every protein have a *backbone,* given by repetitions of the basic block $-\mathrm{N}-\mathrm{C}_\alpha-(\mathrm{CO})-$. To every C_α there corresponds a side chain. See Figure 1.2 for a schematic view of a polypeptide chain. Because we have an amino group at one end of the backbone and a carboxy group at the other end, we can distinguish both ends of a polypeptide chain and thus give it a direction. The convention is that polypeptides begin at the amino group (*N-terminal*) and end at the carboxy group (*C-terminal*).

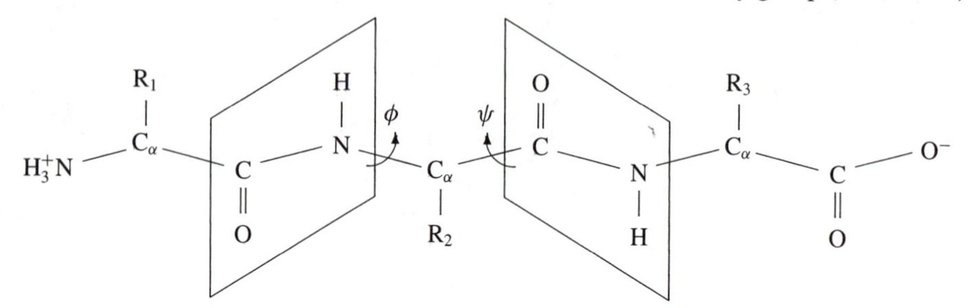

FIGURE 1.2

A polypeptide chain. The R_i *side chains identify the component amino acids. Atoms inside each quadrilateral are on the same plane, which can rotate according to angles* ϕ *and* ψ.

A protein is not just a linear sequences of residues. This sequence is known as its **primary structure**. Proteins actually fold in three dimensions, presenting **secondary**, **tertiary**, and **quaternary** structures. A protein's secondary structure is formed through interactions between backbone atoms only and results in "local" structures such as helices. Tertiary structures are the result of secondary structure packing on a more global level. Yet another level of packing, or a group of different proteins packed together, receives the name of quaternary structure. Figure 1.3 depicts these structures schematically.

Proteins can fold in three dimensions because the plane of the bond between the C_α atom and the nitrogen atom may rotate, as can the plane between the C_α atom and the other C atom. These rotation angles are known as ϕ and ψ, respectively, and are illustrated in Figure1.2. Side chains can also move, but it is a secondary movement with respect to the backbone rotation. Thus if we specify the values of all $\phi - \psi$ pairs in a protein, we know its exact folding. Determining the folding, or three-dimensional structure, of a protein is one of the main research areas in molecular biology, for three reasons. First, the three-dimensional shape of a protein is related to its function. Second, the fact that a protein can be made out of 20 different kinds of amino acids makes the resulting three-dimensional structure in many cases very complex and without symmetry. Third, no simple and accurate method for determining the three-dimensional structure is known. These reasons motivate Chapter 8, where we discuss some molecular structure prediction methods. These methods try to predict a molecule's structure from its primary sequence.

The three-dimensional shape of a protein determines its function in the following way. A folded protein has an irregular shape. This means that it has varied nooks and

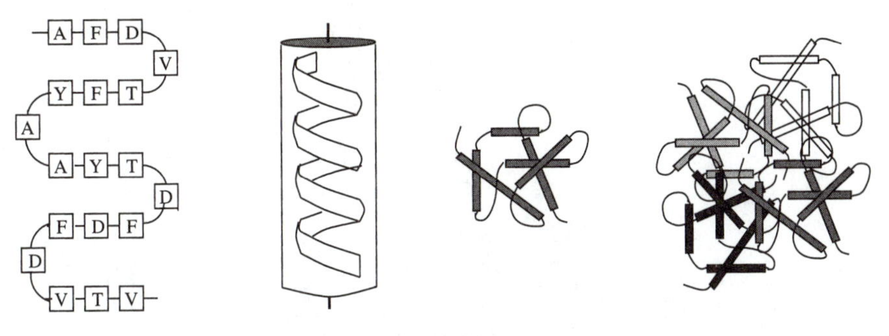

FIGURE 1.3

Primary, secondary, tertiary, and quaternary structures of proteins. (Based on a figure from [28].)

bulges, and such shapes enable the protein to come in closer contact with, or **bind to,** some other specific molecules. The kinds of molecules a protein can bind to depend on its shape. For example, the shape of a protein can be such that it is able to bind with several identical copies of itself, building, say, a thread of hair. Or the shape can be such that molecules A and B bind to the protein and thereby start exchanging atoms. In other words, a reaction takes place between A and B, and the protein is fulfilling its role as a catalyst.

But how do we get our proteins? Proteins are produced in a cell structure called *ribosome*. In a ribosome the component amino acids of a protein are assembled one by one thanks to information contained in an important molecule called **messenger ribonucleic acid**. To explain how this happens, we need to explain what nucleic acids are.

1.3 NUCLEIC ACIDS

Living organisms contain two kinds of nucleic acids: **ribonucleic acid,** abbreviated by RNA, and **deoxyribonucleic acid**, or DNA. We describe DNA first.

1.3.1 DNA

Like a protein, a molecule of DNA is a chain of simpler molecules. Actually it is a *double* chain, but let us first understand the structure of one simple chain, called **strand**. It has a backbone consisting of repetitions of the same basic unit. This unit is formed by a sugar molecule called $2'$-deoxyribose attached to a phosphate residue. The sugar molecule contains five carbon atoms, and they are labeled $1'$ through $5'$ (see Figure 1.4). The bond that creates the backbone is between the $3'$ carbon of one unit, the phosphate residue, and the $5'$ carbon of the next unit. For this reason, DNA molecules also have an

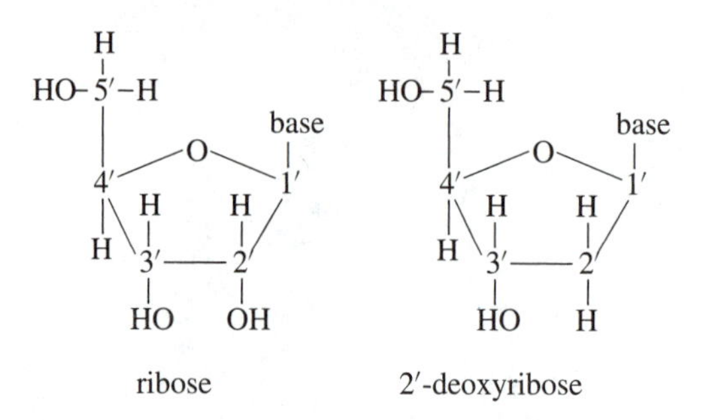

FIGURE 1.4

Sugars present in nucleic acids. Symbols 1′ *through* 5′ *represent carbon atoms. The only difference between the two sugars is the oxygen in carbon* 2′*. Ribose is present in RNA and* 2′*-deoxyribose is found in DNA.*

orientation, which by convention, starts at the 5′ end and finishes at the 3′ end. When we see a single stranded DNA sequence in a technical paper, book, or a sequence database file, it is always written in this canonical, $5' \rightarrow 3'$ direction, unless otherwise stated.

Attached to each 1′ carbon in the backbone are other molecules called **bases**. There are four kinds of bases: adenine (A), guanine (G), cytosine (C), and thymine (T). In Figure 1.5 we show the schematic molecular structure of each base, and in Figure 1.6 we show a schematic view of the single DNA strand described so far. Bases A and G belong to a larger group of substances called *purines,* whereas C and T belong to the *pyrimidines*. When we see the basic unit of a DNA molecule as consisting of the sugar, the phosphate, and its base, we call it a **nucleotide**. Thus, although bases and nucleotides are not the same thing, we can speak of a DNA molecule having 200 bases or 200 nucleotides. A DNA molecule having a few (tens of) nucleotides is referred to as an *oligonucleotide*. DNA molecules in nature are very long, much longer than proteins. In a human cell, DNA molecules have hundreds of millions of nucleotides.

As already mentioned, DNA molecules are double strands. The two strands are tied together in a helical structure, the famous double helix discovered by James Watson and Francis Crick in 1953. How can the two strands hold together? Because each base in one strand is paired with (or *bonds to*) a base in the other strand. Base A is always paired with base T, and C is always paired with G, as shown in Figures 1.5 and 1.7. Bases A and T are said to be the **complement** of each other, or a pair of **complementary bases**. Similarly, C and G are complementary bases. These pairs are known as *Watson–Crick base pairs.* Base pairs provide the unit of length most used when referring to DNA molecules, abbreviated to **bp**. So we say that a certain piece of DNA is 100,000 bp long, or 100 kbp.

In this book we will generally consider DNA as string of letters, each letter representing a base. Figure 1.8 presents this "string-view" of DNA, showing that we represent the double strand by placing one of the strings on top of the other. Notice the base-pairing. Even though the strands are linked, each one preserves its own orientation, and the two orientations are opposite. Figure 1.8 illustrates this fact. Notice that the 3′ end of one strand corresponds to the 5′ end of the other strand. This property is sometimes expressed by saying that the two strands are *antiparallel*. The fundamental consequence of

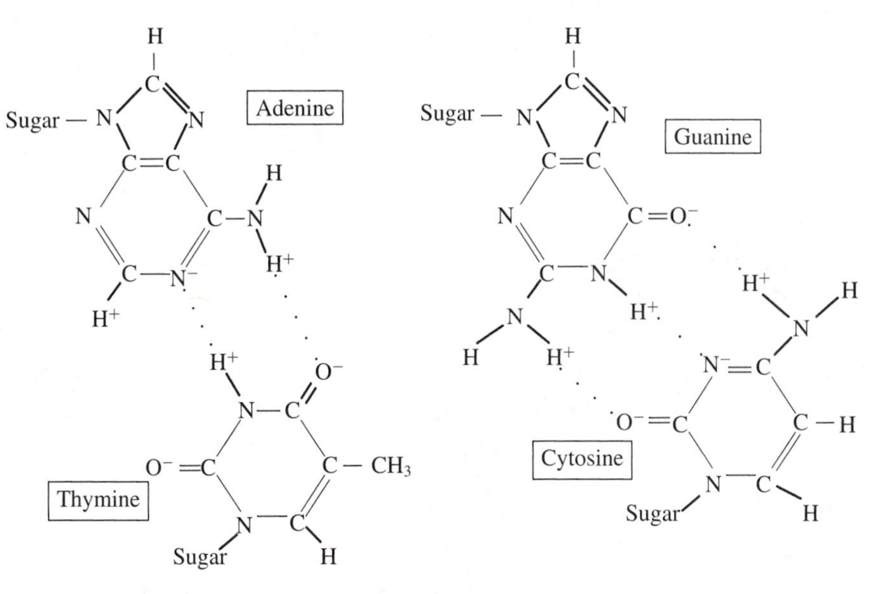

FIGURE 1.5

Nitrogenated bases present in DNA. Notice the bonds that can form between adenine and thymine and between guanine and cytosine, indicated by the dotted lines.

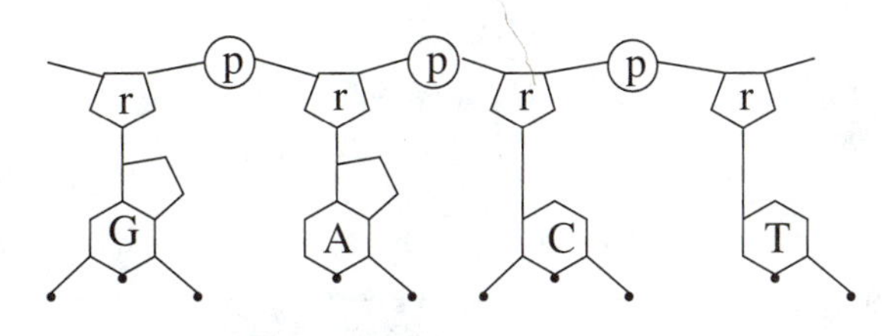

FIGURE 1.6

A schematic molecular structure view of one DNA strand.

this structure is that it is possible to infer the sequence of one strand given the other. The operation that enables us to do that is called **reverse complementation**. For example, given strand s = AGACGT in the canonical direction, we do the following to obtain its reverse complement: First we reverse s, obtaining s' = TGCAGA, and then we replace each base by its complement, obtaining $\overline{s}$ = ACGTCT. (Note that we use the bar over the s to denote the reverse complement of strand s.) It is precisely this mechanism that allows DNA in a cell to *replicate,* therefore allowing an organism that starts its life as one cell to grow into billions of other cells, each one carrying copies of the DNA molecules from the original cell.

In organisms whose cells do not have a nucleus, DNA is found free-floating inside each cell. In higher organisms, DNA is found inside the nucleus and in cell organelles called *mitochondria* (animals and plants) and *chloroplasts* (plants only).

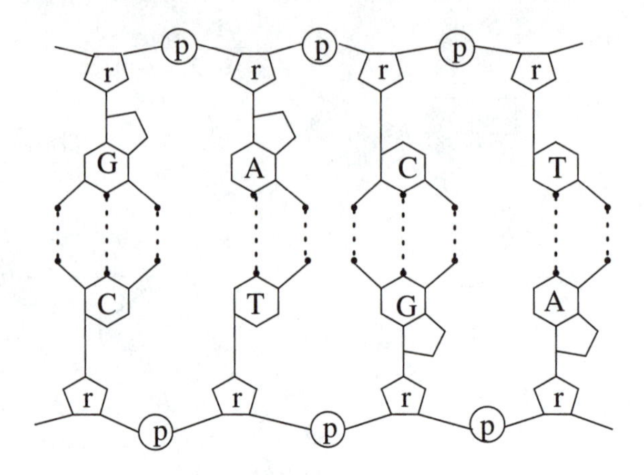

FIGURE 1.7

A schematic molecular structure view of a double strand of DNA.

5′	···	TACTGAA	···	3′
3′	···	ATGACTT	···	5′

FIGURE 1.8

A double-stranded DNA sequence represented by strings of letters.

1.3.2 RNA

RNA molecules are much like DNA molecules, with the following basic compositional and structural differences:

- In RNA the sugar is ribose instead of 2′-deoxyribose (see Figure 1.4).
- In RNA we do not find thymine (T); instead, uracil (U) is present. Uracil also binds with adenine like thymine does.
- RNA does not form a double helix. Sometimes we see RNA-DNA hybrid helices; also, parts of an RNA molecule may bind to other parts of the same molecule by complementarity. The three-dimensional structure of RNA is far more varied than that of DNA.

Another difference between DNA and RNA is that while DNA performs essentially one function (that of encoding information), we will see shortly that there are different kinds of RNAs in the cell, performing different functions.

THE MECHANISMS OF MOLECULAR GENETICS 1.4

The importance of DNA molecules is that the information necessary to build each protein or RNA found in an organism is encoded in DNA molecules. For this reason, DNA is sometimes referred to as "the blueprint of life." In this section we will describe this encoding and how a protein is built out of DNA (the process of *protein synthesis*). We will see also how the information in DNA, or genetic information, is passed along from a parent to its offspring.

1.4.1 GENES AND THE GENETIC CODE

Each cell of an organism has a few very long DNA molecules. Each such molecule is called a **chromosome**. We will have more to say about chromosomes later, so for the moment let us examine the encoding of genetic information from the point of view of only one very long DNA molecule, which we will simply call "the DNA." The first important thing to know about this DNA is that certain contiguous stretches along it encode information for building proteins, but others do not. The second important thing is that to each different kind of protein in an organism there usually corresponds one and only one contiguous stretch along the DNA. This stretch is known as a **gene**. Because some genes originate RNA products, it is more correct to say that a gene is a contiguous stretch of DNA that contains the information necessary to build a protein or an RNA molecule. Gene lengths vary, but in the case of humans a gene may have something like 10,000 bp. Certain cell mechanisms are capable of recognizing in the DNA the precise points at which a gene starts and at which it ends.

A protein, as we have seen, is a chain of amino acids. Therefore, to "specify" a protein all you have to do is to specify each amino acid it contains. And that is precisely what the DNA in a gene does, using triplets of nucleotides to specify each amino acid. Each nucleotide triplet is called a **codon**. The table that gives the correspondence between each possible triplet and each amino acid is the so-called **genetic code**, seen in Table 1.2. In the table you will notice that nucleotide triplets are given using RNA bases rather than DNA bases. The reason is that it is RNA molecules that provide the link between the DNA and actual protein synthesis, in a process to be detailed shortly. Before that let us study the genetic code in more detail.

Notice that there are 64 possible nucleotide triplets, but there are only 20 amino acids to specify. The consequence is that different triplets correspond to the same amino acid. For example, both AAG and AAA code for lysine. On the other hand, three of the possible codons do not code for any amino acid and are used instead to signal the end of a gene. These special termination codons are identified in Table 1.2 with the word STOP written in the corresponding entry. Finally, we remark that the genetic code shown above is used by the vast majority of living organisms, but some organisms use a slightly modified code.

TABLE 1.2

The genetic code mapping codons to amino acids.

First position	*Second position* G	A	C	U	*Third position*
G	Gly	Glu	Ala	Val	G
	Gly	Glu	Ala	Val	A
	Gly	Asp	Ala	Val	C
	Gly	Asp	Ala	Val	U
A	Arg	Lys	Thr	Met	G
	Arg	Lys	Thr	Ile	A
	Ser	Asn	Thr	Ile	C
	Ser	Asn	Thr	Ile	U
C	Arg	Gln	Pro	Leu	G
	Arg	Gln	Pro	Leu	A
	Arg	His	Pro	Leu	C
	Arg	His	Pro	Leu	U
U	Trp	STOP	Ser	Leu	G
	STOP	STOP	Ser	Leu	A
	Cys	Tyr	Ser	Phe	C
	Cys	Tyr	Ser	Phe	U

1.4.2 TRANSCRIPTION, TRANSLATION, AND PROTEIN SYNTHESIS

Now let us describe in some detail how the information in the DNA results in proteins. A cell mechanism recognizes the beginning of a gene or gene cluster thanks to a *promoter*. The promoter is a region before each gene in the DNA that serves as an indication to the cellular mechanism that a gene is ahead. The codon AUG (which codes for methionine) also signals the start of a gene. Having recognized the beginning of a gene or gene cluster, a copy of the gene is made on an RNA molecule. This resulting RNA is the *messenger RNA*, or mRNA for short, and will have exactly the same sequence as one of the strands of the gene but substituting U for T. This process is called **transcription**. The mRNA will then be used in cellular structures called ribosomes to manufacture a protein.

Because RNA is single-stranded and DNA is double-stranded, the mRNA produced is identical in sequence to only one of the gene strands, being complementary to the other strand — keeping in mind that T is replaced by U in RNA. The strand that looks like the mRNA product is called the *antisense* or *coding* strand, and the other one is the *sense* or *anticoding* or else *template* strand. The template strand is the one that is actually transcribed, because the mRNA is composed by binding together ribonucleotides complementary to this strand. The process always builds mRNA molecules from their 5′ end to their 3′ end, whereas the template strand is read from 3′ to 5′. Notice also that it is not the case that the template strand for genes is always the same; for example, the template strand for a certain gene *A* may be one of the strands, and the template strand for another gene *B* may be the other strand. For a given gene, the cell can recognize the corresponding template strand thanks to a promoter. Even though the reverse complement of the promoter appears in the other strand, this reverse complement is *not* a promoter and thus will not be recognized as such. One important consequence of this fact is that genes from the same chromosome have an *orientation* with respect to each other: Given two genes, if they appear in the same strand they have the same orientation; otherwise they have opposite orientation. This is a fundamental fact for Chapter 7. We finally note that the terms *upstream* and *downstream* are used to indicate positions in the DNA in reference to the orientation of the coding strand, with the promoter being upstream from its gene.

Transcription as described is valid for organisms categorized as *prokaryotes*. These organisms have their DNA free in the cell, as they lack a nuclear membrane. Examples of prokaryotes are bacteria and blue algae. All other organisms, categorized as *eukaryotes,* have a nucleus separated from the rest of the cell by a nuclear membrane, and their DNA is kept inside the nucleus. In these organisms genetic transcription is more complex. Many eukaryotic genes are composed of alternating parts called *introns* and *exons*. After transcription, the introns are spliced out from the mRNA. This means that introns are parts of a gene that are *not* used in protein synthesis. An example of exon-intron distribution is given by the gene for bovine atrial naturietric peptide, which has 1082 base pairs. Exons are located at positions 1 to 120, 219 to 545, and 1071 to 1082. Introns occupy positions 121 to 218 and 546 to 1070. Thus, the mRNA coding regions has just 459 bases, and the corresponding protein has 153 residues. After introns are spliced out, the shortened mRNA, containing copies of only the exons plus regulatory regions in the beginning and end, leaves the nucleus, because ribosomes are outside the nucleus.

Because of the intron/exon phenomenon, we use different names to refer to the entire gene as found in the chromosome and to the spliced sequence consisting of exons only. The former is called *genomic* DNA and the latter *complementary* DNA or cDNA. Scientists can manufacture cDNA without knowing its genomic counterpart. They first capture the mRNA outside the nucleus on its way to the ribosomes. Then, in a process called *reverse transcription,* they produce DNA molecules using the mRNA as a template. Because the mRNA contains only exons, this is also the composition of the DNA produced. Thus, they can obtain cDNA without even looking at the chromosomes. Both transcription and reverse transcription are complex processes that need the help of enzymes. *Transcriptase* and *reverse transcriptase* are the enzymes that catalyze these processes in the cell. There is also a phenomenon called *alternative splicing*. This occurs when the same genomic DNA can give rise to two or more different mRNA molecules, by choosing the introns and exons in different ways. They will in general produce different proteins.

Now let us go back to mRNA and protein synthesis. In this process two other kinds of RNA molecules play very important roles. As we have already mentioned, protein synthesis takes place inside cellular structures called ribosomes. Ribosomes are made of proteins and a form of RNA called *ribosomal* RNA, or rRNA. The ribosome functions like an assembly line in a factory using as "inputs" an mRNA molecule and another kind of RNA molecule called *transfer* RNA, or tRNA.

Transfer RNAs are the molecules that actually implement the genetic code in a pro-

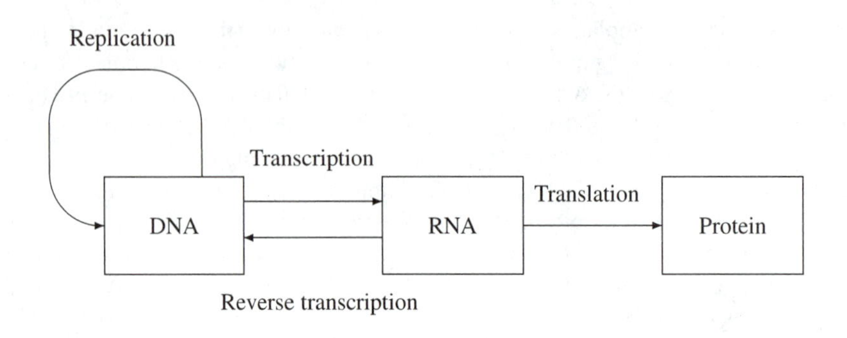

FIGURE 1.9

Genetic information flow in a cell: the so-called central dogma of molecular biology.

cess called **translation**. They make the connection between a codon and the specific amino acid this codon codes for. Each tRNA molecule has, on one side, a conformation that has high affinity for a specific codon and, on the other side, a conformation that binds easily to the corresponding amino acid. As the messenger RNA passes through the interior of the ribosome, a tRNA matching the current codon — the codon in the mRNA currently inside the ribosome — binds to it, bringing along the corresponding amino acid (a generous supply of amino acids is always "floating around" in the cell). The three-dimensional position of all these molecules in this moment is such that, as the tRNA binds to its codon, its attached amino acid falls in place just next to the previous amino acid in the protein chain being formed. A suitable enzyme then catalyzes the addition of this current amino acid to the protein chain, releasing it from the tRNA. A protein is constructed residue by residue in this fashion. When a STOP codon appears, no tRNA associates with it, and the synthesis ends. The messenger RNA is released and degraded by cell mechanisms into ribonucleotides, which will be then recycled to make other RNA.

One might think that there are as many tRNAs as there are codons, but this is not true. The actual number of tRNAs varies among species. The bacterium *E. coli*, for instance, has about 40 tRNAs. Some codons are not represented, and some tRNAs can bind to more than one codon.

Figure 1.9 summarizes the processes we have just described. The expression *central dogma* is generally used to denote our current synthetic view of genetic information transfer in cells.

1.4.3 JUNK DNA AND READING FRAMES

In this section we provide some additional details regarding the processes described in previous sections.

As mentioned, genes are certain contiguous regions of the chromosome, but they do not cover the entire molecule. Each gene, or group of related genes, is flanked by regulatory regions that play a role in controlling gene transcription and other related processes, but otherwise intergenic regions have no known function. They are called "junk DNA" because they appear to be there for no particular reason. Moreover, they accumulate mutations, as a change not affecting genes or their regulatory regions is often not lethal and is therefore propagated to the progeny. Recent research has shown, however, that junk DNA has more information content than previously believed. The amount of junk DNA varies from species to species. Prokaryotes tend to have little of it — their chromosomes are almost all covered by genes. In contrast, eukaryotes have plenty of junk DNA. In human beings it is estimated that as much as 90% of the DNA in chromosomes is composed of junk DNA.

An aspect of the transcription process that is important to know is the concept of *reading frame*. A reading frame is one of the three possible ways of grouping bases to form codons in a DNA or RNA sequence. For instance, consider the sequence

`TAATCGAATGGGC.`

One reading frame would be to take as codons `TAA`, `TCG`, `AAT`, `GGG`, leaving out the last `C`. Another reading frame would be to ignore the first `T` and get codons `AAT`, `CGA`, `ATG`, `GGC`. Yet another reading frame would yield codons `ATC`, `GAA`, `TGG`, leaving out two bases at the beginning (`TA`) and two bases at the end (`GC`).

Notice that the three reading frames start at positions 1, 2, and 3 in the given sequence, respectively. If we were to consider a reading frame starting at position 4, the codons obtained would be a subset of the ones for starting position 1, so this is actually the same reading frame starting at a different position. In general, if we take starting positions i and j where the difference $j - i$ is a multiple of three, we are in fact considering the same reading frame.

Sometimes we talk about six, not three, different reading frames in a sequence. In this case, what we have is a DNA sequence and we are looking at the opposite strand as well. We have three reading frames in one strand plus another three in the complementary strand, giving a total of six. It is common to do that when we have newly sequenced DNA and want to compare it to a protein database. We have to translate the DNA sequence into a protein sequence, but there are six ways of doing that, each one taking a different reading frame. The fact that we lose one or two bases at the extremities of the sequence is not important; these sequences are long enough to yield a meaningful comparison even with a few missing residues.

An *open reading frame,* or ORF, in a DNA sequence is a contiguous stretch of this sequence beginning at the start codon, having an integral number of codons (its length is a multiple of three), and such that none of its codons is a STOP codon. The presence of additional regulatory regions upstream from the start codon is also used to characterize an ORF.

1.4.4 CHROMOSOMES

In this section we briefly describe the process of genetic information transmission at the chromosome level. First we note that the complete set of chromosomes inside a cell is called a **genome**. The number of chromosomes in a genome is characteristic of a species. For instance, every cell in a human being has 46 chromosomes, whereas in mice this number is 40. Table 1.3 gives the number of chromosomes and genome size in base pairs for selected species.

TABLE 1.3

Genome sizes of certain species. Apart from our own species, the organisms listed are important in molecular biology and genetics research.

Species	*Number of Chromosomes (diploid)*	*Genome Size (base pairs)*
Bacteriophage λ (virus)	1	5×10^4
Escherichia coli (bacterium)	1	5×10^6
Saccharomyces cerevisiae (yeast)	32	1×10^7
Caenorhabditis elegans (worm)	12	1×10^8
Drosophila melanogaster (fruit fly)	8	2×10^8
Homo sapiens (human)	46	3×10^9

Prokaryotes generally have just one chromosome, which is sometimes a circular DNA molecule. On the other hand, in eukaryotes chromosomes appear in pairs (and for this reason the cells that carry them are called *diploid*). In humans, for example, there are 23 pairs. Each member of a pair was inherited from each parent. The two chromosomes that form a pair are called *homologous* and a gene in one member of the pair corresponds to a gene in the other. Certain genes are exactly the same in both the paternal and the maternal copies. One example is the gene that codes for hemoglobin, a protein that carries oxygen in the blood. Other genes may appear in different forms, which are called *alleles*. A typical example is the gene that codes for blood type in humans. It appears in three forms: A, B, and O. As is well known, if a person receives, say, the A allele from the mother and the B allele from the father, this person's blood type will be AB.

Cells that carry only one member of each pair of chromosomes are called *haploid*. These are the cells that are used in sexual reproduction. When a haploid cell from the mother is merged with a haploid cell from the father we have an *egg cell,* which is again diploid. Haploid cells are formed through a process called *meiosis,* in which a cell divides in two and each daughter cell gets one member of each pair of chromosomes.

It is interesting to note that despite the fact that all genes are present in all cells, only a portion of the genes are normally used (*expressed,* in biological jargon) by any specific cell. For instance, liver cells express a different set of genes than do skin cells. The mechanisms through which the cells in an organism differentiate into liver cells, skin cells, and so on, are still largely unclear.

1.4.5 IS THE GENOME LIKE A COMPUTER PROGRAM?

Having reviewed the basic mechanisms through which proteins are produced, it is tempting to view them in light of the so-called "genetic program metaphor." In this metaphor, the genome of an organism is seen as a computer program that completely specifies the organism, and the cell machinery is simply an interpreter of this program. The biological functions performed by proteins would be the execution of this "program."

This metaphor is overly simplistic, due in part to the following two facts:

- The "DNA program" in fact undergoes changes during transcription and translation, so that one cannot simply apply the genetic code to a stretch of DNA known to contain a gene in order to know what protein corresponds to that gene.
- Gene expression is a complex process that may depend on spatial and temporal context. For example, not all genes in a genome are expressed in an organism's lifetime, whereas others are expressed over and over again; some genes are expressed only when the organism is subject to certain outside phenomena, such as a virus invasion. The opposite is also true: Genes that are normally expressed may be repressed because of outside stimuli. It is true that the expression of certain genes is essentially context-free, and this is what makes biotechnology possible. But this is by no means true for all genes. If we view gene expression inside a cell as a "computing process," we can say that, in the case of the human genome, there are more than 10^{18} such processes occurring and interacting simultaneously.

In view of these observations, it seems better to view an organism not as determined by its genome but rather as the result of a very complex network of simultaneous interactions, in which the genome sequence is one of several contributing factors.

1.5 HOW THE GENOME IS STUDIED

In the scientific study of a genome, the first thing to notice is the different orders of magnitude that we have to deal with. Let us use the human genome as an example. The basic information we want to extract from any piece of DNA is its base-pair sequence. The process of obtaining this information is called **sequencing**. A human chromosome has around 10^8 base pairs. On the other hand, the largest pieces of DNA that can be sequenced in the laboratory are 700 bp long. This means that there is a gap of some 10^5 between the scale of what we can actually sequence and a chromosome size. This gap is at the heart of many problems in computational biology, in particular those studied in Chapters 4 (fragment assembly) and 5 (physical mapping). In this section we briefly describe some of the lab techniques that underlie those problems.

1.5.1 MAPS AND SEQUENCES

One particular piece of information that is very important is the location of genes in chromosomes. The term *locus* (plural *loci*) is used to denote the location of a gene in a chromosome. (Sometimes the word locus is used as a synonym for the word gene.) The simplest question in this context is, given two genes, are they in the same homologous pair? This can be answered without resorting to molecular techniques, as long as the genes in question are ones that affect visible characteristics, such as eye color or wing shape. We have to test whether the characteristics are being inherited independently. We will say that they are, or more technically, that they *assort* or *segregate* independently if an offspring has about a 50% chance of inheriting both characteristics from the same parent. If the characteristics do assort independently, then the genes are probably not linked; that is, they probably belong to different chromosomes. Two genes carried in the same pair should segregate together, and the offspring will probably inherit both characteristics from the same parent.

In fact, as is usually the case in biology, things are not so clear-cut: 100% or 50% segregation does not always happen. Any percentage in between can occur, due to *crossing over*. When cells divide to produce other cells that will form the progeny, new gene arrangements can form. We say that *recombination* occurs. Recombination can happen because homologous chromosomes can "cross" and exchange their end parts before segregating. There are an enormous number of recombination possibilities, so what we really see is that rates of recombination vary a great deal. These rates in turn give information on how far apart the genes are in the chromosome. If they are close together, there is a small chance of separation due to crossing over. If they are far apart, the chances of separation increase, to the point that they appear to assort independently.

The first genetic maps were constructed by producing successive generations of certain organisms and analyzing the observed segregation percentages of certain characteristics. A *genetic linkage map* of a chromosome is a picture showing the order and relative distance among genes using such information. Genetic maps constructed from recombination percentages are important, but they have two drawbacks: (1) They do not tell the actual distance in base pairs or other linear unit of length along the chromosome; and (2) If genes are very close, one cannot resolve their order, because the probability of separation is so small that observed recombinant frequencies are all zero.

Maps that reflect the actual distance in base pairs are called **physical maps**. To construct physical maps, we need completely different techniques. In particular we have to work with pieces of DNA much smaller than a chromosome but still too large to be sequenced directly. A physical map can tell the location of certain *markers*, which are precisely known small sequences, within 10^4 base pairs or so. Computational problems associated with physical map construction are studied in Chapter 5.

Finally, for pieces of DNA that are on the order of 10^3 base pairs, we can use still other techniques and obtain the whole sequence. We have mentioned that current lab techniques can sequence DNA pieces of at most 700 bp; to sequence a 20,000-bp piece (in what is known as **large-scale sequencing**), the basic idea is to break apart several copies of the piece in different ways, sequence the (small) fragments directly, and then put the fragments together again by using computational techniques that are studied in Chapter 4.

Figure 1.10 illustrates the different scales at which the human genome is studied.

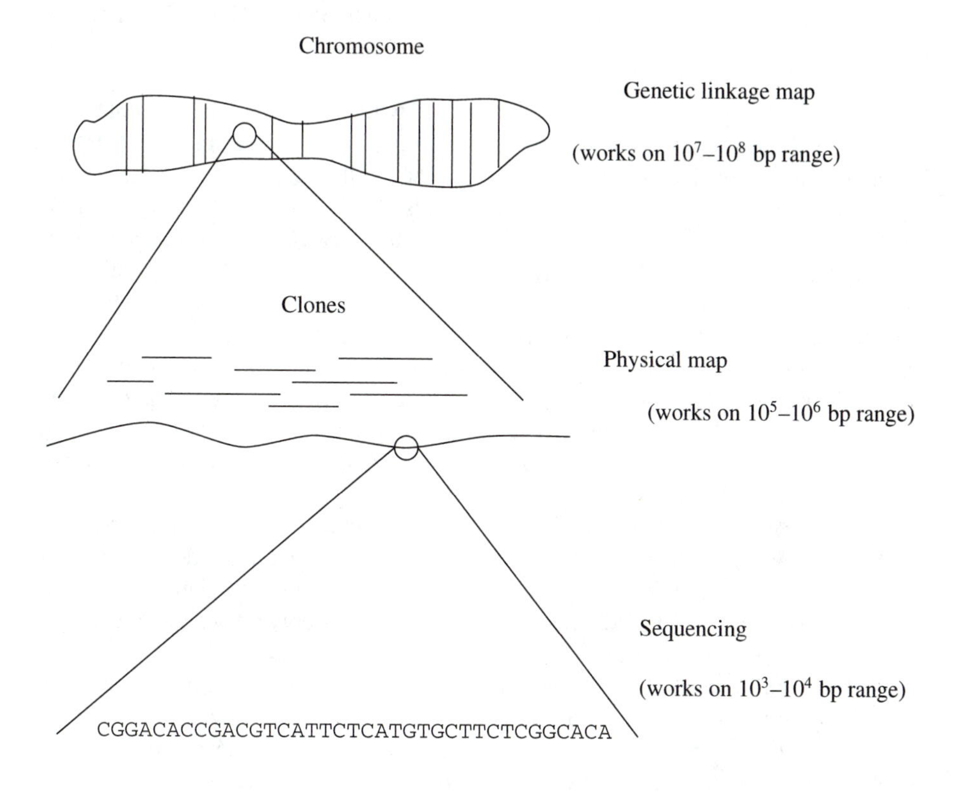

FIGURE 1.10

The different levels at which a genome is studied. (Based on a figure from [19].)

1.5.2 SPECIFIC TECHNIQUES

Some special laboratory techniques must be used to obtain maps and sequences. In this section we give an overview of such techniques.

It is important to realize that lab techniques in molecular biology nearly always produce data that have errors. For this reason, most algorithms for problems in computational biology are useful only to the extent that they can handle errors. Such concern will appear throughout this book.

Viruses and Bacteria

We begin by briefly describing the organisms most used in genetics research: viruses and bacteria.

Viruses are parasites at the molecular level. Viruses can hardly be considered a life form, yet they can reproduce when infecting suitable cells, called *hosts*. Viruses do not exhibit any metabolism — no biochemical reactions occur in them. Instead, viruses rely on the host's metabolism to replicate, and it is this fact that is exploited in laboratory experiments.

Most viruses consist of a protein cap (a *capsid*) with genetic material (either DNA or RNA) inside. Viral DNA is much smaller than the DNA in chromosomes, and therefore is much easier to manipulate. When viruses infect a cell, the genetic material is introduced in the cytoplasm. This genetic material is mistakenly interpreted by the cell mechanism as its own DNA, and for this reason the cell starts producing virus-coded proteins as though they were the cell's own. These proteins promote viral DNA replication and formation of new capsids, so that a large number of virus particles are assembled inside the infected cell. Still other viral proteins break the cell membrane and release all the new virus particles in the environment, where they can attack other cells. Certain viruses do not kill their hosts at once. Instead, the viral DNA gets inserted into the host genome and can stay there for a long time without any noticeable change in cell life. Under certain conditions the dormant virus can be activated, whereupon it detaches itself from the host genome and starts its replication activity.

Viruses are highly specific; they are capable of infecting one type of cell only. Thus, for instance, virus T2 infects only *E. coli* cells; HIV, the human immunodeficiency virus, infects only a certain kind of human cell involved in defending the organism against intruders in general; TMV, the tobacco mosaic virus, infects only tobacco leaf cells; and so on. *Bacteriophages,* or just *phages,* are viruses that infect bacteria.

A bacterium is a single-cell organism having just one chromosome. Bacteria can multiply by simple DNA replication, and they can do this in a very short period of time, which makes them very useful in genetic research. The bacterium most commonly used in labs is the already mentioned *Escherichia coli,* which can divide in 20 minutes. As a result of their size and speed of reproduction, millions of bacteria can be easily generated and handled in a laboratory.

Cutting and Breaking DNA

Because a DNA molecule is so long, some tool to cut it at specific points (like a pair of scissors) or to break it apart in some other way is needed. We review basic techniques for these processes in this section.

The pair of scissors is represented by **restriction enzymes**. They are proteins that catalyze the hydrolysis of DNA (molecule breaking by adding water) at certain specific points called **restriction sites** that are determined by their local base sequence. In other words, they cut DNA molecules in all places where a certain sequence appears. For instance, *Eco*RI is a restriction enzyme that cuts DNA wherever the sequence `GAATTC` is found. Notice that this sequence is its own reverse complement, that is, $\overline{\mathtt{GAATTC}} = \mathtt{GAATTC}$. Sequences that are equal to their reverse complement are called *palindromes*. So, every time this sequence appears in one strand it appears in the other strand as well. The cuts are made in both strands between the `G` and the first `A`. Therefore, the remaining DNA pieces will have "sticky" ends, that is, their 5′ end in the cut point will be four bases shorter than the 3′ end; see Figure 1.11. This favors relinking with another DNA piece cut with the same enzyme, providing a kind of "DNA cut and paste" technique very useful in genetic engineering for the production of recombinant DNA.

```
...ATCCAG|AATTCTCGGA...      ...ATCCAG        AATTCTCGGA...
...TAGGTCTTAA|GAGCCT...      ...TAGGTCTTAA        GAGCCT...
```

DNA before cutting DNA after cutting

FIGURE 1.11

Enzyme that cuts DNA, leaving "sticky" ends.

Some common types of restriction enzymes are 4-cutters, 6-cutters, and 8-cutters. It is rare to see an odd-cutter because sequences of odd length cannot be palindromes. Restriction enzymes are also called *endonucleases* because they break DNA in an internal point. *Exonucleases* are enzymes that degrade DNA from the ends inwards.

In bacteria, restriction enzymes may act as a defense against virus attacks. These enzymes can cut the viral DNA before it starts its damage. Bacterial DNA on the other hand protects itself from restriction enzymes by adding a methyl group to some of its bases.

DNA molecules can be broken apart by the *shotgun method,* which is sometimes used in DNA sequencing. A solution containing purified DNA — a large quantity of identical molecules — is subjected to some breaking process, such as submitting it to high vibration levels. Each individual molecule breaks down at several random places, and then some of the fragments are filtered and selected for further processing, in particular for copying or *cloning* (see below). We then get a collection of cloned fragments that correspond to random contiguous pieces of the purified DNA sequence. Sequencing these pieces and assembling the resulting sequences is a standard way of determining the purified DNA's sequence. This method can also be used to construct a cloning library — a collection of clones covering a certain long DNA molecule.

Copying DNA

Another very important tool needed in molecular biology research is a DNA copying process (also called DNA *amplification*). We are dealing here with molecules, which are microscopic objects. The more copies we have of a molecule, the easier it is to study it. Fortunately, several techniques have been developed for this purpose.

DNA Cloning: To do any experiment with DNA in the lab, one needs a minimum quantity of material; one molecule is clearly not enough. Yet one molecule is sometimes all we have to start with. In addition, the material should be stored in a way that permits essentially unlimited production of new material for repetition of the experiment or for new experiments as needed. DNA cloning is the name of the general technique that allows these goals to be attained.

Given a piece of DNA, one way of obtaining further copies is to use nature itself:

We insert this piece into the genome of an organism, a *host* or *vector,* and then let the organism multiply itself. Upon host multiplication, the inserted piece (the *insert*) gets multiplied along with the original DNA. We can then kill the host and dispose of the rest, keeping only the inserts in the desired quantity. DNA produced in this way is called *recombinant*. Popular vectors include *plasmids, cosmids, phages,* and *YACs*.

A plasmid is a piece of circular DNA that exists in bacteria. It is separated from and much smaller than the bacterial chromosome. It does get replicated when the cell divides, though, and each daughter cell keeps one copy of the plasmid. Plasmids make good vectors, but they place a limitation on insert size: about 15 kbp. Inserts much larger than this limit will make very big plasmids, and big plasmids tend to get shortened when replicated.

Phages are viruses often used as vectors. One example is phage λ, which infects the bacteria *E. coli*. Inserts in phage DNA get replicated when the virus infects a host colony. Phage λ normally has a DNA of 48 kbp, and inserts of up to 25 kbp are well tolerated. Inserts larger than this limit are impossible, though, because the resulting DNA will not fit in the phage protein capsule. However, if the entire phage DNA is replaced by an insert plus some minimum replicative apparatus, inserts of up to 50 kbp are feasible. These are called *cosmids*.

For very large inserts, on the range of a million base pairs, a YAC (yeast artificial chromosome) can be used. A YAC is an extra, artificially-made chromosome that can be built by adding yeast control chromosomal regions to the insert and making it look like an additional chromosome to the yeast replication mechanism.

Polymerase Chain Reaction: A way of producing many copies of a DNA molecule without cloning it is afforded by the polymerase chain reaction (PCR). DNA Polymerase is an enzyme that catalyzes elongation of a single strand of DNA, provided there is a template DNA to which this single strand is attached. Nucleotides complementary to the ones in the template strand are added until both strands have the same size and form a normal double strand of DNA. The small stretch of double stranded DNA at the beginning needed for polymerase to start its job is called a **primer**.

PCR consists basically of an alternating repetition of two phases: A phase in which double stranded DNA is separated into two single strands by heat and a phase in which each single strand thus obtained is converted into a double strand by addition of a primer and polymerase action. Each repetition doubles the number of molecules. After enough repetitions, the quantity of material produced is large enough to perform further experiments, thanks to the exponential growth of this doubling procedure.

A curious footnote here is that Kary B. Mullis, the inventor of PCR (in 1983), realized it was a good idea because he "had been spending a lot of time writing computer programs" and was thus familiar with iterative functions and thereby with exponential growth processes.

Reading and Measuring DNA

How do we actually "read" the base pairs of a DNA sequence? Reading is done with a technique known as *gel electrophoresis,* which is based on separation of molecules by their size. The process involves a gel medium and a strong electric field. DNA or

RNA molecules are charged in aqueous solution and will move to a definite direction by action of the electric field. The gel medium makes them move slowly, with speed inversely proportional to their size. All molecules are initially placed at one extremity in a gel block. After a few hours, the smaller molecules will have migrated to the other end of the block, whereas larger molecules will stay behind, near the starting place. By interpolation, the relative sizes of molecules can be calculated with good approximation.

In this kind of experiment, DNA molecules can be labeled with radioactive isotopes so that the gel can be photographed, producing a graphic record of the positions at the end of a run. This process is used in DNA sequencing, determination of restriction fragment lengths, and so on. An alternative is to use fluorescent dyes instead of radioactive isotopes. A laser beam can trace the dyes and send the information directly to a computer, avoiding the photographic process altogether. Sequencing machines have been built using this technique.

DNA or RNA bases can be read using this process with the following technique. Given a DNA molecule, it is possible to obtain all fragments from it that end at every position where an `A` appears. Similarly all fragments that end in `T`, in `G`, or in `C` can be obtained. We thus get four different test tubes, one for each base. The fragments in each tube will have differing lengths. For example, suppose the original DNA piece is the following:

`GACTTAGATCAGGAAACT`

The fragments that end in `T` are `GACT`, `GACTT`, `GACTTAGAT`, and the whole sequence itself. Thus if we separate these fragments by size, and we do it simultaneously but separately for all four test tubes, we will know the precise base composition of the original DNA sequence. This is illustrated in Figure 1.12.

Some errors may occur in reading a gel film, because sometimes marks are blurred, especially near the film borders. One other limitation is the size of DNA fragments that can be read in this way, which is about 700 bp.

1.6 THE HUMAN GENOME PROJECT

The Human Genome Project is a multinational effort, begun in 1988, whose aim is to produce a complete physical map of all human chromosomes, as well as the entire human DNA sequence. As part of the project, genomes of other organisms such as bacteria, yeast, flies, and mice are also being studied.

This is no easy task, since the human genome is so large. So far, many virus genomes have been entirely sequenced, but their sizes are generally in the 1 kbp to 10 kbp range. The first free-living organism to be totally sequenced was the bacterium *Haemophilus influenzae,* containing a 1800 kbp genome. In 1996 the whole sequence of the yeast genome — a 10 million bp sequence — was also determined. This was an important milestone, given that yeast is a free-living eukaryote. Mapping the human genome still seems remote, as it is 100 times bigger than the largest genome sequenced so far.

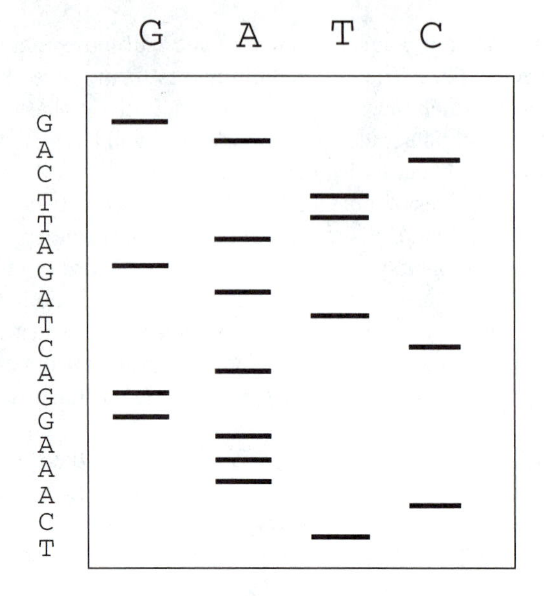

FIGURE 1.12

Schematic view of film produced by gel electrophoresis. Individual DNA bases can be identified in each of the four columns. Shorter fragments leave their mark near the top, whereas longer fragments leave their mark near the bottom.

One of the reasons why other organisms are part of the project is to perfect sequencing methods so that they can then be applied to the human genome. The cost of sequencing is also expected to drop as a result of new, improved technology. Another reason lies in research benefits. All species targeted are largely employed in genetic and molecular research.

Some lessons have been learned. A large effort like this cannot be entertained by a single lab. Rather, a consortium of first-rate labs working together is a more efficient way — and perhaps the only way — of getting the work done. Coordinating these tasks is in itself a challenge. On the computer science side, databases with updated and consistent information have to be maintained, and fast access to the data has to be provided. A major concern is with possible errors in the sequences obtained. The current target is to obtain DNA sequences with at most one error for every 10,000 bases.

Another problem is that of the "average genome." Different individuals have different genomes (this is what makes *DNA fingerprinting* possible). A gene may have many alleles in the population, and the sequence of intergenic regions (which do not code for products) may vary from person to person. It is estimated that the genomes of two distinct humans differ on average in one in every 500 bases. So the question is: Whose genome is going to be sequenced? Even if one individual is somehow chosen and her or his DNA is taken as the standard, there is the problem of transposable elements. It is now known that certain parts of the genome keep moving from one place to another, so that at best what we will get after sequencing is a "snapshot" of the genome in a given instant.

Once the entire sequence is obtained, we will face the difficult task of analyzing it.

We have to recognize the genes in it and determine the function of the proteins produced. But gene recognition is still in its infancy, and protein function determination is still a very laborious procedure. Treatment of genetic diseases based on data produced by the Human Genome Project is still far ahead, although encouraging pioneering efforts have already yielded results.

1.7 SEQUENCE DATABASES

Thanks in part to the techniques described in previous sections, large numbers of DNA, RNA, and protein sequences have been determined in the past decades. Some institutional sequence databases have been set up to harbor these sequences as well as a wealth of associated data. The rate at which new sequences are being added to these databases is exponential. Computational techniques have been developed to allow fast search on these databases, some of which are described in Chapter 3. Here we give a brief description of some representative sequence databases.

GenBank: Maintained by the National Center for Biotechnology Information (NCBI), USA, GenBank contains hundreds of thousands of DNA sequences. It is divided into several sections with sequences grouped according to species, including:

- ***PLN:*** Plant sequences
- ***PRI:*** Primate sequences
- ***ROD:*** Rodent sequences
- ***MAM:*** Other mammalian sequences
- ***VRT:*** Other vertebrate sequences
- ***IVN:*** Invertebrate sequences
- ***BCT:*** Bacterial sequences
- ***PHG:*** Phage sequences
- ***VRL:*** Other viral sequences
- ***SYN:*** Synthetic sequences
- ***UNA:*** Unannotated sequences
- ***PAT:*** Patent sequences
- ***NEW:*** New sequences

Searches can be made by keywords or by sequence. A typical entry is shown in Figure 1.13. The entry is divided into *fields,* with each field composed by a field identifier, which is a word describing the contents of the field, and the information per se. The entries are just plain text. One important field is *accession no.,* the accession number, which is a code that is unique to this entry and can be used for faster access to it. In the example, the accession number is `M12174`. Some entries have several accession numbers as a result of combination of several related but once distinct entries into one comprehensive entry. Other fields are self-explanatory for the most part.

This database is part of an international collaboration effort, which also includes the DNA DataBank of Japan (DDBJ) and the European Molecular Biology Laboratory (EMBL). GenBank can be reached through the following locator:

http://www.ncbi.nlm.nih.gov/

EMBL: The European Molecular Biology Laboratory is an institution that maintains several sequence repositories, including a DNA database called the Nucleotide Sequence Database. Its organization is similar to that of GenBank, with the entries having roughly the same fields. In the EMBL database, entries are identified by two-letter codes (see Figure 1.14). The accession number, for instance, is identified by the letters `AC`. Code `XX` indicates blank lines. Both GenBank and EMBL separate the sequences in blocks of ten characters, with six blocks per line. This scheme makes it easy to find specific positions within the sequence. EMBL can be reached through the following locator:

http://www.embl-heidelberg.de/

PIR: The Protein Identification Resource (PIR) is a database of protein sequences cooperatively maintained and distributed by three institutions: the National Biomedical Research Foundation (in the USA), the Martinsried Institute for Protein Sequences (in Europe), and the Japan International Protein Information Database (in Japan). A typical entry appears in Figure 1.15. Several web sites provide interfaces to this database, including the following:

http://www.gdb.org/
http://www.mips.biochem.mpg.de/

PDB: The Protein Data Bank is a repository of three-dimensional structures of proteins. For each protein represented, a general information header is provided followed by a list of all atoms present in the structure, with three spatial coordinates for each atom that indicate their position with three decimal places. An example is given in Figures 1.16 and 1.17. This repository is maintained in Brookhaven, USA. Access to it can be gained through

http://www.pdb.bnl.gov/

Other Databases: Apart from those mentioned above, many other databases have been created to keep molecular biology information. Among them we cite ACEDB, a powerful platform prepared for the *C. elegans* sequencing project but adaptable to similar projects; Flybase, a database of fly sequences; and databases for restriction enzymes, codon preferential usage, and so on.

```
LOCUS       HUMRHOA      539 bp    mRNA          PRI    04-AUG-1986
DEFINITION  Human ras-related rho mRNA (clone 6), partial cds.
ACCESSION   M12174
KEYWORDS    c-myc proto-oncogene; ras oncogene; rho gene.
SOURCE      Human peripheral T-cell, cDNA to mRNA, clone 6.
  ORGANISM  Homo sapiens
            Eukaryota; Animalia; Chordata; Vertebrata; Mammalia;
            Theria; Eutheria; Primates; Haplorhini; Catarrhini;
            Hominidae.
REFERENCE   1  (bases 1 to 539)
  AUTHORS   Madaule,P.
  JOURNAL   Unpublished (1985) Columbia U, 701 W 168th St,
            New York, NY 10032
REFERENCE   2  (bases 1 to 539)
  AUTHORS   Madaule,P. and Axel,R.
  TITLE     A novel ras-related gene family
  JOURNAL   Cell 41, 31-40 (1985)
  MEDLINE   85201682
COMMENT     [2] has found and sequenced a family of highly
            evolutionarily conserved genes with homology to the
            ras family (H-ras, K-ras, N-ras) of oncogenes.
            [2] named this family rho (for ras homology).
            In humans at least three distinct rho genes are
            present. A draft entry and computer-readable copy of
            this sequence were kindly
            provided by P.Madaule (07-OCT-1985).

            NCBI gi: 337392
FEATURES             Location/Qualifiers
     source          1..539
                     /organism="Homo sapiens"
     CDS             <1..509
                     /note="rho protein;  NCBI gi: 337393"
                     /codon_start=2
BASE COUNT      105 a    180 c    172 g     82 t
ORIGIN      185 bp upstream of HinfI site.
        1 cgagttcccc gaggtgtacg tgc .... tatgtggccg acattgaggt
       61 ggacggcaag caggtggagc tgg .... ggccaggagg actacgaccg
      121 cctgcggccg ctctcctacc cgg .... atgtgcttct cggtggacag
      181 cccggactcg ctggagaaca tcc .... gaggtgaagc acttctgtcc
      ...
      421 cgaggtcttc gagacggcca cgc .... cgctacggct cccagaacgg
      481 ctgcatcaac tgctgcaagg tgc .... cgcgcctgcc cctgccggc
//
```

FIGURE 1.13

Typical GenBank entry. The actual file was edited to fit the page. In a real entry each sequence line has 60 characters, with the possible exception of the last line.

```
ID   ECTRGA      standard; RNA; PRO; 75 BP.
XX
AC   M24860;
XX
DT   24-APR-1990 (Rel. 23, Created)
DT   31-MAR-1992 (Rel. 31, Last updated, Version 3)
XX
DE   E.coli Gly-tRNA.
XX
KW   transfer RNA-Gly.
XX
OS   Escherichia coli
OC   Prokaryota; Bacteria; Gracilicutes; Scotobacteria;
OC   Facultatively anaerobic rods; Enterobacteriaceae;
OC   Escherichia.
XX
RN   [1]
RP   1-75
RA   Carbon J., Chang S., Kirk L.L.;
RT   "Clustered tRNA genes in Escherichia coli: Transcription
RT   and processing";
RL   Brookhaven Symp. Biol. 26:26-36(1975).
XX
FH   Key             Location/Qualifiers
FH
FT   tRNA            1..75
FT                   /note="Gly-tRNA"
XX
SQ   Sequence 75 BP; 13 A; 24 C; 19 G; 19 T; 0 other;
     gcgggcatcg tataatggct attacctcag ... tgatgatgcg ggttcgattc
     ccgctgcccg ctcca
//
```

FIGURE 1.14

A typical EMBL entry, edited to fit the page. Actual entries have 60 characters per line in the sequence section, with the possible exception of the last line.

```
PIR1:CCHP
cytochrome c - hippopotamus

Species: Hippopotamus amphibius (hippopotamus)

Date: 19-Feb-1984 #sequence_revision 19-Feb-1984 #text_change
      05-Aug-1994

Accession: A00008

Thompson, R.B.; Borden, D.; Tarr, G.E.; Margoliash, E.
    J. Biol. Chem. 253, 8957-8961, 1978
    Title: Heterogeneity of amino acid sequence in hippopotamus
      cytochrome c.
    Reference number: A00008; MUID:79067782
    Accession: A00008
    Molecule type: protein
    Residues: 1-104 <THO>
    Note: 3-Ile was also found

Superfamily: cytochrome c; cytochrome c homology

Keywords: acetylated amino end; electron transfer; heme;
    mitochondrion; oxidative phosphorylation; respiratory chain

Residues                Feature
1                       Modified site: acetylated amino end (Gly)
                          #status predicted
14,17                   Binding site: heme (Cys) (covalent) #status
                          predicted
18,80                   Binding site: heme iron (His, Met) (axial
                          ligands) #status predicted

                                Composition
           6 Ala  A       4 Gln  Q       6 Leu  L       2 Ser  S
           2 Arg  R       8 Glu  E      17 Lys  K       8 Thr  T
           5 Asn  N      14 Gly  G       2 Met  M       1 Trp  W
           3 Asp  D       3 His  H       4 Phe  F       4 Tyr  Y
           2 Cys  C       6 Ile  I       4 Pro  P       3 Val  V
     Mol. wt. unmod. chain = 11,530     Number of residues = 104

                5         10        15        20        25        30
   1 G D V E K G K K I F V Q K C A Q C H T V E K G G K H K T G P
  31 N L H G L F G R K T G Q S P G F S Y T D A N K N K G I T W G
  61 E E T L M E Y L E N P K K Y I P G T K M I F A G I K K K G E
  91 R A D L I A Y L K Q A T N E
```

FIGURE 1.15

A typical PIR entry, edited to fit the page.

```
HEADER    MYOGLOBIN (CARBONMONOXY)                  19-JUL-95   1MCY
TITLE     SPERM WHALE MYOGLOBIN (MUTANT WITH INITIATOR MET AND
TITLE    2 WITH HIS 64 REPLACED BY GLN, LEU 29 REPLACED BY PHE
COMPND    MOL_ID: 1;
COMPND   2 MOLECULE: MYOGLOBIN (CARBONMONOXY);
COMPND   3 CHAIN: NULL;
COMPND   4 ENGINEERED: YES;
COMPND   5 MUTATION: INS(MET 0), F29L, Q64H, N122D
SOURCE    MOL_ID: 1;
SOURCE   2 SYNTHETIC: SYNTHETIC GENE;
SOURCE   3 ORGANISM_SCIENTIFIC: PHYSETER CATODON;
SOURCE   4 ORGANISM_COMMON: SPERM WHALE;
SOURCE   5 EXPRESSION_SYSTEM: ESCHERICHIA COLI
KEYWDS    HEME, OXYGEN TRANSPORT, RESPIRATORY PROTEIN
EXPDTA    X-RAY DIFFRACTION
AUTHOR    T.LI,G.N.PHILLIPS JUNIOR
REVDAT   1   07-DEC-95 1MCY    0
JRNL        AUTH   X.ZHAO,K.VYAS,B.D.NGUYEN,K.RAJARATHNAM,
JRNL        AUTH 2 G.N.LAMAR,T.LI,G.N.PHILLIPS JUNIOR,R.EICH,
JRNL        AUTH 3 J.S.OLSON,J.LING,D.F.BOCIAN
JRNL        TITL   A DOUBLE MUTANT OF SPERM WHALE MYOGLOBIN
JRNL        TITL 2 MIMICS THE STRUCTURE AND FUNCTION OF
JRNL        TITL 3 ELEPHANT MYOGLOBIN
JRNL        REF    J.BIOL.CHEM.                 V. 270 20763 1995
JRNL        REFN   ASTM JBCHA3  US ISSN 0021-9258          0071
REMARK   1
REMARK   2
REMARK   2 RESOLUTION. 1.7  ANGSTROMS.
REMARK   3
REMARK   3 REFINEMENT.
REMARK   3   PROGRAM                      X-PLOR
REMARK   3   AUTHORS                      BRUNGER
REMARK   3   R VALUE                      0.182
REMARK   3   RMSD BOND DISTANCES          0.020  ANGSTROMS
REMARK   3   RMSD BOND ANGLES             1.82   DEGREES
REMARK   3
REMARK   3   NUMBER OF REFLECTIONS        23187
REMARK   3   RESOLUTION RANGE         5.0 - 1.7  ANGSTROMS
REMARK   3   DATA CUTOFF                  0.0    SIGMA(F)
REMARK   3
REMARK   3  DATA COLLECTION.
REMARK   3   NUMBER OF UNIQUE REFLECTIONS        25787
REMARK   3   RESOLUTION RANGE  INFINITY - 1.7    ANGSTROMS
REMARK   3   COMPLETENESS OF DATA                95.8   %
```

FIGURE 1.16

A typical PDB entry, partial header. The last columns containing the characters 1MCY *i, where i is the line number, are omitted. Some other editing was done to fit the page.*

```
ATOM      1  N    MET   0    24.486   8.308   -9.406  1.00 37.00
ATOM      2  CA   MET   0    24.542   9.777   -9.621  1.00 36.40
ATOM      3  C    MET   0    25.882  10.156  -10.209  1.00 34.30
ATOM      4  O    MET   0    26.833   9.391  -10.078  1.00 34.80
ATOM      5  CB   MET   0    24.399  10.484   -8.303  1.00 39.00
ATOM      6  CG   MET   0    24.756   9.581   -7.138  1.00 41.80
ATOM      7  SD   MET   0    24.017  10.289   -5.719  1.00 44.80
ATOM      8  CE   MET   0    24.761  12.009   -5.824  1.00 41.00
ATOM      9  N    VAL   1    25.951  11.334  -10.816  1.00 31.10
ATOM     10  CA   VAL   1    27.185  11.834  -11.382  1.00 28.40
ATOM     11  C    VAL   1    27.330  13.341  -11.124  1.00 26.30
ATOM     12  O    VAL   1    26.444  14.135  -11.452  1.00 26.60
ATOM     13  CB   VAL   1    27.270  11.547  -12.912  1.00 29.30
ATOM     14  CG1  VAL   1    28.532  12.207  -13.526  1.00 28.60
ATOM     15  CG2  VAL   1    27.275  10.038  -13.163  1.00 29.70
ATOM     16  N    LEU   2    28.435  13.739  -10.500  1.00 23.00
ATOM     17  CA   LEU   2    28.691  15.142  -10.318  1.00 20.80
ATOM     18  C    LEU   2    29.289  15.737  -11.599  1.00 20.10
ATOM     19  O    LEU   2    30.129  15.110  -12.276  1.00 19.50
ATOM     20  CB   LEU   2    29.661  15.356   -9.134  1.00 20.90
ATOM     21  CG   LEU   2    29.036  15.387   -7.726  1.00 20.40
ATOM     22  CD1  LEU   2    28.556  13.983   -7.402  1.00 19.60
ATOM     23  CD2  LEU   2    30.058  15.904   -6.689  1.00 19.50
ATOM     24  N    SER   3    28.996  17.003  -11.826  1.00 18.80
ATOM     25  CA   SER   3    29.696  17.733  -12.852  1.00 19.40
ATOM     26  C    SER   3    31.096  18.141  -12.385  1.00 19.50
ATOM     27  O    SER   3    31.397  18.121  -11.174  1.00 19.10
ATOM     28  CB   SER   3    28.861  18.954  -13.223  1.00 20.60
ATOM     29  OG   SER   3    29.019  19.969  -12.261  1.00 22.10
ATOM     30  N    GLU   4    31.947  18.561  -13.310  1.00 18.40
ATOM     31  CA   GLU   4    33.293  18.956  -12.937  1.00 19.00
ATOM     32  C    GLU   4    33.173  20.215  -12.047  1.00 19.20
ATOM     33  O    GLU   4    34.026  20.457  -11.206  1.00 19.10
ATOM     34  CB   GLU   4    34.135  19.270  -14.198  1.00 19.60
ATOM     35  CG   GLU   4    35.491  19.937  -13.932  1.00 21.10
ATOM     36  CD   GLU   4    36.537  19.020  -13.295  1.00 22.60
ATOM     37  OE1  GLU   4    36.355  17.787  -13.230  1.00 23.90
ATOM     38  OE2  GLU   4    37.569  19.550  -12.840  1.00 24.60
ATOM     39  N    GLY   5    32.182  21.062  -12.313  1.00 18.30
ATOM     40  CA   GLY   5    32.039  22.287  -11.532  1.00 19.10
ATOM     41  C    GLY   5    31.669  22.025  -10.056  1.00 18.20
ATOM     42  O    GLY   5    32.251  22.638   -9.140  1.00 18.60
```

FIGURE 1.17

A typical PDB entry, coordinates. The last columns, containing the characters 1MCY *i, where i is the line number, are omitted. A few blank characters in each line were removed.*

EXERCISES

1. Consider the sequence TAATCGAATGGGC. Derive the six possible protein sequences translated from it.
2. Given the fictitious "gene" below, find
   a. the sequence of the corresponding mRNA.
   b. the sequence of the resulting protein (use the standard genetic code).

   ```
   ATGATACCGACGTACGGCATTTAA
   TACTATGGCTGCATGCCGTAAATT
   ```
3. Given the protein sequence LMK, how many DNA sequences could possibly have given rise to it? *Hint:* One of them is CTGATGAAG.
4. Find the cut patterns of restriction enzymes *Bam*HI and *Hind*III.
5. Suppose we have a DNA molecule of length 40,000 and digest it with a 4-cutter restriction enzyme. Assuming a random distribution of bases in the molecule, how many pieces can we expect to get?
6. Browse through the GenBank viral sequences section and find a disease-causing agent you have never heard of before.

BIBLIOGRAPHIC NOTES

Readers completely new to molecular biology may wish to take a look at the introductory "comic-book" by Rosenfeld, Ziff, and van Loon [165]. Readers who wish to go beyond the overview we have given should look at the standard references in molecular biology, some of which are:

- Watson, Hopkins, Roberts, Steitz, and Weiner [203, 204].
- Alberts, Bray, Lewis, Raff, Roberts, and Watson [7].
- Lewin [124].

The reader should keep in mind that such books are constantly being updated, due to the rapid progress in molecular biology research.

Other textbooks that we found useful in preparing this chapter were as follows. For biochemistry, Mathews and van Holde [131]; for genetics, Tamarin [183]; for protein structures, Branden and Tooze [28].

The quote from R. F. Doolittle on page 2 is from [50], which is a good review of proteins for the "educated layperson," and appears on a special issue dedicated to the "molecules of life." An entertaining account of the discovery of PCR by Mullis appears in [141]. The quote on page 20 was taken from that article.

Robbins [162] and Frenkel [66] provide articles discussing the difficulties of the Human Genome Project from the point of view of computer science. Robbins has emphasized the view of an organism as the result of a complex interaction of many "cell processes." Lewontin [125] presents a sobering critique of the Human Genome Project.

The problems with the "genetic program metaphor" were pointed out by many people; our source was the paper by Atlan and Koppel [17]. On the other hand, the interested reader may wish to take a look at the book by Hofstadter [96], which contains an interesting discussion of parallels between molecular biology and mathematical logic.

CHAPTER

2

STRINGS, GRAPHS, AND ALGORITHMS

This chapter presents notation, conventions, and a brief review of the main mathematical and computer science concepts used in this book. It is not intended to be an introduction to the subjects treated. To the reader who is not trained in these topics we suggest several books in the bibliographic notes.

STRINGS

Strings are the basic type of data we deal with in this book. A **string** is an ordered succession of **characters** or **symbols** drawn from a finite set called the **alphabet**. The term **sequence** is used throughout as a synonym of string. Most of our examples use either the DNA alphabet {A, C, G, T} of nucleotides or the 20-character alphabet of amino acids (see Chapter 1). But we note that some of the results presented make sense and are valid for all alphabets. In particular, the definitions we give here are valid in general.

Strings may have repeated characters, for example, s = AATGCA. The **length** of a string s, denoted by $|s|$, is the number of characters in it. In the previous example, $|s| = 6$. The character occupying position i in a string s is denoted by $s[i]$. Character indices start from 1 and go up to $|s|$. Using the same example again we have $s[1]$ = A, $s[2]$ = A, $s[3]$ = T, and so on. There is a string of length zero, called the **empty string**. We denote the empty string by the special symbol ϵ.

Whereas the terms *string* and *sequence* have the same meaning, the terms *substring* and *subsequence* represent distinct concepts. A **subsequence** of s is a sequence that can be obtained from s by removal of some characters. For instance, TTT is a subsequence of ATATAT, but TAAA is not. The empty string is a subsequence of every sequence. When sequence t is a subsequence of sequence s, we say that s is a **supersequence** of t.

In contrast to the subsequence concept, a **substring** of s is a string formed by consecutive characters of s, in the same order as they appear in s. For instance, TAC is a substring of AGTACA, but not of TTGAC. Being a substring of some string means being a subsequence as well, but not all subsequences of a sequence s are substrings of s.

When sequence t is a substring of sequence s we say that s is a **superstring** of t.

Sometimes a substring w of u appears several times inside u, as when $w = \mathtt{TT}$ and $u = \mathtt{CTTTAGCATTAA}$. In certain cases, it is important to distinguish these occurrences, and for this we use the concept of interval. An **interval** of a string s is a set of consecutive indices $[i..j]$ such that $1 \leq i \leq j+1 \leq |s|+1$. The interval contains all indices between i and j, including i and j themselves. For $[i..j]$ an interval of s, $s[i..j]$ denotes the substring $s[i]s[i+1]\ldots s[j]$ of s if $i \leq j$, and the empty string when $i = j+1$. Therefore, for any substring t of s there is at least one interval $[i..j]$ of s with $t = s[i..j]$.

The **concatenation** of two strings s and t is denoted by st and is formed by appending all characters of t after s, in the order they appear in t. For instance, if $s = \mathtt{GGCTA}$ and $t = \mathtt{CAAC}$, then $st = \mathtt{GGCTACAAC}$. The length of st is $|s| + |t|$. The concatenation of several copies of the same string s is indicated by raising s to a suitable power, as in $s^3 = sss$.

A **prefix** of s is any substring of s of the form $s[1..j]$ for $0 \leq j \leq |s|$. We admit $j = 0$ and define $s[1..0]$ as being the empty string, which is a prefix of s as well. Note that t is a prefix of s if and only if there is another string u such that $s = tu$. Sometimes we need to refer to the prefix of s with exactly k characters, with $0 \leq k \leq |s|$, and we use the notation *prefix*(s, k) to denote this string.

Analogously, a **suffix** of s is a substring of the form $s[i..|s|]$ for a certain i such that $1 \leq i \leq |s|+1$. We admit $i = |s|+1$, in which case $s[|s|+1..|s|]$ denotes the empty string. A string t is a suffix of s if and only if there is u such that $s = ut$. The notation *suffix*(s, k) denotes the unique suffix of s with k characters, for $0 \leq k \leq |s|$.

The Killer Agent All strings s satisfy $|s| \geq 0$. To simplify certain arguments, we postulate the existence of a special string κ such that

$$|\kappa| = -1.$$

This string functions as a **killer agent** that destroys characters it comes in contact with by concatenation, shortening strings in the process. For instance,

$$\kappa\mathtt{GCTAGT} = \mathtt{CTAGT}.$$

Notice that the result is a suffix of the original string. In general, concatenation with powers of κ can be used to indicate prefixes, suffixes, and substrings in general.

$$\begin{aligned} \mathit{prefix}(s,k) &= s\kappa^{|s|-k}, \\ \mathit{suffix}(s,k) &= \kappa^{|s|-k}s, \\ s[i..j] &= \kappa^{i-1}s\kappa^{|s|-j}. \end{aligned}$$

There is one drawback in using the killer agent. Concatenation is no longer associative when κ is present. For instance,

$$(\mathtt{ATC}\kappa)\mathtt{GTC} = \mathtt{ATGTC}$$

whereas

$$\mathtt{ATC}(\kappa\mathtt{GTC}) = \mathtt{ATCTC}.$$

Hence, the meaning of $\mathtt{ATC}\kappa\mathtt{GTC}$ is not uniquely defined without parentheses. However, we still have the equality

$$|st| = |s| + |t|,$$

where s and t are general string expressions, possibly involving κ.

GRAPHS 2.2

A **graph** can be described by two sets. One of them is V, the set of *nodes,* or *vertices;* the other is E, the set of *edges,* which is a set of distinct pairs of vertices. A graph is denoted by $G = (V, E)$, vertices by letters such as u, v, or w, and edges by letters such as e or by pairs of vertices in parentheses, as in (u, v). A graph can be *undirected,* in which case an edge is an unordered pair of vertices [so that $(u, v) = (v, u)$], or *directed,* in which case an edge is an ordered pair of vertices [so that $(u, v) \neq (v, u)$]. When we mention graphs in the rest of this book without qualifications, we mean an undirected graph; in this section, the unqualified term "graph" means that the statement applies to both directed an undirected graphs. Although graphs are abstract mathematical objects, they are usually depicted by drawings such as those in Figure 2.1. Unless otherwise stated, all graphs in this book are *simple graphs,* meaning that they do not contain loops, which are edges of the form (u, u), or multiple edges between the same pair of vertices [but note that in a directed graph the edges (u, v) and (v, u) are not considered to be multiple]. We denote the number of vertices of a graph by $|V|$ or n; and the number of edges by $|E|$ or m.

Given edge (u, v) we say that u and v are its *endpoints*. We also say that u and v are *incident* to (u, v) and that (u, v) is *incident* to u and v. If (u, v) is a directed edge, u is the *tail* of this edge and v is its *head.* When u and v are the endpoints of an undirected edge, we say that u and v are *adjacent*. The *degree* of a vertex v in an undirected graph is the number of vertices adjacent to it. In the case of directed graphs a vertex v has an *outdegree,* which is the number of edges of the form (v, x), and an *indegree,* which is the number of edges of the form (x, v). For example, the degree of vertex v_2 in Figure 2.1 is three; the degree of vertex v_5 is one. The indegree of vertex u_3 is one, and its outdegree is two.

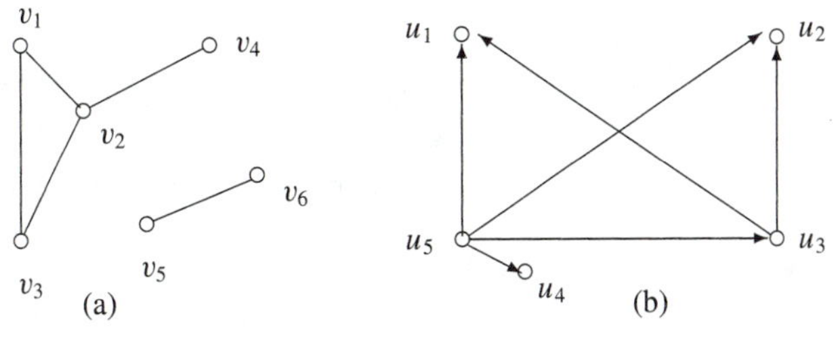

FIGURE 2.1

Examples of graphs. (a) An undirected graph; (b) a directed graph.

In a *weighted* graph we associate a real number with each edge. Depending on context, the weight of an edge (u, v) is called its *cost,* the *distance* between u and v, or just its *weight.*

Given graphs $G = (V, E)$ and $G' = (V', E')$, we say that G' is a *subgraph* of G when $V' \subseteq V$ and $E' \subseteq E$. If G' is a subgraph of G but $G' \neq G$, we say that G' is a *proper subgraph* of G. If G' is a subgraph of G and $V = V'$ we say that G' is a *spanning subgraph* of G. If V' is the set of all vertices incident to some edge of E', for a given E', we say that G' is the graph *induced* by E'. Conversely, if E' is the set of edges whose endpoints are both in V', for a given V', we say that G' is the graph *induced* by V'.

A *path* in a graph is a list of distinct vertices $(v_1, v_2, \ldots, v_k)$ such that (v_i, v_{i+1}) is an edge in the graph for $1 \leq i < k$. A *cycle* in a directed graph is a path such that $k > 1$ and $v_1 = v_k$. A cycle in an undirected graph is a path such that $v_1 = v_k$ and no edge is repeated. Unless otherwise stated, all cycles are *simple,* which means that all vertices except the first and last are distinct. An example of a (simple) cycle is (v_1, v_2, v_3, v_1) in the undirected graph of Figure 2.1. We say that vertex v is *reachable* from vertex u when there is a path between u and v. In weighted graphs the *weight of a path* is the sum of the weights of its edges.

In an undirected graph G, if every vertex is reachable from every other vertex, we say that the graph is *connected.* If a graph is not connected, we can find its *connected components*. The notion of connected component is intuitive. For example, Figure 2.1(a) shows an undirected graph with two connected components. More formally, the set of connected components of a graph G is the set of all connected subgraphs of G such that no element of the set is a subgraph of another element of the set. In the case of directed graphs, we say that it can be *strongly connected, weakly connected,* or not connected. In a strongly connected graph, every vertex can be reached from every other vertex. In a weakly connected graph this is true only if we disregard edge directions (thus the *underlying undirected graph* is connected); such is the case of the directed graph of Figure 2.1(b). If neither of the two properties is true, the graph is not connected.

Graph theory classifies graphs in classes according to certain properties. Some important examples are:

- *The acyclic graph:* a graph without cycles.
- *The complete graph:* a graph where for every pair of vertices v and w, $(v, w) \in E$ [and $(w, v) \in E$ as well, in the case of directed graphs].
- *The bipartite graph:* a graph where vertices are separated into two disjoint subsets U and V, such that any edge has one of its endpoints in U and the other in V.
- *The tree:* an acyclic and connected graph. A graph whose connected components are trees is called a *forest*.

Trees are an important class of graphs and have their own terminology. A *leaf* is a node of the tree with degree 1. All other nodes are *interior nodes*. A tree can be *rooted,* which means that one of its nodes is distinguished and called the *root* (denoted by r). Given a node u and another node v on the path from u to r, we say that v is an *ancestor* of u and that u is a *descendant* of v; if u and v are adjacent, then v is the *parent* of u and u is the *child* of v. Leaves are nodes without children; the root has no parent. The *depth* of a node v is the number of edges on the path from v to r. The *lowest common ancestor* of two nodes u and v is the deepest node that is ancestor of both u and v.

Another important class of graphs in computational biology is the class of *interval graphs*. An interval graph $G = (V, E)$ is an undirected graph obtained from a collection C of intervals on the real line. To each interval in C there corresponds a vertex in G; we place an edge between vertices u and v if and only if their intervals have a nonempty intersection.

Below we list a few important problems on graphs that are especially relevant in this book:

1. Given a directed or undirected graph, find a cycle such that every edge of the graph is in the cycle but each edge appears exactly once (vertices may be repeated). Graphs for which this can be done are called *Eulerian graphs*. A version of this problem requires that we find a *path* in the graph fulfilling the same condition (the graph is said to contain an *Eulerian path*).
2. Given a directed or undirected graph, find a cycle in the graph such that every vertex of the graph is in the cycle but each vertex appears exactly once (except for the first and last vertices). Graphs for which this can be done are called *Hamiltonian graphs*. A version of this problem requires that we find a *path* in the graph such that every vertex of the graph is in the path but each vertex appears exactly once (the graph is said to contain a *Hamiltonian path*). Still another version of this problem is defined on weighted graphs: Find a Hamiltonian cycle of minimum weight. This is also known as *the traveling salesman problem,* since cities can be modeled as vertices and roads as edges, and traveling salesmen presumably are interested in minimizing the distance they travel when visiting many cities.
3. Given an undirected weighted connected graph, find a spanning tree such that the sum of the weights of its edges is minimum among all spanning trees. This is known as the *minimum spanning tree problem.*
4. Given an undirected graph, find the minimum number of colors necessary to obtain a *coloring*. A *vertex-coloring* (or simply *coloring*) of a graph is a function that assigns integers (the *colors*) to vertices so that no two adjacent vertices have the same color.
5. Given an undirected graph, find a *maximum cardinality matching*. A matching is a subset M of the edges such that no two edges in M share an endpoint. The weighted version of this problem asks for a matching of maximum total weight.

Graphs can be represented in computer programs in a variety of ways. One of the most common is the *adjacency matrix representation*. Given an $n \times n$ matrix M, we set $M_{ij} = 1$ if $(i, j) \in E$ and $M_{ij} = 0$ otherwise. If the graph is weighted, we replace the 1s in the matrix by the weight of each corresponding edge. This representation uses on the order of n^2 storage space. Another way of representing graphs is through the *adjacency list representation.* In this case, we store all vertices in a list, and for every vertex in the list we create another list containing its adjacent vertices. In the case of directed graphs, we may wish to keep two lists per vertex v: one for edges that go out of v and another for edges that go into v. This representation uses on the order of $n + m$ storage space. Adjacency lists have the advantage of using less storage space than adjacency matrices when the number of edges in the graph is relatively small. Such graphs are called *sparse*. When m is much larger than n we say that the graph is *dense*.

ALGORITHMS 2.3

In a general sense an algorithm is a well-defined and finite sequence of steps used to solve a well-defined problem. An example of an algorithm is the sequence of steps we use to multiply two integers. Algorithms are meant to be executed by persons or machines. It is important to specify who or what is going to execute the algorithm, because that is what gives meaning to "well defined" in the preceding definition of algorithm. In this book we are interested primarily in algorithms that can be understood by people but that could (with details added) be executed by machines. The machine we have in mind is not a concrete machine but an abstraction called the *random access machine model.* Briefly, this model has the following components: one processor, which can execute the steps sequentially; a finite memory (a collection of *words*); and a finite set of registers (another collection of words). In each word a single integer or real number can be stored. The processor executes the steps of a *program,* which is the algorithm coded using the machine's *instructions.* In this book we abuse the term *machine instruction* and allow *informal instructions,* which are meant to be understood by people but require considerable expansion before they can be actually implemented on a real computer.

Algorithms in this book are expressed both in informal English sentences and in pseudo programming language code. One of the goals in presenting such pseudo-code is to help readers who wish to implement the algorithms described. We do not formally define this code, but simply explain it with reference to well-known programming languages such as C or Pascal. First we describe the notation conventions:

1. Algorithms are headed by the keyword **Algorithm**.

2. The **input:** and **output:** keywords are used to show the algorithm's input and output, respectively.

3. Keywords are typeset in **boldface**.

4. Variable, procedure, and function names are typeset in *italic.*

5. Constants are typeset using SMALL CAPS.

6. Comments begin with the symbol // , extend to the rest of the line, and are typeset in *slanted font.*

7. Informal statements and conditions are typeset in normal type.

8. Different statements generally appear in different lines; semicolons are used only to separate statements in the same line. When an informal statement needs more than one line, indentation is used from the second line onward.

9. Block structure is denoted by indentation, without **begin**s or **end**s. Occasionally we use comments to signal the end of a long block.

10. Indexing of arrays is done by square brackets (as in $V[i]$) and indexing of records is done by the dot notation (as in $R.field$).

11. The parameters in a procedure or function call appear in parentheses, separated by commas.

Below we describe the formal statements we use in our algorithms. Most of them are similar to those occurring in any modern procedural computer language.

1. The assignment statement, as in $a \leftarrow a + 1$.

2. The **for** iterative statement, as in **for** $i \leftarrow 1$ **to** n **do**. Variations include the keyword **downto** and expressing the iteration in informal terms (as in **for** each element of set S **do**).

3. The conditional statement **if** *<condition>* **then** ... **else** ....

4. The iterative statement **while** *<condition>* **do** ....

5. The iterative statement **repeat** ... **until** *<condition>*.

6. The **break** statement, which stops execution of the innermost enclosing iterative statement.

7. The **return** statement, which stops execution of the function where it appears and returns a value or structure to whomever called the function.

The *<condition>* in statements above can be expressed either formally or informally.

Algorithms are meant to solve all instances of the problems for which they were designed; when they do, they are said to be *correct*. Hence for any algorithm a *proof of correctness* is essential. Another important aspect of algorithms is their *running time*. The running time of an algorithm is the number of machine instructions it executes when the algorithm is run on a particular instance. Presenting a proof of correctness and an estimate of the running time are aspects of the *analysis* of the algorithm.

We estimate the running time of algorithms rather than count the exact number of machine instructions because we want to have an idea of the algorithm's behavior on *any* instance of the problem. To obtain such information two rules are necessary. The first is that the running time be given for the *worst case* instance of the problem. The second is that we make the running time depend on the size of the instance being solved. Hence, instead of being a number, the running time is a *function*. For example, we could say that the running time of algorithm A is $5n^2 + 3n + 72$, where n measures the problem size. Running time functions are simplified by dropping all constants and lower-order terms. In the example above, $3n$ and 72 are lower-order terms with respect to n^2. Hence, the running time of A becomes $O(n^2)$, where the O (big-oh) notation is used precisely to indicate that we are not caring for constants or lower-order terms. A consequence of this is that we will be able to compare algorithms only in terms of their *asymptotic behavior*. Thus an $O(n^2)$ algorithm may be faster than an $O(n)$ algorithm for small values of n; however, we know that for a sufficiently large value of n the $O(n)$ algorithm will be faster. Incidentally, algorithms whose running times are $O(n)$ are said to be *linear-time;* when the running time is $O(n^2)$ they are *quadratic*.

From the discussion in the previous paragraph it is clear that in order to compare the running time of algorithms expressed with the O notation we must be able to compare the rate of growth of functions. That is, given two functions that take integers to reals, which one grows faster than the other? We assume the reader is capable of comparing simple functions using the following formal definitions. Given two functions $f(n)$ and $g(n)$, from the integers to the reals, we say that $g(n) = O(f(n))$ when there exist constants c and n_0 such that for all $n > n_0$, $g(n) \leq cf(n)$. In other words, if $g(n) = O(f(n))$ then we know that function $g(n)$ grows no faster than function $f(n)$; f is an *upper bound* for g. For example, this is true if $g(n) = \log n$ and $f(n) = \sqrt{n}$. Given that we do not care for constants, the base of logarithms (when it is a constant) is irrelevant inside the O, because $\log_a n$ and $\log_b n$ are related by a multiplicative constant when a and b are constants.

Sometimes it is desirable to describe a function $g(n)$ as growing *at least* as fast as some other function $f(n)$. In this case we say that $g(n) = \Omega(f(n))$. Formally, this means that there exist constants c and n_0 such that for all $n > n_0$, $g(n) \geq cf(n)$. Function f provides a *lower bound* for g. When $g(n)$ is both $O(f(n))$ and $\Omega(f(n))$, we say that $g(n) = \Theta(f(n))$.

In this book we are primarily interested in *efficient* algorithms. An algorithm is efficient for a problem of size n when its running time is bounded by a polynomial $p(n)$. An algorithm whose running time is $\Omega(2^n)$ is *not* efficient, since the function 2^n grows faster than any polynomial on n.

One of the main tasks of theoretical computer scientists is to classify problems in two classes: those for which there exist efficient algorithms (which belong to class P), and those for which such algorithms do not exist (this task is known as *complexity analysis*). Unfortunately, there is a very important class of problems for which the classification is as yet undecided. Nobody has been able to find efficient algorithms for them, but neither has it been shown that no such algorithms exist. These are the **NP-complete problems**. These problems belong to class NP, the class of problems whose solution, once found, can be checked in polynomial time. This class also includes the class P as a subset. In a very precise sense, NP-complete problems are the hardest problems in class NP. This means that any instance of any problem in class NP can be transformed in polynomial time to an instance of an NP-complete problem. A particular case of this statement is that all NP-complete problems are equivalent under polynomial transformations. This in turn means that if a polynomial-time algorithm is found for just one NP-complete problem, then all NP-complete problems can be solved in polynomial time.

The undecided question alluded to above is known as the P = NP problem. In words it means: Can all problems of the class NP be solved in polynomial time? If P = NP, then there are efficient algorithms for every problem in NP, and in particular for the NP-complete problems; if P $\neq$ NP, then NP-complete problems are all computationally hard, and P is a proper subset of NP. Most computer scientists believe the latter is true.

NP-complete problems appear frequently in this book. Another related term used is **NP-hard** problem. To explain the difference between the two, we have to make more precise our statements about problem classes made above. Strictly speaking, classes P and NP include only *decision* problems: The answer must be yes or no. A typical example is the problem of deciding whether a graph is Hamiltonian. However, most problems we will meet in this book are *optimization* problems, meaning that there is some function associated with the problem and we want to find the solution that minimizes or maximizes the value of this function. A typical example is the traveling salesman problem, mentioned in Section 2.2. An optimization problem can be transformed into a decision problem by including as part of the input a parameter K and asking whether there is a solution whose value is $\leq K$ (in the case of minimization problems) or $\geq K$ (in the case of maximization problems). This trick enables us to speak of the *decision version* of every optimization problem and hence try to classify it as being in class P or in class NP. Thus it can be shown that the decision version of the traveling salesman problem is an

NP-complete problem. But how about the traveling salesman problem itself? It does not belong to class NP, since it is not a decision problem; but clearly it is at least as hard as its decision version. Problems that are at least as hard as an NP-complete problem but which do not belong to class NP are called NP-hard. All optimization problems whose decision versions are NP-complete are themselves NP-hard.

For the problems in Section 2.2, the following is known. To decide whether a graph is Eulerian is a problem that belongs to class P, given that it can be done in $O(n + m)$ time. To decide whether a graph is Hamiltonian is an NP-complete problem. As already mentioned, the traveling salesman problem is NP-hard. A minimum spanning tree can be found in time $O(n^2)$, and hence (its decision version) is in P. Vertex coloring is another NP-hard problem. Finally, the matching problems can be solved in polynomial time.

When we first meet a problem X of unknown complexity, we have two choices: We can try to find an algorithm that solves X efficiently (that is, we try to prove that the problem belongs to class P), or we can try to prove that X is NP-complete. A very brief description of how one goes about doing the latter is as follows. We first must show that the problem belongs to class NP (can its solution, once found, be checked in polynomial time?). Then we must show that a known NP-complete problem Y reduces to X. This is done by showing that if we did have a polynomial-time algorithm for solving X, then we could use it to solve Y also in polynomial time. Finding the NP-complete problem Y that allows the easiest reduction is usually the hardest part in NP-completeness proofs. The fact that more than a thousand problems have been shown to be NP-complete has made this task considerably easier.

What are we to do when we face an NP-complete or NP-hard problem? Here are some possibilities:

1. Verify whether the instances to be solved are really general. In some cases, solving an NP-hard problem on a restricted class of inputs can be done in polynomial time. A classical example is graph coloring. Finding a minimum coloring for general graphs is an NP-hard problem, but if the inputs are restricted to be bipartite graphs then it can be done in polynomial time (any bipartite graph can be colored using at most two colors).

2. Devise an exhaustive search algorithm (that is, an algorithm that enumerates all possible solutions and picks the best one). This is generally practical only for small instances of the problem, since the running time of such an algorithm is exponential.

3. Devise a polynomial-time approximation algorithm. Such an algorithm finds a solution that is guaranteed to be close to optimal, but it is not necessarily an optimal solution. The guarantee can be expressed in terms of the ratio between the value of the solution found by the algorithm and the optimal solution. For example, there exists an approximation algorithm for one version of the traveling salesman problem that is guaranteed to give a solution whose weight is no larger than 1.5 times the optimal value. This guarantee may or may not have practical significance, depending on the guarantee itself and the problem being solved.

4. Devise heuristics. These are polynomial-time algorithms with no guarantee whatsoever on the solution's value. In many cases heuristics have satisfactory performance in practice and are widely used.

Let us now leave the NP-complete problems and describe a few important problems for which efficient algorithms are known. The first of these is to verify whether an undirected graph is connected. This problem can be solved using well-known *graph traversal* techniques. One of them is *depth-first search:* Beginning at an arbitrary vertex v, visit it, then go to one of the unvisited adjacent vertices. We repeat this process until all vertices have been visited or if we get stuck (all reachable vertices have been visited, but there are still unvisited vertices). In the first case the graph is connected, and in the second it is not. Another technique is *breadth-first search,* in which we use a *queue* to process vertices. We visit all neighbors of the first vertex in the queue, then put them all in the queue for further exploration. We then get the next vertex from the queue, and so on. Both algorithms run in time $O(n+m)$. These same techniques can be used in many other graph problems; depth-first search, in particular, can be used to verify whether a graph is acyclic. Trees can also be traversed using one of these techniques. A tree is of course connected, but the techniques provide a systematic and useful way of visiting its nodes.

The minimum spanning tree (MST) problem can be solved efficiently, although with techniques that are more elaborate than simple traversals. On a graph with n vertices, we can find an MST with a simple algorithm in time $O(n^2)$. The idea is to start at an arbitrary vertex, and from it grow the tree by adding the smallest weight edge that connects a vertex in the tree to a vertex not yet in the tree. It is surprising that such a simple idea works, because we are choosing edges based purely on what appears "locally" to be the best choice. Algorithms that use such techniques are called *greedy*.

Another problem in the class P is sorting. Given a vector with n distinct numbers, sort it in increasing or decreasing order. It can be shown that sorting requires $\Theta(n \log n)$ time, when the numbers are arbitrary. If the numbers are bounded by some constant, or even by a fixed-degree polynomial (such as n^3), then the problem can be solved in linear time.

We now close this section mentioning a few ideas and tools for the design of algorithms. One powerful idea is that of *induction*. In broad terms, the idea is this. Given a problem whose instances have size n, we begin by finding a way to solve very small instances of the problem (e.g., how to sort 1 number) and by assuming that we know how to solve instances of size $n-1$. Then we see what must be done to extend the solution of instances of size $n-1$ to instances of size n. In many cases this extension is simple, and by mathematical induction it can be shown that we have solved the problem in general. When designing algorithms using this technique, the result is frequently expressed using *recursion*. In programming language terms, a recursive function is a function that calls itself (directly or indirectly). These ideas are depicted in the example of Figure 2.2.

A particularly important algorithm design technique for this book is called *dynamic programming*. The basic idea is that we obtain a solution for a problem by building it up from solutions of many smaller but similar subproblems. It can be seen as a special case of design by induction. Concrete examples are given in Chapter 3. Note that there is no relation between this technique and computer programming, in the modern sense of the word. The technique has this name for historical reasons.

Data structures are another kind of tool used in algorithm design. These are structures that store data and are designed to support certain operations like retrieval or insertion in a very efficient manner, usually in constant time or $O(\log n)$ time, where n is the number of items stored. The types of data structures used vary depending on the problem.

```
Algorithm RSort
    input: array A with n distinct elements
    output: array A sorted in increasing order
    if n = 1 then
        return A
    else
        x ← min(A)   // min returns the smallest element
        Remove x from A
        RSort(A)   // recursive call
        return x : A   // we concatenate x to the beginning of A
```

FIGURE 2.2

A recursive algorithm to sort n elements.

Among the best known data structures are special kinds of lists called *stacks* and *queues, hash tables,* and *binary search trees*. Readers unfamiliar with these structures may wish to consult books cited in the bibliographic notes. One structure used in this book but not so well known as those above is the *disjoint-set forest* data structure. It supports operations on a dynamic collection of disjoint sets, such that initially each data item is in a set by itself, and as the algorithm progresses the sets are gradually joined to each other. This data structure supports three basic operations: *MakeSet*(a), which puts element a in a set by itself; *FindSet*(a), which returns the set to which element a currently belongs; and *Union*(S, T), which joins sets S and T (for this reason, data structures that support such operations are also known as *union-find data structures*). If we have n total elements, a disjoint-set forest coupled with two other techniques allows a series of $m > n$ *FindSet* and/or *Union* operations to be done in time $O(m\alpha(m, n))$, where $\alpha(m, n)$ is a function that grows so slowly that for all practical purposes it can be considered a constant.

EXERCISES

1. How many distinct substrings does ACTAC have? how many distinct subsequences? how many intervals?
2. Design and analyze an algorithm that checks whether a string p is a substring of another string t. Can you design an algorithm that runs in time $O(|p| + |t|)$?
3. Write a computer program using your favorite programming language that accepts as input a DNA sequence of *any* size and returns its reverse complement.
4. How many edges does an n-vertex complete directed graph have? how many if the graph is undirected? How many edges does an n-node tree have?
5. Show that in an undirected graph $G = (V, E)$, $\sum_{v \in V} \delta(v) = 2|E|$, where $\delta(v)$ is the degree of vertex v.
6. Give adjacency matrix and adjacency list representations for the graphs depicted in Figure 2.1.

7. Explain how an $O(n^2)$ algorithm can be faster than an $O(n)$ algorithm for small values of n.

8. Say you have two algorithms A_1 and A_2 for the same problem, whose size is defined by parameter n. If A_1 runs in time $O(n/\log n)$ and A_2 runs in time $O(\sqrt{n})$, which one is the fastest asymptotically?

9. Leonhard Euler showed that a graph is Eulerian if and only if it is connected and all vertices have even degree. Based on this observation, propose an algorithm that finds an Eulerian cycle in a graph if such a cycle exists.

10. Show that in a weighted undirected graph where all edge costs are distinct the minimum spanning tree is unique.

11. Completely design an algorithm for the minimum spanning tree problem based on the information given in the text. Analyze your algorithm.

12. A *topological ordering* for the vertices of a directed graph G is such that for all edges (u, v), u appears before v in the ordering. Topological orderings are well defined only for directed acyclic graphs. Design and analyze an algorithm that determines a topological ordering of a directed acyclic graph G.

13. Assume you have a disjoint-set forest data structure. Use it to design an algorithm that will determine the connected components of an undirected graph G. Compare your algorithm with another based on breadth-first or depth-first search.

14. Suppose you have a problem X whose complexity is unknown. You succeed in reducing it to a known NP-complete problem. What can you say about X's complexity?

15. The *vertex cover problem* is defined over an undirected graph $G = (V, E)$ and asks for a minimum set of vertices $W \subseteq V$ such that for any edge $e \in E$, at least one of its endpoints belongs to W. Write out the decision version of this problem.

16. A *clique* in a graph G is a subgraph of G that is complete. The *clique problem* is defined over an undirected graph $G = (V, E)$ and asks for a maximum set of vertices $C \subseteq V$ such that C is a clique. Given that the vertex cover problem defined in the previous exercise is NP-complete, show that the clique problem is also NP-complete.

17. Given that the Hamiltonian cycle problem is NP-complete, show that the Hamiltonian path problem between two given vertices u and v is NP-complete.

18. An approximation algorithm for the vertex cover problem (defined in Exercise 15) is the following. Begin with an empty cover W. Build up a vertex cover by choosing any edge (u, v) and adding both u and v to W. Then delete u and v from G (that is, remove any other edges incident to them), and go on repeating these steps until no more edges are left. Show that this algorithm finds a vertex cover whose size is at most twice the size of the minimum vertex cover.

19. Design and analyze an exhaustive search algorithm for the Hamiltonian cycle problem.

20. The *knapsack problem* is the following. Given a set of n objects, object i having weight w_i, and a positive integer K (the knapsack's capacity), determine whether there exists a subset of the objects whose total weight is exactly K. This is an NP-complete problem, but it is possible to design an algorithm for it that runs in time $O(nK)$. There is no contradiction here, because the value of K may be arbitrarily large (for example, it could be 2^n), and hence this algorithm is not strictly polynomial. Design such an algorithm using the dynamic programming technique mentioned in the text.

BIBLIOGRAPHIC NOTES

Two books are devoted solely to algorithms on strings, one by Stephen [180] and the other by Crochemore and Rytter [39]. A useful survey is given by Aho [6]. The forthcoming book by Gusfield [85] should be especially relevant for computational biology.

Readers unfamiliar with graph theory might wish to read a chapter on this subject in a college textbook on discrete mathematics. One such book, by Rosen [163], contains additional references on books devoted to graph theory. Our notation and exposition of graph theory concepts were based on Chapter 1 of Tarjan's book [184].

Two excellent books on algorithms are by Manber [129] and by Cormen, Leiserson, and Rivest [37]. Manber stresses the use of induction in the design of algorithms, while Cormen et al. provide a more encyclopedic coverage. Sedgewick [170] and Gonnet and Baeza-Yates [76] are handy references on algorithms for a variety of problems.

The classical reference on NP-complete problems and the general theory of NP-completeness is the book by Garey and Johnson [68]. A more recent book that covers computational complexity as a whole is by Papadimitriou [151].

CHAPTER

3

SEQUENCE COMPARISON AND DATABASE SEARCH

In this chapter we present some of the most practical and widely used methods for sequence comparison and database search. Sequence comparison is undoubtedly the most basic operation in computational biology. It appears explicitly or implicitly in all subsequent chapters of this book. It has also many applications in other subareas of computer science.

3.1 BIOLOGICAL BACKGROUND

Why compare sequences? Why use a computer to do that? How to compare sequences using computers? These are the main questions that will concern us in this chapter. We will answer the first two in this section, and devote the rest of the chapter to the last one.

Sequence comparison is the most important primitive operation in computational biology, serving as a basis for many other, more complex, manipulations. Roughly speaking, this operation consists of finding which parts of the sequences are alike and which parts differ. However, behind this apparently simple concept, a great variety of truly distinct problems exist, with diverse formalizations and sometimes requiring completely different data structures and algorithms for an efficient solution.

As examples, we will give a list of problems that often appear in computational biology. In these examples we use two notions that will be precisely defined in later sections. One is the *similarity* of two sequences, which gives a measure of how similar the sequences are. The other is the *alignment* of two sequences, which is a way of placing one sequence above the other in order to make clear the correspondence between similar characters or substrings from the sequences. The examples (and the whole chapter) also use basic concepts about strings, which are defined in Section 2.1. Here are the examples.

1. We have two sequences over the same alphabet, both about the same length (tens of thousands of characters). We know that the sequences are almost equal, with only a few isolated differences such as insertions, deletions, and substitutions of characters. The average frequency of these differences is low, say, one each hundred characters. We want to find the places where the differences occur.

2. We have two sequences over the same alphabet with a few hundred characters each. We want to know whether there is a prefix of one which is similar to a suffix of the other. If the answer is yes, the prefix and the suffix involved must be produced.

3. We have the same problem as in (2), but now we have several hundred sequences that must be compared (each one against all). In addition, we know that the great majority of sequence pairs are unrelated, that is, they will not have the required degree of similarity.

4. We have two sequences over the same alphabet with a few hundred characters each. We want to know whether there are two substrings, one from each sequence, that are similar.

5. We have the same problem as in (4), but instead of two sequences we have one sequence that must be compared to thousands of others.

Problems like (1) appear when, for instance, the same gene is sequenced by two different labs and they want to compare the results; or even when the same long sequence is typed twice into the computer and we are looking for typing errors. Problems like (2) and (3) appear in the context of fragment assembly in programs to help large-scale DNA sequencing. Problems like (4) and (5) occur in the context of searches for local similarities using large biosequence databases.

We will see in this chapter that a single basic algorithmic idea can be used to solve all of the above problems. However, this may not be the most efficient solution. Sometimes less general but faster methods are better suited to each task.

The use of computers hardly needs justification when we deal with large quantities of data, as in searches in large databases. Yet, even in cases where we could conceivably do comparisons "by hand," the use of computers is safer and more convenient, as the following examples, taken from a paper by Smith and coworkers [176], illustrate.

In a 1979 paper, Sims and colleagues examined similar regions from the DNA of two bacteriophages [173]. They presented an alignment between regions of the H-gene from phages St-1 and G4 containing 11 matches (that is, 11 columns in which there are equal characters in both sequences). The procedure they used to obtain good alignments was described as the "insertion of occasional gaps to maximize homology [number of identities]." Smith and colleagues [176], however, produced 12 matches in an alignment found by computer, which was certainly overlooked by the first group of researchers.

In another 1979 paper, Rosenberg and Court completed a study aligning 46 promoter sequences from various organisms [164]. The multiple alignment was constructed starting from the alignment of certain regions reputed relatively invariant and then extending this "seed" alignment in both directions without introducing gaps. Although the resulting alignment was good as a whole, the pairwise alignments obtained from it were relatively poor. This happens fairly frequently in multiple alignments, but in this case Smith and colleagues found, with the help of a computer, an alignment between two substrings of the sequences involved that had 44 identical characters over a total of 45. This is certainly an astonishingly good local alignment, and the fact that it is not even mentioned by Rosenberg and Court must mean again that they were not able to find it by hand.

These two examples show that the use of computers can reveal intriguing similarities that might go unnoticed otherwise.

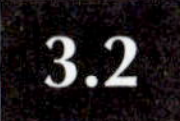

COMPARING TWO SEQUENCES 3.2

In this section we study methods for comparing two sequences. More specifically, we are interested in finding the best alignments between these two sequences. Several versions of this problem occur in practice, depending on whether we are interested in alignments involving the entire sequences or just substrings of them. This leads to the definition of global and local comparisons. There is also a third kind of comparison in which we are interested in aligning not arbitrary substrings, but prefixes and suffixes of the given sequences. We call this third kind semiglobal comparison. All the problems mentioned can be solved efficiently by dynamic programming, as we will see in the sequel.

3.2.1 GLOBAL COMPARISON — THE BASIC ALGORITHM

Consider the following DNA sequences: `GACGGATTAG` and `GATCGGAATAG`. We cannot help but notice that they actually look very much alike, a fact that becomes more obvious when we align them one above the other as follows.

```
GA-CGGATTAG
GATCGGAATAG
```
(3.1)

The only differences are an extra `T` in the second sequence and a change from `A` to `T` in the fourth position from right to left. Observe that we had to introduce a space (indicated by a dash above) in the first sequence to let equal bases before and after the space in the two sequences align perfectly.

Our goal in this section is to present an efficient algorithm that takes two sequences and determines the best alignment between them as we did above. Of course, we must define what the "best" alignment is before approaching the problem. To simplify the discussion, we will adopt a simple formalism; later, we will see possible generalizations.

To begin, let us be precise about what we mean by an *alignment* between two sequences. The sequences may have different sizes. As we saw in the recent example, alignments may contain spaces in any of the sequences. Thus, we define an alignment as the insertion of spaces in arbitrary locations along the sequences so that they end up with the same size. Having the same size, the augmented sequences can now be placed one over the other, creating a correspondence between characters or spaces in the first sequence and characters or spaces in the second sequence. In addition, we require that no space in one sequence be aligned with a space in the other. Spaces can be inserted even in the beginning or end of sequences.

Given an alignment between two sequences, we can assign a *score* to it as follows. Each column of the alignment will receive a certain value depending on its contents and the total score for the alignment will be the sum of the values assigned to its columns. If a column has two identical characters, it will receive value $+1$ (a *match*). Different characters will give the column value -1 (a *mismatch*). Finally, a space in a column drops down its value to -2. The best alignment will be the one with maximum total score. This maximum score will be called the *similarity* between the two sequences and will be denoted by $\text{sim}(s, t)$ for sequences s and t. In general, there may be many alignments with maximum score.

To exercise these definitions, let us compute the score of alignment (3.1). There are nine columns with identical characters, one column with distinct characters, and one column with a space, giving a total score of

$$9 \times 1 + 1 \times (-1) + 1 \times (-2) = 6.$$

Why did we choose these particular values ($+1$, -1, and -2)? This scoring system is often used in practice. We reward matches and penalize mismatches and spaces. In Section 3.6.2 we discuss in more detail the choice of these parameters.

One approach to computing the similarity between two sequences would be to generate all possible alignments and then pick the best one. However, the number of alignments between two sequences is exponential, and such an approach would result in an intolerably slow algorithm. Fortunately, a much faster algorithm exists, which we will now describe.

The algorithm uses a technique known as *dynamic programming*. It basically consists of solving an instance of a problem by taking advantage of already computed solutions for smaller instances of the same problem. Given two sequences s and t, instead of determining the similarity between s and t as whole sequences only, we build up the solution by determining all similarities between arbitrary *prefixes* of the two sequences. We start with the shorter prefixes and use previously computed results to solve the problem for larger prefixes.

Let m be the size of s and n the size of t. There are $m + 1$ possible prefixes of s and $n + 1$ prefixes of t, including the empty string. Thus, we can arrange our calculations in an $(m + 1) \times (n + 1)$ array where entry (i, j) contains the similarity between $s[1..i]$ and $t[1..j]$.

Figure 3.1 shows the array corresponding to $s =$ AAAC and $t =$ AGC. We placed s along the left margin and t along the top to indicate the prefixes more easily. Notice that the first row and the first column are initialized with multiples of the space penalty (-2 in our case). This is because there is only one alignment possible if one of the sequences is empty: Just add as many spaces as there are characters in the other sequence. The score of this alignment is $-2k$, where k is the length of the nonempty sequence. Hence, filling the first row and column is easy.

Now let us concentrate on the other entries. The key observation here is that we can compute the value for entry (i, j) looking at just three previous entries: those for $(i - 1, j)$, $(i - 1, j - 1)$, and $(i, j - 1)$. The reason is that there are just three ways of obtaining an alignment between $s[1..i]$ and $t[1..j]$, and each one uses one of these previous values. In fact, to get an alignment for $s[1..i]$ and $t[1..j]$, we have the following three choices:

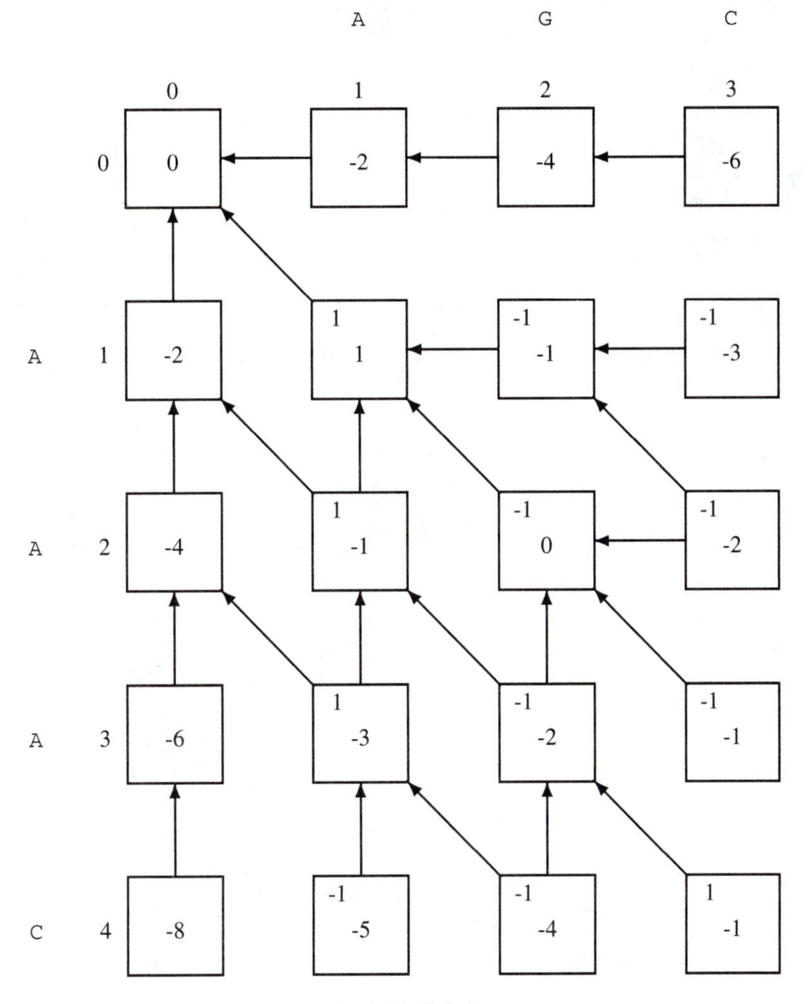

FIGURE 3.1

Bidimensional array for computing optimal alignments. The value in the upper left corner of cell (i, j) indicates whether $s[i] = t[j]$. Indices of rows and columns start at zero.

- Align $s[1..i]$ with $t[1..j-1]$ and match a space with $t[j]$, or
- Align $s[1..i-1]$ with $t[1..j-1]$ and match $s[i]$ with $t[j]$, or
- Align $s[1..i-1]$ with $t[1..j]$ and match $s[i]$ with a space.

These possibilities are exhaustive because we cannot have two spaces paired in the last column of the alignment. Scores of the best alignments between smaller prefixes are already stored in the array if we choose an appropriate order in which to compute the entries. As a consequence, the similarity sought can be determined by the formula

$$\mathrm{sim}(s[1..i], t[1..j]) = \max \begin{cases} \mathrm{sim}(s[1..i], t[1..j-1]) - 2 \\ \mathrm{sim}(s[1..i-1], t[1..j-1]) + p(i, j) \\ \mathrm{sim}(s[1..i-1], t[1..j]) - 2, \end{cases} \tag{3.2}$$

where $p(i, j)$ is $+1$ if $s[i] = t[j]$ and -1 if $s[i] \neq t[j]$. The values of $p(i, j)$ are written in the upper left corners of the boxes in Figure 3.1. If we denote the array by a, this equation can be rewritten as follows:

$$a[i, j] = \max \begin{cases} a[i, j-1] - 2 \\ a[i-1, j-1] + p(i, j) \\ a[i-1, j] - 2. \end{cases} \tag{3.3}$$

As we mentioned earlier, a good computing order must be followed. This is easy to accomplish. Filling in the array either row by row, left to right on each row, or column by column, top to bottom on each column, suffices. Any other order that makes sure $a[i, j-1]$, $a[i-1, j-1]$, and $a[i-1, j]$ are available when $a[i, j]$ must be computed is fine, too.

Finally, we drew arrows in Figure 3.1 to indicate where the maximum value comes from according to Equation (3.3). For instance, the value of $a[1, 2]$ was taken as the maximum among the following figures.

$$a[1, 1] - 2 = -1$$
$$a[0, 1] - 1 = -3$$
$$a[0, 2] - 2 = -6.$$

Therefore, there is only one way of getting this maximum value, namely, coming from entry $(1, 1)$, and that is what the arrows show.

Figure 3.2 presents an algorithm for filling in an array as explained. This algorithm computes entries row by row. It depends on a parameter g that specifies the space penalty

```
Algorithm Similarity
    input: sequences s and t
    output: similarity between s and t
    m ← |s|
    n ← |t|
    for i ← 0 to m do
        a[i, 0] ← i × g
    for j ← 0 to n do
        a[0, j] ← j × g
    for i ← 1 to m do
        for j ← 1 to n do
            a[i, j] ← max(a[i − 1, j] + g,
                          a[i − 1, j − 1] + p(i, j),
                          a[i, j − 1] + g)
    return a[m, n]
```

FIGURE 3.2

Basic dynamic programming algorithm for comparison of two sequences.

(usually $g < 0$) and on a scoring function p for pairs of characters. We have used $g = -2$, $p(a, b) = 1$ if $a = b$, and $p(a, b) = -1$ if $a \neq b$ in Figure 3.1.

Optimal Alignments

We saw how to compute the similarity between two sequences. Here we will see how to construct an optimal alignment between them. The arrows in Figure 3.1 will be useful in this respect. All we need to do is start at entry (m, n) and follow the arrows until we get to $(0, 0)$. Each arrow used will give us one column of the alignment. In fact, consider an arrow leaving entry (i, j). If this arrow is horizontal, it corresponds to a column with a space in s matched with $t[j]$; if it is vertical, it corresponds to $s[i]$ matched with a space in t; finally, a diagonal arrow means $s[i]$ matched with $t[j]$. Observe that the first sequence, s, is always placed along the vertical edge. Thus, an optimal alignment can be easily constructed from right to left if we have the matrix a computed by the basic algorithm. It is not necessary to implement the arrows explicitly — a simple test can be used to choose the next entry to visit.

Figure 3.3 shows a recursive algorithm for determining an optimal alignment given the matrix a and sequences s and t. The call *Align*(m, n, len) will construct an optimal alignment. The answer will be given in a pair of vectors *align-s* and *align-t* that will hold in the positions $1..len$ the aligned characters, which can be either spaces or symbols from the sequences. The variables *align-s* and *align-t* are treated as globals in the code. The

Algorithm *Align*
  input: indices i, j, array a given by algorithm *Similarity*
  output: alignment in *align-s*, *align-t*, and length in *len*
  if $i = 0$ **and** $j = 0$ **then**
    $len \leftarrow 0$
  else if $i > 0$ **and** $a[i, j] = a[i-1, j] + g$ **then**
    $Align(i-1, j, len)$
    $len \leftarrow len + 1$
    $align\text{-}s[len] \leftarrow s[i]$
    $align\text{-}t[len] \leftarrow$ -
  else if $i > 0$ **and** $j > 0$ **and** $a[i, j] = a[i-1, j-1] + p(i, j)$ **then**
    $Align(i-1, j-1, len)$
    $len \leftarrow len + 1$
    $align\text{-}s[len] \leftarrow s[i]$
    $align\text{-}t[len] \leftarrow t[j]$
  else // has to be $j > 0$ and $a[i, j] = a[i, j-1] + g$
    $Align(i, j-1, len)$
    $len \leftarrow len + 1$
    $align\text{-}s[len] \leftarrow$ -
    $align\text{-}t[len] \leftarrow t[j]$

FIGURE 3.3

Recursive algorithm for optimal alignment.

length of the alignment is also returned by the algorithm in the parameter *len*. Note that $\max(|s|, |t|) \leq len \leq m + n$.

As we said previously, many optimal alignments may exist for a given pair of sequences. The algorithm of Figure 3.3 returns just one of them, giving preference to the edges leaving (i, j) in counterclockwise order (see Figure 3.4).

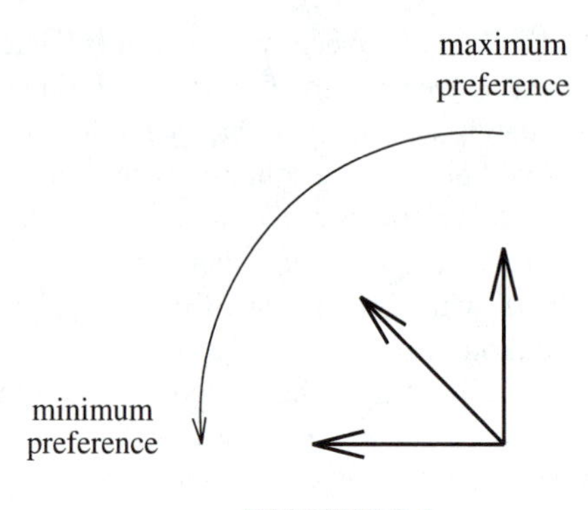

FIGURE 3.4

Arrow preference.

As a result, the optimal alignment returned by this algorithm has the following general characteristic: When there is choice, a column with a space in t has precedence over a column with two symbols, which in turn has precedence over a column with a space in s. For instance, when aligning $s =$ `ATAT` with $t =$ `TATA`, we get

```
-ATAT
TATA-
```

rather than

```
ATAT-
-TATA
```

which is the other optimal alignment for these two sequences. Similarly, when aligning $s =$ `AA` with $t =$ `AAAA`, we get

```
--AA
AAAA
```

although there are five other optimal alignments. This alignment is sometimes referred to as the *upmost* alignment because it uses the arrows higher up in the matrix. To reverse these preferences, we would reverse the order of the **if** statements in the code, obtaining the *downmost* alignment in this case. A column appearing in both the upmost and downmost alignments will be present in all optimal alignments between the two sequences considered.

It is possible to modify the algorithm to produce *all* optimal alignments between s and t. We need to keep a stack with the points at which there are options and backtrack to them to explore all possibilities of reaching (0, 0) through the arrows. However, there might be a very large number of optimal alignments. In these cases, it is often advisable to keep some sort of short representation of all or part of these alignments, for instance, the upmost and downmost ones.

Let us now determine the complexity of the algorithms described in this section. The basic algorithm of Figure 3.2 has four loops. The first two do the initialization and consume time $O(m)$ and $O(n)$, respectively. The last two loops are nested and fill the rest of the matrix. The number of operations performed depends essentially on the number of entries that must be computed, that is, the size of the matrix. Thus, we spend time $O(mn)$ in this part and this is the dominant term in the time complexity. The space used is also proportional to the size of the matrix. Hence, the complexity of the basic algorithm is $O(mn)$ both for time and space. If the sequences have the same or nearly the same length, say n, we get $O(n^2)$. That is why we say that these algorithms have quadratic complexity.

The construction of the alignment — given the already filled matrix — is done in time $O(len)$, where *len* is the size of the returned alignment, which is $O(m + n)$.

3.2.2 LOCAL COMPARISON

A *local alignment* between s and t is an alignment between a substring of s and a substring of t. In this section we present an algorithm to find the highest scoring local alignments between two sequences.

This algorithm is a variation of the basic algorithm. The main data structure is, as before, an $(m + 1) \times (n + 1)$ array. Only this time the interpretation of the array values is different. Each entry (i, j) will hold the highest score of an alignment between a *suffix* of $s[1..i]$ and a *suffix* of $t[1..j]$. The first row and the first column are initialized with zeros.

For any entry (i, j), there is always the alignment between the empty suffixes of $s[1..i]$ and $t[1..j]$, which has score zero; therefore the array will have all entries greater than or equal to zero. This explains in part the initialization above.

Following initialization, the array can be filled in the usual way, with $a[i, j]$ depending on the value of three previously computed entries. The resulting recurrence is

$$a[i, j] = \max \begin{cases} a[i, j-1] + g \\ a[i-1, j-1] + p(i, j) \\ a[i-1, j] + g \\ 0, \end{cases}$$

that is, the same as in the basic algorithm, except that now we have a fourth possibility, not available in the global case, of an empty alignment.

In the end, it suffices to find the maximum entry in the whole array. This will be the score of an optimal local alignment. Any entry containing this value can be used as a starting point to get such an alignment. The rest of the alignment is obtained tracing back as usual, but stopping as soon as we reach an entry with no arrow going out. Alternatively, we can stop as soon as we reach an entry with value zero.

In general, when doing local comparison, we are interested not only in the optimal alignments, but also in near optimal alignments with scores above a certain threshold. References to methods for retrieving near optimal alignments are given in the bibliographic notes.

3.2.3 SEMIGLOBAL COMPARISON

In a semiglobal comparison, we score alignments ignoring some of the *end spaces* in the sequences. An interesting characteristic of the basic dynamic programming algorithm is that we can control the penalty associated with end spaces by doing very simple modifications to the original scheme.

Let us begin by defining precisely what we mean by *end spaces* and why it might be better to let them be included for free in certain situations. End spaces are those that appear before the first or after the last character in a sequence. For instance, all the spaces in the second sequence in the alignment below are end spaces, while the single space in the first sequence is not an end space.

```
CAGCA-CTTGGATTCTCGG
---CAGCGTGG--------
```
(3.4)

Notice that the lengths of these two sequences differ considerably. One has size 8, and the other has 18 characters. When this happens, there will be many spaces in any alignment, giving a large negative contribution to the score. Nevertheless, if we ignore end spaces, the alignment is pretty good, with 6 matches, 1 mismatch, and 1 space.

Observe that this is not the best alignment between these sequences. In alignment (3.5) below we present another alignment with a higher score (-12 against -19 of the previous one) according to the scoring system we have been using so far.

```
CAGCACTTGGATTCTCGG
CAGC-----G-T----GG
```
(3.5)

In spite of having scored higher and having matched all characters of the second sequence with identical characters in the first one, this alignment is not so interesting from the point of view of finding similar regions in the sequences. The second sequence was simply torn apart brutally by the spaces just for the sake of matching exactly its characters. If we are looking for regions of the longer sequence that are approximately the same as the shorter sequence, then undoubtedly the first alignment (3.4) is more to the point. This is reflected in the scores obtained when we disregard (that is, not charge for) end spaces: (3.4) gets 3 points against the same -12 for (3.5).

Let us now describe a variation of the basic algorithm that will ignore end spaces. Consider initially the case where we do not want to charge for spaces after the last character of s. Take an optimal alignment in this case. The spaces after the end of s are matched to a suffix of t. If we remove this final part of the alignment, the remaining is an alignment between s and a prefix of t, with score equal to the original alignment. Therefore, to get the score of the optimal alignment between s and t without charge for spaces after the end of s, all we need to do is to find the best similarity between s and a prefix of t. But we saw in Section 3.2.1 that the entry (i, j) of matrix a contains the similarity between $s[1..i]$ and $t[1..j]$. Hence, it suffices to take the maximum value in the last row of a, that is,

$$\mathrm{sim}(s, t) = \max_{j=1}^{n} a[m, j].$$

Notice that in this section we have changed the definition of $\mathrm{sim}(s, t)$ so that it now indicates similarity, ignoring final spaces in s. The expression above gives the score of the best alignment. To recover the alignment itself, we proceed just as in the basic algorithm, but starting at (m, k) where k is such that $\text{sim}(s, t) = a[m, k]$.

An analogous argument solves the case in which we do not charge for final spaces in t. We take the maximum along the last column of a in this case. We can even combine the two ideas and seek the best alignment without charging for final spaces in either sequence. The answer will be found by taking the maximum along the border of the matrix formed by the union of the last row and the last column. In all cases, to recover an optimal alignment, we start at an array entry that contains the similarity value and follow the arrows until we reach (0, 0). Each arrow will give one column of the alignment as in the basic case.

Now let us turn our attention to the case of initial spaces. Suppose that we want the best alignment that does not charge for initial spaces in s. This is equivalent to the best alignment between s and a suffix of t. To get the desired answer, we use an $(m + 1) \times (n + 1)$ array just as in the basic algorithm, but with a slight difference. Each entry (i, j) now will contain the highest similarity between $s[1..i]$ and a suffix of a prefix, which is to say a suffix of $t[1..j]$.

Doing that, it is clear that $a[m, n]$ will be the answer. What is less clear, but nevertheless true, is that the array can be filled in using exactly the same formula as in the basic algorithm! That is Equation (3.3). The initialization will be different, however. The first row must be initialized with zeros instead of multiples of the space penalty because of the new meaning of the entries. We leave it to the reader to verify that Equation (3.3) works in this case.

We can apply the same trick and initialize the first column with zeros, and by doing this we will be forgiving spaces before the beginning of t. If in addition we initialize both the first row and the first column with zeros, and proceed with Equation (3.3) for the other entries, we will be computing in each entry the highest similarity between a suffix of s and a suffix of t. To find an optimal alignment, we follow the arrows from the maximum value entry until we reach one of the borders initialized with zeros, and then follow the border back to the origin.

Table 3.1 summarizes these variations. There are four places where we may not want to charge for spaces: beginning or end of s, and beginning or end of t. We can combine these conditions independently in any way and use the variations above to find the similarity. The only things that change are the initialization and where to look for the maximum value. Forgiving initial spaces translates into initializing certain positions with zero. Forgiving final spaces means looking for the maximum along certain positions. But filling in the array is always the same process, using Equation (3.3).

TABLE 3.1

Summary of end space charging procedures.

Place where spaces are not charged for	*Action*
Beginning of first sequence	Initialize first row with zeros
End of first sequence	Look for maximum in last row
Beginning of second sequence	Initialize first column with zeros
End of second sequence	Look for maximum in last column

3.3 EXTENSIONS TO THE BASIC ALGORITHMS

The algorithms presented in Section 3.2 are generally adequate for most applications. Sometimes, however, we have a special situation and we need a better algorithm. In this section we study a few techniques that can be employed in some of these cases.

One kind of improvement is related to the problem's computational complexity. We show that it is possible to reduce the space requirements of the algorithms from quadratic (mn) to linear ($m+n$), at a cost of roughly doubling computation time. On the other hand, we show a way of reducing also the time complexity, but this only works for similar sequences and for a certain family of scoring parameters.

Another improvement has to do with the biological interpretation of alignments. From the biological point of view, it is more realistic to consider a series of consecutive spaces instead of individual spaces. We study variants of the basic algorithms adapted to this point of view.

3.3.1 SAVING SPACE

The quadratic complexity of the basic algorithms makes them unattractive in some applications involving very long sequences or repeated comparison of several sequences. No algorithm is known that uses asymptotically less time and has the same generality, although faster algorithms exist if we restrict ourselves to particular choices of the parameters.

With respect to space, however, it is possible to improve complexity from quadratic to linear and keep the same generality. The price to pay is an increase in processing time, which will roughly double. Nevertheless, the asymptotic time complexity is still the same, and in many cases space and not time is the limiting factor, so this improvement is of great practical value. In this section we describe this elegant space-saving technique.

We begin by noticing that computing $\text{sim}(s, t)$ can be easily done in linear space. Each row of the matrix depends only on the preceding one, and it is possible to perform the calculations keeping only one vector in memory, which will hold partly the new row being computed and partly the previous row. This is done in the code shown in Figure 3.5. Obviously, the same is valid for columns, and if $m < n$, using this trick with the columns uses less space. Notice that at the end of each iteration in the loop on i in Figure 3.5 the vector a contains the similarities between $s[1..i]$ and all prefixes of t. This fact will be used later.

The hard part is to get an optimal alignment in linear space. The algorithm we saw earlier depends on the whole matrix to do its job. To remove this difficulty, we use a *divide and conquer* strategy, that is, we divide the problem into two smaller subproblems and later combine their solutions to obtain a solution for the whole problem.

The key idea is the following. Fix an optimal alignment and a position i in s, and consider what can possibly be matched with $s[i]$ in this alignment. There are only two possibilities:

```
Algorithm BestScore
    input: sequences s and t
    output: vector a
    m ← |s|
    n ← |t|
    for j ← 0 to n do
        a[j] ← j × g
    for i ← 1 to m do
        old ← a[0]
        a[0] ← i × g
        for j ← 1 to n do
            temp ← a[j]
            a[j] ← max(a[j] + g,
                       old + p(i, j),
                       a[j − 1] + g)
            old ← temp
```

FIGURE 3.5

Algorithm for similarity in linear space. In the end, $a[n]$ *contains* $\mathrm{sim}(s, t)$.

1. The symbol $t[j]$ will match $s[i]$, for some j in $1..n$.
2. A space between $t[j]$ and $t[j+1]$ will match $s[i]$, for some j in $0..n$.

In the second case the index j varies between 0 and n because there is always one more position for spaces than for symbols in a sequence. We also abused notation when $j = 0$ or $j = n$. What we mean in these cases is that the space will be before $t[1]$ or after $t[n]$, respectively.

Let

$$Optimal\begin{pmatrix} x \\ y \end{pmatrix}$$

denote an optimal alignment between x and y. Every alignment between s and t, optimal or not, satisfies (1) or (2). In particular, our fixed optimal alignment must satisfy one of these as well. If it satisfies (1), to obtain all of it we must concatenate

$$Optimal\begin{pmatrix} s[1..i-1] \\ t[1..j-1] \end{pmatrix} + \begin{matrix} s[i] \\ t[j] \end{matrix} + Optimal\begin{pmatrix} s[i+1..m] \\ t[j+1..n] \end{pmatrix}, \quad (3.6)$$

while in case (2) we must concatenate

$$Optimal\begin{pmatrix} s[1..i-1] \\ t[1..j] \end{pmatrix} + \begin{matrix} s[i] \\ - \end{matrix} + Optimal\begin{pmatrix} s[i+1..m] \\ t[j+1..n]) \end{pmatrix}. \quad (3.7)$$

These considerations give us a recursive method to compute an optimal alignment, as long as we can determine, for a given i, which one of the cases (1) or (2) occurs and what is the corresponding value of j.

This can be done as follows. According to Equations (3.6) and (3.7) we need, for fixed i, the similarities between $s[1..i-1]$ and an arbitrary prefix of t, and also the similarities between $s[i+1..m]$ and an arbitrary suffix of t. If we had these values, we could explicitly compute the scores of the j alignments represented in (3.6) and of the $j+1$ alignments represented in (3.7). By choosing the best among these, we will have the information necessary to proceed in the recursion.

As we saw earlier, it is possible to compute in linear space the best scores between a given prefix of s and all prefixes of t (see Figure 3.5). A similar algorithm exists for suffixes. Hence, our problem is almost solved. The only thing left is to decide which value of i to use in each recursive call. The best choice is to pick i as close as possible to the middle of the sequence. The complete code appears in Figure 3.6. In this code, the call

$$BestScore(s[a..i-1], t[c..d], pref\text{-}sim)$$

returns in *pref-sim* the similarities between $s[a..i-1]$ and $t[c..j]$ for all j in $c-1..d$. Analogously, the call

$$BestScoreRev(s[i+1..b], t[c..d], suff\text{-}sim)$$

returns in *suff-sim* the similarities between $s[i+1..b]$ and $t[j+1..d]$ for all j in $c-1..d$. The call

$$Align(1, m, 1, n, 1, len)$$

will return an optimal alignment in the global variables *align-s* and *align-t*, and the size of this alignment in *len*.

One last concern remains: Can the processing time go up too much with these additional calculations? Not really. In fact, the time roughly doubles, as we show below.

Let $T(m,n)$ be the number of times a maximum is computed in the internal loop of *BestScore* or *BestScoreRev* as a result of a call *Align*$(a, b, c, d, start, end)$ where $m = b-a+1$ and $n = c-d+1$. It is easy to see that the total processing time will be proportional to $T(m,n)$ plus linear terms due to control and initializations. We claim that $T(m,n) \le 2mn$.

A proof can be developed by induction on m. For $m = 1$ no maximum computations will occur, so obviously $T(1,n) \le 2n$. For $m > 1$ we will have a call to *BestScore* with at most $mn/2$ maximum computations, another such amount for *BestScoreRev*, and two recursive calls to *Align*, producing at most $T(m/2, j)$ and $T(m/2, n-j)$ maximum computations. Adding this all up, we have

$$\begin{aligned} T(m,n) &\le \frac{mn}{2} + \frac{mn}{2} + T(m/2, j) + T(m/2, n-j) \\ &\le mn + mj + m(n-j) \\ &= 2mn, \end{aligned}$$

proving the claim.

3.3.2 GENERAL GAP PENALTY FUNCTIONS

Let us define a **gap** as being a consecutive number $k > 1$ of spaces. It is generally accepted that, when mutations are involved, the occurrence of a gap with k spaces is more probable than the occurrence of k isolated spaces. This is because a gap may be due to a

```
Algorithm Align
    input: sequences s and t, indices a, b, c, d, start position start
    output: optimal alignment between s[a..b] and t[c..d] placed in vectors align-s
        and align-t beginning at position start and ending at end
    if s[a..b] empty or t[c..d] empty then
        // Base case: s[a..b] empty or t[c..d] empty
        Align the nonempty sequence with spaces
        end ← start + max(|s|, |t|)
    else
        // General case
        i ← ⌊(a + b)/2⌋
        BestScore(s[a..(i − 1)], t[c..d], pref-sim)
        BestScoreRev(s[(i + 1)..b], t[c..d], suff-sim)

        posmax ← c − 1
        typemax ← SPACE
        vmax ← pref-sim[c − 1] + g + suff-sim[c − 1]
        for j ← c to d do
            if pref-sim[j − 1] + p(i, j) + suff-sim[j] > vmax then
                posmax ← j
                typemax ← SYMBOL
                vmax ← pref-sim[j − 1] + p(i, j) + suff-sim[j]

            if pref-sim[j] + g + suff-sim[j] > vmax then
                posmax ← j
                typemax ← SPACE
                vmax ← pref-sim[j] + g + suff-sim[j]

        if typemax = SPACE then
            Align(a, i − 1, c, posmax, start, middle)
            align-s[middle] ← s[i]
            align-t[middle] ← SPACE
            Align(i + 1, b, posmax + 1, d, middle + 1, end)
        else // typemax = SYMBOL
            Align(a, i − 1, c, posmax − 1, start, middle)
            align-s[middle] ← s[i]
            align-t[middle] ← t[posmax]
            Align(i + 1, b, posmax + 1, d, middle + 1, end)
```

FIGURE 3.6

An optimal alignment algorithm that uses linear space.

single mutational event that removed a whole stretch of residues, while separated spaces are most probably due to distinct events, and the occurrence of one event is more common than the occurrence of several events.

Up to now, we have not made any distinction between clustered or isolated spaces. This means that a gap is penalized through a linear function. Denoting by $w(k)$, for $k \geq 1$, the penalty associated with a gap with k spaces, we have

$$w(k) = bk,$$

where b is the absolute value of the score associated with a space.

In this section we present an algorithm that computes similarities with respect to general gap penalty functions w. The algorithm has time complexity $O(n^3)$ for sequences of length n and is therefore slower than the basic algorithm. The algorithm still has a lot in common with the basic algorithm. The main difference is the fact that the scoring scheme is not *additive,* in the sense that we cannot break an alignment in two parts and expect the total score to be the sum of the partial scores. In particular, we cannot separate the last column of an alignment and expect the alignment score to be the sum of the score for this last column and the score for the remaining prefix of the alignment. However, score additivity is still valid if we break the alignment in block boundaries. Every alignment can be uniquely decomposed into a number of consecutive *blocks.* There are three kinds of blocks, enumerated below.

1. Two aligned characters from Σ
2. A maximal series of consecutive characters in t aligned with spaces in s
3. A maximal series of consecutive characters in s aligned with spaces in t

The term *maximal* means that it cannot be extended further. Figure 3.7 shows an alignment and its blocks. Blocks in the first category above receive score $p(a, b)$, where a and b are the two aligned characters. Blocks in categories (2) and (3) receive score $-w(k)$, where k is the length of the gap series.

s_1	AAC---AATTCCGACTAC
s_2	ACTACCT------CGC--

s_1	A	A	C	---	A	TATCCG	A	C	T	AC
s_2	A	C	T	ACC	T	------	C	G	C	--

FIGURE 3.7

An alignment and its blocks.

The scoring of an alignment is not done at the column level now but rather at the block level. Accordingly, only if we break in block boundaries can we expect score additivity to hold. This implies some significant changes in our dynamic programming algorithm to compute the similarity under a general gap penalty function. Instead of reasoning on the last column of the alignment, we must reason on the last block. Furthermore, blocks cannot follow other blocks arbitrarily. A block of type 2 or 3 above cannot follow another block of the same type. This requires that we keep, for each pair (i, j), not only the best score of an alignment between prefixes $s[1..i]$ and $t[1..j]$, but rather the best score of these prefixes that ends in a particular type of block.

To compare sequence s of length m to sequence t of length n, we use three arrays of size $(m+1) \times (n+1)$, one for each type of ending block. Array a is used for alignments ending in character-character blocks; b is used for alignments ending in spaces in s; and c is used for alignments ending with spaces in t.

Initialization of the first row and column is done as follows, according to the meaning of the arrays:

$$\begin{aligned} a[0,0] &= 0 \\ b[0,j] &= -w(j) \\ c[i,0] &= -w(i). \end{aligned}$$

All other values in the initial row and column should be set to $-\infty$ to make them harmless in the maximum computations that use them.

We now ask the standard question — What type of block terminates our optimal alignment? The answer determines which array, a, b, or c, will be updated. The recurrence relations are as follows:

$$a[i,j] = p(i,j) + \max \begin{cases} a[i-1,j-1] \\ b[i-1,j-1] \\ c[i-1,j-1] \end{cases}$$

$$b[i,j] = \max \begin{cases} a[i,j-k] - w(k), & \text{for } 1 \le k \le j \\ c[i,j-k] - w(k), & \text{for } 1 \le k \le j \end{cases}$$

$$c[i,j] = \max \begin{cases} a[i-k,j] - w(k), & \text{for } 1 \le k \le i \\ b[i-k,j] - w(k), & \text{for } 1 \le k \le i. \end{cases}$$

As usual, $p(i,j)$ indicates the score of a matching between $s[i]$ and $t[j]$.

Notice that entries in arrays b and c depend on more than one earlier value, because the last block can have variable length. Also, when computing an entry of b, we do not look at previous b entries, because type 2 blocks cannot immediately follow type 2 blocks. Likewise, c entries do not depend directly on earlier c entries. To obtain the final answer, that is, the value of $\text{sim}(s,t)$, we take the maximum among $a[m,n]$, $b[m,n]$, and $c[m,n]$.

The time complexity of the algorithm is $O(mn^2+m^2n)$. To see this, count the number of times some entry of some array is read, based on the given formulas. This is the dominant term in the time complexity. It is easy to see that to compute $a[i,j]$, $b[i,j]$, and $c[i,j]$, we need to perform

$$3 + 2j + 2i$$

accesses to previous array entries, which sum up to

$$\sum_{i=1}^{m}\sum_{j=1}^{n}(3 + 2j + 2i)$$

accesses for the entire arrays. The sum above can be computed in closed form as follows.

$$\sum_{j=1}^{n}(2i + 2j + 3) = 2ni + n(n+1) + 3n = 2ni + n^2 + 4n$$

$$\sum_{i=1}^{m}(2ni + n^2 + 4n) = nm(m+1) + mn^2 + 4mn = m^2n + 5mn + mn^2.$$

Getting an optimal alignment is straightforward. The same ideas used in the previous algorithms work. We trace back in the arrays the entries that contributed to the maximum computations, at the same time keeping track of which array we used.

Thus, we see that it is possible to use any gap penalty function. However, the price to pay for this generality is a considerable increase in computing time and in storage space (the three auxiliary arrays). This can be critical when sequence comparison by dynamic programming is the bottleneck of a larger computation.

3.3.3 AFFINE GAP PENALTY FUNCTIONS

We have seen that the basic algorithm with a linear gap function runs in $O(n^2)$ time, for sequences of length n; we have also seen that allowing general gap functions makes the running time go up to $O(n^3)$. The question we now pose is: If we use a less general gap penalty function but one that still charges less for a gap with k spaces than for k isolated spaces, can we still get an $O(n^2)$ algorithm? In this section we show that this is in fact possible.

If we want to introduce the idea that k spaces together are more probable than k isolated spaces (or at least equally probable), we must have

$$w(k) \leq kw(1)$$

or, in general,

$$w(k_1 + k_2 + \cdots + k_n) \leq w(k_1) + w(k_2) + \cdots + w(k_n). \tag{3.8}$$

A function w that satisfies Equation (3.8) is called a *subadditive function*. An *affine function* is a function w of the form $w(k) = h + gk$, $k \geq 1$, with $w(0) = 0$. It will be subadditive if $h, g > 0$.

Another way of thinking about a function w as above is by saying that the first space in a gap costs $h + g$, while the other spaces cost g. This cost is based on the difference $\Delta w(k) = w(k) - w(k-1)$ of the penalty from $k-1$ spaces to k spaces.

Let us now describe the dynamic programming algorithm for this case. The main difference between the basic algorithm and this one is that here we need to make a distinction between the first space in a gap and the others, so that they can be penalized accordingly. This is accomplished by the use of three arrays, a, b, and c, as was done for the general gap function case. The entries in each one of these arrays have the following meaning:

$a[i, j]$ = maximum score of an alignment between $s[1..i]$ and $t[1..j]$ that ends in $s[i]$ matched with $t[j]$.
$b[i, j]$ = maximum score of an alignment between $s[1..i]$ and $t[1..j]$ that ends in a space matched with $t[j]$.
$c[i, j]$ = maximum score of an alignment between $s[1..i]$ and $t[1..j]$ that ends in $s[i]$ matched with a space.

The entries (i, j) of these arrays depend on previous entries according to the following formulas, valid for $1 \leq i \leq m$ and $1 \leq j \leq n$:

$$a[i, j] = p(i, j) + \max \begin{cases} a[i-1, j-1] \\ b[i-1, j-1] \\ c[i-1, j-1] \end{cases}$$

$$b[i, j] = \max \begin{cases} -(h+g) + a[i, j-1] \\ -g + b[i, j-1] \\ -(h+g) + c[i, j-1] \end{cases}$$

$$c[i, j] = \max \begin{cases} -(h+g) + a[i-1, j] \\ -(h+g) + b[i-1, j] \\ -g + c[i-1, j]. \end{cases}$$

As before, $p(i, j)$ indicates the score of a matching between $s[i]$ and $t[j]$.

Let us understand the preceding formulas. The formula for $a[i, j]$ includes the term $p(i, j)$, which is the score of the last column, plus the best score of an alignment between the prefixes $s[1..i-1]$ and $t[1..j-1]$.

For $b[i, j]$, we know that the last column will contain a space. We still have to check whether this is the first space of a gap or a continuation space, so that we can penalize it correctly. Furthermore, we need to read from the arrays the values of the best scores for the prefixes involved, namely, $s[1..i]$ and $t[1..j-1]$. That is why we always look at entries $(i, j-1)$ in the formulas for $b[i, j]$. Alignments corresponding to $a[i, j-1]$ and $c[i, j-1]$ do not end with a space in s, hence the final space must be penalized as the first of a gap, that is, subtracting $h+g$. Alignments corresponding to $b[i, j-1]$ already have a space in s at the end, so the space matched with $t[j]$ must be penalized as a continuation space, that is, subtracting g. These three possibilities are exhaustive, so taking the maximum among them gives the correct value. A similar argument explains the formula for $c[i, j]$.

The initialization of the arrays requires some care. Again, it will depend on whether we want to charge for initial spaces. To make it concrete, let us assume that we want to charge for all spaces, leaving to the reader the task of adapting the algorithm to the other cases.

The entries that need initialization are those with indices of the form $(i, 0)$ for $0 \le i \le m$ or $(0, j)$ for $0 \le j \le n$. The initialization has to be done according to the contents of each array as given in the definition. Thus, for array b, for instance, we have

$$b[i, 0] = -\infty \quad \text{for} \quad 0 \le i \le m \qquad \text{and}$$
$$b[0, j] = -(h + gj) \quad \text{for} \quad 1 \le j \le n.$$

We assign the value $-\infty$ to $b[i, 0]$ to be consistent with the definition, because there is no alignment between $s[1..i]$ and $t[1..0]$ (an empty sequence) in which a space is matched with $t[0]$, since there is no $t[0]$. Hence, if there are no alignments, the maximum value must be $-\infty$, which is the identity for maximum computations. On the other hand, there is exactly one alignment ending with a space in s between $s[1..0]$ (empty sequence again) and $t[1..j]$, and this alignment has score $-(h + gj)$.

In an analogous manner we initialize c letting

$$c[i, 0] = -(h + gi) \quad \text{for} \quad 1 \le i \le m \qquad \text{and}$$
$$c[0, j] = -\infty \quad \text{for} \quad 0 \le j \le n,$$

and, finally,

$$a[0, 0] = 0,$$
$$a[i, 0] = -\infty \quad \text{for} \quad 1 \le i \le m \qquad \text{and}$$
$$a[0, j] = -\infty \quad \text{for} \quad 1 \le j \le n.$$

In the initialization of a we used alignments *that do not end in spaces,* generalizing the given definition. To get the final result, that is, the similarity, it suffices to take the maximum among $a[m, n]$, $b[m, n]$, and $c[m, n]$, given that this covers all possibilities.

We can construct an optimal alignment in a manner analogous to the case of general gap functions. Just trace back from the final position (m, n) the entries that were chosen as maxima in the process of filling the arrays until the initial position $(0, 0)$ is reached. During this traceback, it is necessary to remember not only the current position but also which array (a, b, or c) it belongs to.

We leave to the reader the task of verifying that the time complexity of this algorithm is indeed $O(mn)$. The version just described has space complexity $O(mn)$ as well, but linear-space versions have been developed (see the bibliographic notes).

3.3.4 COMPARING SIMILAR SEQUENCES

Two sequences are similar when they "look alike" in some sense. We have already developed our intuition about the concept of similarity and have formalized it through alignments and scores. So, for us, two sequences are similar when the scores of the optimal alignments between them are very close to the maximum possible (where "maximum possible" will be made precise). This section is about faster algorithms to find good alignments in this case. We analyze global alignments only.

Let us first treat the case where the two sequences of interest, s and t, have the same length n. This is a fair assumption, given that we are focusing on similar sequences. If s and t have the same length, the dynamic programming matrix is a square matrix and its main diagonal runs from position $(0, 0)$ to (n, n). Following this diagonal corresponds to the unique alignment without spaces between s and t. If this is not an optimal alignment, we need to insert some spaces in the sequences to obtain a better score. Notice that spaces will always be inserted in pairs, one in s and one in t.

As we insert these space pairs, the alignment is thrown off the main diagonal. Consider for a moment the following two sequences:

$$s = \texttt{GCGCATGGATTGAGCGA}$$
$$t = \texttt{TGCGCCATGGATGAGCA}.$$

The optimal alignments correspond to paths that reach up to the diagonal twice removed from the main one, as in Figure 3.8. The best alignments include two pairs of spaces. In this case, the number of space pairs is equal to how far the alignment departed from the main diagonal, but this does not always happen. One thing is certain, though: The number of space pairs is greater than or equal to the maximum departure.

The basic idea then goes as follows. If the sequences are similar, the best alignments have their paths near the main diagonal. To compute the optimal score and alignments, it is not necessary to fill in the entire matrix. A narrow band around the main diagonal should suffice.

The algorithm *KBand* in Figure 3.9 performs the matrix fill-in in a band of horizontal (or vertical) width $2k + 1$ around the main diagonal. At the end, the entry $a[n, n]$ contains the highest score of an alignment confined to that band. This algorithm runs in time $O(kn)$, which is a big win over the usual $O(n^2)$ if k is small compared to n.

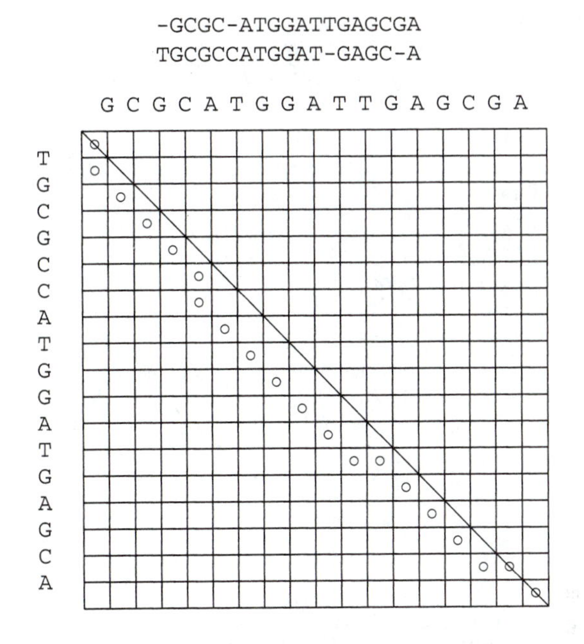

FIGURE 3.8

An optimal alignment and its corresponding path in the dynamic programming matrix. A line is drawn along the main diagonal.

Observe that we do not use entries outside of the k-strip at all. We do not initialize them, and we do not consult them for maximum computations. To test whether a certain position (i, j) is inside the k-strip, we use the following criterion:

$$\textit{InsideStrip}(i, j, k) = (-k \leq i - j \leq k).$$

As in the other dynamic programming algorithms, each entry $a[i, j]$ depends on $a[i - 1, j]$, $a[i-1, j-1]$, and $a[i, j-1]$. In the code, we need not test position $(i-1, j-1)$, because it is in the same diagonal as (i, j), so it will always be inside the strip. Tests must be performed for $(i - 1, j)$ and $(i, j - 1)$, which may be outside the strip when (i, j) is in the border.

How can we use algorithm *KBand*? We choose a value for k and run the algorithm. If $a[n, n]$ is greater than or equal to the best score that could possibly come from an alignment with $k+1$ or more space pairs, we are lucky: We have found an optimal alignment with just $O(kn)$ steps. The best possible score, given that we have at least $k + 1$ space pairs, is

$$M(n - k - 1) + 2(k + 1)g, \tag{3.9}$$

which is computed assuming that there are exactly $k + 1$ space pairs and that the other pairs are matches. Here M is the score of a match, and g is added for each space. We will assume that $M > 0$ and $g \leq 0$.

```
Algorithm KBand
    input: sequences s and t of equal length n, integer k
    output: best score of alignment at most k diagonals away from main diagonal
    n ← |s|
    for i ← 0 to k do
        a[i, 0] ← i × g
    for j ← 0 to k do
        a[0, j] ← j × g
    for i ← 1 to n do
        for d ← −k to k do
            j ← i + d
            if 1 ≤ j ≤ n do
                // compute maximum among predecessors
                a[i, j] ← a[i − 1, j − 1] + p(i, j)
                if InsideStrip(i − 1, j, k) then
                    a[i, j] ← max(a[i, j], a[i − 1, j] + g)
                if InsideStrip(i, j − 1, k) then
                    a[i, j] ← max(a[i, j], a[i, j − 1] + g)
    return a[n, n]
```

FIGURE 3.9

Algorithm for k-strip around main diagonal.

If $a[n, n]$ is smaller than the quantity (3.9), we double k and run the algorithm again. If the initial value of k is 1, further values will be powers of two. The stop condition becomes

$$a_k[n, n] \geq M(n - k - 1) + 2(k + 1)g,$$

which is equivalent to

$$k \geq \frac{Mn - a_k[n, n]}{M - 2g} - 1.$$

At this point, we have already run the algorithm many times, each time with a larger k, so that the total complexity is

$$n + 2n + 4n + \cdots + kn \leq 2kn,$$

assuming we use powers of two. To bound this total complexity, we need an upper bound on k. Now, we did not stop earlier, so

$$\frac{k}{2} < \frac{Mn - a_{k/2}[n, n]}{M - 2g} - 1.$$

If $a_k[n, n] = a_{k/2}[n, n]$, then this is the optimum score $\text{sim}(s, t)$, and our bound is

$$k < 2\left(\frac{Mn - \text{sim}(s, t)}{M - 2g} - 1\right).$$

If $a_k[n, n] > a_{k/2}[n, n]$, then optimal alignments have more than $k/2$ pairs of spaces, and therefore

$$\mathrm{sim}(s, t) \leq M\left(n - \frac{k}{2} - 1\right) + 2\left(\frac{k}{2} + 1\right)g,$$

which leads to

$$k \leq 2\left(\frac{Mn - \mathrm{sim}(s, t)}{M - 2g} - 1\right),$$

almost the same bound as above.

Observe that $M - 2g$ is a constant. The time complexity is then $O(dn)$, where d is the difference between the maximum possible score Mn — the score of two identical sequences — and the optimal score. Thus, the higher the similarity, the faster the answer.

It is straightforward to extend this method to general sequences, not necessarily with the same length. Space-saving versions can also be easily derived.

3.4 COMPARING MULTIPLE SEQUENCES

So far in this chapter we have concentrated on the comparison between a pair of sequences. However we often are given several sequences that we have to align simultaneously in the best possible way. This happens, for example, when we have the sequences for certain proteins that have similar function in a number of different species. We may want to know which parts of these sequences are similar and which parts are different. To get this information, we need to build a **multiple alignment** for these sequences, and that is the topic of this section.

The notion of multiple alignment is a natural generalization of the two-sequence case. Let $s_1, \ldots, s_k$ be a set of sequences over the same alphabet. A multiple alignment involving $s_1, \ldots, s_k$ is obtained by inserting spaces in the sequences in such a way as to make them all of the same size. It is customary to place the extended sequences in a vertical list so that characters — or spaces — in corresponding positions occupy the same column. We further require that no column be made exclusively of spaces. Figure 3.10 shows a multiple alignment involving four short amino acid sequences. (Multiple alignments are more common with proteins, so in some of our examples in this section we use sequences of amino acids.)

One important issue to decide in multiple alignment is the precise definition of the

```
MQPILLL
MLR-LL-
MK-ILLL
MPPVLIL
```

FIGURE 3.10

Multiple alignment of four amino acid sequences.

quality of an alignment. We will next study one way of scoring a multiple alignment based on pairwise alignments. In addition, scientists also look at multiple alignments by placing the sequences in a tree structure rather than piling them up. This leads to different measures of quality that we discuss in subsequent sections.

3.4.1 THE SP MEASURE

Scoring a multiple alignment is more complex than its pairwise counterpart. We restrict ourselves to purely additive functions here; that is, the alignment score is the sum of column scores. Therefore we need a way to assign a score to each column and then add them up to get the alignment score. However, to score a column, we want a function with k arguments, where k is the number of sequences. Each one of these arguments is a character or a space. One way to do that would be to have a k-dimensional array that could be indexed with the arguments and return the value. The problem with this approach is that it is necessary to specify a value for each possible combination of arguments, and there could be as many as $2^k - 1$ such combinations. A typical value for k is 10, and that results in more than 1000 possibilities. Some more manageable methods to define such a function are necessary.

Such methods can be obtained by determining "reasonable" properties that such a function should have. First, the function must be independent of the order of arguments. For instance, if a column has `I, -, I, V` and another has `V, I, I, -`, they should both receive the same score. Second, it should be a function that rewards the presence of many equal or strongly related residues and penalizes unrelated residues and spaces. A solution that satisfies these properties is the so-called **sum-of-pairs** (SP) function. It is defined as the sum of pairwise scores of all pairs of symbols in the column. For instance, the score of a column with the above content is

$$\begin{aligned} \textit{SP-score}(\mathtt{I}, \mathtt{-}, \mathtt{I}, \mathtt{V}) = {} & p(\mathtt{I}, \mathtt{-}) + p(\mathtt{I}, \mathtt{I}) + p(\mathtt{I}, \mathtt{V}) + \\ & p(\mathtt{-}, \mathtt{I}) + p(\mathtt{-}, \mathtt{V}) + p(\mathtt{I}, \mathtt{V}), \end{aligned}$$

where $p(a, b)$ is the pairwise score of symbols a and b. Notice that this may include a space penalty specification when either a or b is a space. This is a very convenient scheme, because it relies on pairwise scores like the ones we use for two-sequence comparison. The SP scoring system is widely used due to its simplicity and effectiveness.

A small but important detail needs to be addressed to complete the definition. Although no column can be composed exclusively by spaces, it is possible to have two or more spaces in a given column. When computing the SP score, we need a value for $p(\mathtt{-}, \mathtt{-})$. This value is not specified in two-sequence comparison because it never appears there. However, it is necessary for SP-based multiple comparison. The common practice is to set $p(\mathtt{-}, \mathtt{-}) = 0$. This may seem strange, given that spaces are generally penalized (i.e., when there is a space, the pairwise score is negative), so two spaces should be even more so. Nevertheless, there are good reasons to define $p(\mathtt{-}, \mathtt{-})$ as zero. One of them is related to pairwise alignments again. We often draw conclusions about a multiple alignment by looking at the pairwise alignments it induces. Indeed, in any multiple alignment, we may select two of the sequences and just look at the way they are aligned to each other, forgetting about all the rest. It is not difficult to see that this produces a pairwise alignment, except for the fact that we may have columns with two spaces in them. But then we just remove these columns and derive a true pairwise alignment. An example of this procedure is presented in Figure 3.11. This is what we call the *induced* pairwise alignment, or the *projection* of a multiple alignment.

A very useful fact, which is true only if we have $p(-, -) = 0$, is the following formula for the SP score of a multiple alignment α:

$$SP\text{-}score(\alpha) = \sum_{i<j} score(\alpha_{ij}), \tag{3.10}$$

where α_{ij} is the pairwise alignment induced by α on sequences s_i and s_j. This is true because it reflects two ways of doing the same thing. We may compute the score of each column and then add all column scores, or compute the score for each induced pairwise alignment and then add these scores. In any case, we are adding, for each column c and for each pair (i, j), the score $p(s_i'[c], s_j'[c])$, where s' indicates the extended sequence (with spaces inserted). But this is true only if $p(-, -) = 0$, because these quantities appear in the first computation only.

Multiple alignment

```
1 PEAALYGRFT---IKSDVW
2 PEAALYGRFT---IKSDVW
3 PESLAYNKF---SIKSDVW
4 PEALNYGRY---SSESDVW
5 PEALNYGWY---SSESDVW
6 PEVIRMQDDNPFSFQSDVY
```

Get only sequences 2 and 4

```
PEAALYGRFT---IKSDVW
PEALNYGRY---SSESDVW
```

Remove columns with two spaces

```
PEAALYGRFT-IKSDVW
PEALNYGRY-SSESDVW
```

FIGURE 3.11

Induced pairwise alignment (projection).

Dynamic Programming

Having decided on a measure, or score, for determining the quality of a multiple alignment, we would like to compute the alignments of maximum score, given a set of sequences. These will be our *optimal alignments*.

It is possible to use a dynamic programming approach here, as we did in the two-sequence case. Suppose, for simplicity, that we have k sequences, all of the same length n.

We use a k-dimensional array a of length $n+1$ in each dimension to hold the optimal scores for multiple alignments of prefixes of the sequences. Thus, $a[i_1, \ldots, i_k]$ holds the score of the optimal alignment involving $s_1[1..i_1], \ldots, s_k[1..i_k]$.

After initializing with $a[0, \ldots, 0] \leftarrow 0$, we must fill in this entire array. Just to store it requires $O(n^k)$ space. This is also a lower bound for the computation time, because we have to compute the value of each entry. The actual time complexity is higher for a number of reasons. First, each entry depends on $2^k - 1$ previously computed entries, one for each possible composition of the current column of the alignment. In this composition, each sequence can participate with either a character or a space. Because we have k sequences, we have 2^k compositions. Removing the forbidden composition of all spaces, we obtain the final count of $2^k - 1$. This incorporates a factor of 2^k to the already exponential time complexity.

Then there is the question of accessing the data in the array. Very few programming languages, and certainly none of the most popular ones, will let users define an array with the number of dimensions k set at run-time. The alternative is to implement our own access routines. In any case, with or without language support, we can expect to spend $O(k)$ steps per access.

Another issue is the computation of column scores. The SP method requires $O(k^2)$ steps per column, as there are $k(k-1)/2$ pairwise scores to add up. Simpler schemes — for instance, just count the number of nonspace symbols — require at least $O(k)$ steps, given that we have to look at all arguments.

Finally, there is the actual computing of the value of $a[i_1, \ldots, i_k]$, which involves a maximum operation. Using boldface letters to indicate k-tuples, the command we must perform can be written as

$$a[\,\boldsymbol{i}\,] \leftarrow \max_{\mathbf{b} \neq \mathbf{0}} \{a[\,\boldsymbol{i} - \boldsymbol{b}] + \textit{SP-score}(\textit{Column}(\,\boldsymbol{s},\ \boldsymbol{i},\ \boldsymbol{b}))\},$$

where $\boldsymbol{b}$ ranges over all nonzero binary vectors of k elements, and

$$\textit{Column}(\,\boldsymbol{s},\ \boldsymbol{i},\ \boldsymbol{b}) = (c_j)_{1 \le j \le k}$$

with

$$c_j = \begin{cases} s_j[i_j] & \text{if } b_j = 1 \\ \texttt{-} & \text{if } b_j = 0. \end{cases}$$

The total running time estimate for this first plan for implementation is therefore $O(k^2 2^k n^k)$ if we use SP, or $O(k 2^k n^k)$ if column scores can be computed in $O(k)$. If k is fixed, k nested **for** loops can be used to fill in the array. Optimal alignments can be recovered from this array by a backtracking procedure analogous to the one used in the pairwise case (Section 3.2.1). Straightforward extensions yield a similar method for the case where the sequences do not necessarily have the same length. In any case, the complexity of this algorithm is exponential in the number of input sequences, and the existence of a polynomial algorithm seems unlikely: It has been shown that the multiple alignment problem with the SP measure is NP-complete (see the bibliographic notes).

Saving Time

The exponential complexity of the pure dynamic programming approach makes it unappealing for general use. The main problem is the size of the array. For three sequences we already have a cube of $O(n^3)$ cells, and as the number k of sequences grows we have larger and larger "volumes" to fill in. Clearly, if we could somehow reduce the amount of cells to compute, this would have a direct impact on processing time.

This section describes a heuristic that does exactly that. We will show how to incorporate it into the dynamic programming algorithm to speed up its computation. It is a heuristic because in the worst case all cells will have to be computed; in practice, however, we can expect a good speedup. The heuristic is based on the relationship between a multiple alignment and its projections on two-sequence arrays, and, in particular, it uses Equation (3.10) relating SP scores to pairwise scores. Thus, the method we are about to see works only for the SP measure.

The outline of the method is as follows. We have k sequences of length n_i, for $1 \leq i \leq k$, and we want to compute the optimal alignments according to the SP measure. We will still use dynamic programming, but now we do not want to treat all cells. We just want the cells "relevant" to optimal alignments, in some sense. But exactly which cells will we deem relevant, and why?

The answer is to look at the pairwise projections of the cell. In a preprocessing step, we create conditions that will allow us to perform a test of relevance for arbitrary cells. To take advantage of this test and reduce the number of cells we need to look at, we have to modify the fill-in order as well.

Relevance Test

Let α be an optimal alignment involving $s_1, \ldots, s_k$. The first thing we must know is that even though α is optimal its projections are not necessarily the best ones for the given sequence pair. Figure 3.12 shows a case in which a projection fails to be optimal. It is unfortunate that such cases can happen. Were it not for them, we could easily establish a test for relevant cells: A cell is relevant when each of its pairwise projections is part of an optimal alignment of the two sequences corresponding to the projection.

Before going on, let us remark that it is easy to use the comparison algorithms we have seen for two sequences s and t to produce a matrix in which each entry (i, j) con-

```
AT
A-          A-
-T          -T
AT
AT
```

Optimal multiple alignment — Nonoptimal projection

FIGURE 3.12

Optimal alignment with nonoptimal projection.

tains the highest score of an alignment that includes the *cut* (i, j). A pairwise alignment α contains a cut (i, j) when α can be divided into two subalignments, one aligning $s[1..i]$ with $t[1..j]$ and the other aligning the rest of s with the rest of t.

To obtain the desired values, we add two dynamic programming matrices a and b with the contents,

$$a[i, j] = \text{sim}(s[1..i], t[1..j])$$
$$b[i, j] = \text{sim}(s[i+1..n], t[j+1..m]),$$

where $n = |s|$ and $m = |t|$. The matrix a is just the standard matrix we have been using all along. The matrix b can be computed just like a, but backward. We initialize the last row and column and proceed backward until we reach $b[0, 0]$. We did this in Section 3.3.1, when we discussed linear space implementations of dynamic programming algorithms. The function *BestScoreRev* there does what we want in b. Then, sum $c = a + b$ contains exactly the highest score of an alignment that cuts at (i, j). We call c the matrix of *total scores*, while a and b are the matrices of *prefix* and *suffix* scores, respectively.

The matrix c is more appealing to the eyes than either a or b. We can quickly spot the best alignments just by looking at c, a feat that is considerably more difficult to do with either a or b. Let us explain this with an example. In Figure 3.13 we have the matrices a and c for a certain pair of sequences and a certain scoring system. Can you see exactly where the optimal alignments are by looking at a? It is much easier with c because we merely follow the cells with the highest score.

		G	A	T	T	C
	0	-2	-4	-6	-8	-10
A	-2	-1	-1	-3	-5	-7
T	-4	-3	-2	0	-2	-4
T	-6	-5	-4	-1	1	-1
C	-8	-7	-6	-3	-1	2
G	-10	-7	-8	-5	-3	0
G	-12	-9	-8	-7	-5	-2

		G	A	T	T	C
	-2	-2	-7	-12	-17	-22
A	-7	-4	-2	-7	-12	-17
T	-10	-7	-5	-2	-7	-12
T	-13	-10	-7	-5	-2	-7
C	-14	-13	-10	-5	-4	-2
G	-17	-14	-13	-8	-4	-2
G	-22	-17	-14	-11	-7	-2

FIGURE 3.13

Dynamic programming matrices: prefix scores (left) and total scores (right).

Although projections of optimal alignments may not be optimal themselves, we can establish a lower bound for the projection scores, as long as we have a lower bound for the optimal score. The following result tells us exactly how this works.

THEOREM 3.1 Let α be an optimal alignment involving $s_1, \ldots, s_k$. If *SP-score*$(\alpha) \geq L$, then

$$score(\alpha_{ij}) \geq L_{ij},$$

where

$$L_{ij} = L - \sum_{\substack{x < y \\ (x,y) \neq (i,j)}} (\text{sim}(s_x, s_y)).$$

Proof. We have, successively,

$$\textit{SP-score}(\alpha) \geq L,$$

$$\sum_{x<y} \textit{score}(\alpha_{xy}) \geq L,$$

$$\sum_{\substack{x < y \\ (x,y) \neq (i,j)}} (\textit{score}(\alpha_{xy})) \geq L - \textit{score}(\alpha_{ij}),$$

$$\sum_{\substack{x < y \\ (x,y) \neq (i,j)}} (\text{sim}(s_x, s_y)) \geq L - \textit{score}(\alpha_{ij}),$$

from which the result follows. ■

We are now in a position to test whether a cell with index $\boldsymbol{i} = (i_1, \ldots, i_k)$ is relevant to optimal alignments with respect to lower bound L. Simply put, this cell is relevant if all of its projections satisfy the conditions of the previous theorem. In other words, $\boldsymbol{i}$ is relevant when

$$c_{xy}[i_x, i_y] \geq L_{xy},$$

for all x and y such that $1 \leq x < y \leq k$, where c_{xy} is the matrix of total scores for s_x and s_y. According to the theorem, only relevant cells can participate in optimal alignments, although not *all* relevant cells will participate in optimal alignments. Nevertheless, this affords a reduction in the number of cells that are potential candidates for an optimal alignment. This reduction becomes more significant as L approaches the true optimal score.

To obtain a suitable bound L, we may just choose an arbitrary multiple alignment involving all sequences and take its score as L. Of course this is a lower bound since optimal alignments have the highest possible score. If we already have a reasonably good alignment, we can use its score and improve the alignment to get an optimal one by the method just sketched. An alternative lower bound can be obtained using the results of Section 3.4.2.

Implementation Details

We have to be a bit careful when implementing the heuristic just described. It is not enough to test all cells for relevance and then use only the relevant ones. This will achieve no appreciable time reduction, because we are still looking at all the cells to test them.

We need a way of actually cutting off completely the irrelevant cells, so that they are not even looked at.

Here is one possible strategy. We start with the cell at $\mathbf{0} = (0, 0, \ldots, 0)$, which is always relevant, and expand its influence to dependent relevant cells. Each one of these will in turn expand its influence, and so on, until we reach the final corner cell at $(n_1, \ldots, n_k)$. In the whole process, only relevant cells will be analyzed.

To explain the strategy better, we need a few definitions. A cell $\boldsymbol{i}$ *influences* another cell $\boldsymbol{j}$ if $\boldsymbol{i}$ is one of the cells used in the maximum computation to determine the value of $a[\boldsymbol{j}]$. In this case, we also say that $\boldsymbol{j}$ *depends on* $\boldsymbol{i}$. Another characterization of this fact is that $\boldsymbol{b} = \boldsymbol{j} - \boldsymbol{i}$ is a vector with either 0 or 1 in each component, and $\boldsymbol{i} \neq \boldsymbol{j}$. So, each cell depends on at most $2^k - 1$ others, and influences at most $2^k - 1$ others. We say "at most" because border cells may influence or depend on less than $2^k - 1$ other cells.

We keep a pool of cells to be examined. Initially, only $\mathbf{0}$ is in the pool. The pool contains only relevant cells at all times. When a cell $\boldsymbol{i}$ enters the pool, its value $a[\boldsymbol{i}]$ is initialized. During its stay in the pool, this value is updated. When this cell is removed from the pool, the current value of $a[\boldsymbol{i}]$ is taken as the true value of this cell and used in a propagation to the relevant cells that depend on $\boldsymbol{i}$.

The value of a cell is propagated as follows. Let $\boldsymbol{j}$ be a relevant cell that depends on $\boldsymbol{i}$. If $\boldsymbol{j}$ is not in the pool, we put it there and initialize its value with the command

$$a[\boldsymbol{j}] \leftarrow a[\boldsymbol{i}] + \textit{SP-score}(\textit{Column}(\boldsymbol{s}, \boldsymbol{i}, \boldsymbol{b})).$$

If $\boldsymbol{j}$ is already in the pool, we conditionally update its value with the command

$$a[\boldsymbol{j}] \leftarrow \max(a[\boldsymbol{j}], a[\boldsymbol{i}] + \textit{SP-score}(\textit{Column}(\boldsymbol{s}, \boldsymbol{i}, \boldsymbol{b}))).$$

It is important to make sure that each time some cell is to be removed from the pool, the lexicographically smaller one is selected. This guarantees that this cell has already received the influences due to other relevant cells, and its value need not be further updated. The process stops when we reach cell $(n_1, \ldots, n_k)$, which is necessarily the last cell examined; its value is the SP score sought. The algorithm for computing the score is shown in Figure 3.14. To recover the optimal alignments, we need to keep track of the updates somehow. One way is to construct a dependence graph where the relevant cells are nodes and the edges represent influences that provided the maximum value. Time and space complexity of this algorithm are proportional to the number of relevant cells.

3.4.2 STAR ALIGNMENTS

Computing optimal multiple alignments can take a long time if we use the standard dynamic programming approach, even with the savings sketched in the previous section; so other methods have been developed. These alternative ways are often heuristic, in the sense that they do not yield any guarantee on the quality of the resulting alignment. They are simply faster ways of getting an answer, which in many cases turns out to be a reasonably good answer.

One such method is what has been termed the *star alignment* method. It consists in building a multiple alignment based upon the pairwise alignments between a fixed sequence of the input set and all others. This fixed sequence is the *center* of the star. The

```
Algorithm Multiple-Sequence Alignment
    input: s = (s_1, ..., s_k) and lower bound L
    output: The value of an optimal alignment
    // Compute L_xy, 1 ≤ x < y ≤ k
    for all x and y, 1 ≤ x < y ≤ k do
        Compute c_xy, the total score array for s_x and s_y
    for all x and y, 1 ≤ x < y ≤ k do
        L_xy ← L − Σ_{(p,q)≠(x,y)} sim(s_p, s_q)
    // Compute array a
    pool ← {0}
    while pool not empty do
        i ← the lexicographically smallest cell in the pool
        pool ← pool \ { i}
        if c_xy[i_x, i_y] ≥ L_xy, ∀x, y, 1 ≤ x < y ≤ k, then // Relevance test
            for all j dependent on i do
                if j ∉ pool then
                    pool ← pool ∪ { j}
                    a[ j] ← a[ i] + SP-score(Column( s, i, j − i))
                else
                    a[ j] ← max(a[ j], a[ i] + SP-score(Column( s, i, j − i)))
    return a[n_1, ..., n_k]
```

FIGURE 3.14

Dynamic programming algorithm for multiple-sequence comparison with heuristics for saving time.

alignment α constructed is such that its projections α_{ij} are optimal when either i or j is the index of the center sequence.

Let $s_1, \ldots, s_k$ be k sequences that we want to align. To construct a star alignment, we must first pick one of the sequences as the center. We will postpone the discussion as to how this selection should be done. For the moment, let us just assume that the center sequence has been selected and its index is a number c between 1 and k. Next we need, for each index $i \neq c$, an optimal alignment between s_i and s_c. This can be obtained with standard dynamic programming, so this phase takes $O(kn^2)$ time, assuming all sequences have $O(n)$ length.

We aggregate these pairwise alignments using a technique known as "once a gap, always a gap," applied to the center sequence s_c. The construction starts with one of the pairwise alignments, say the one between s_1 and s_c, and goes on with each pairwise alignment being added to the bunch. During the process, we progressively increase the gaps in s_c to suit further alignments, never removing gaps already present in s_c — once a gap in s_c, always a gap.

Each subsequent pairwise alignment is added using s_c as a guide. We have a multiple alignment involving s_c and some other sequences in one hand, and a pairwise alignment between s_c and a new sequence in the other. We add as few gaps as necessary in both alignments so that the extended copies of s_c agree. Then just include the new extended sequence in the cluster, given that it now has the same length as the other extended sequences.

The time complexity of this joining operation depends on the data structure used to represent alignments, but it should not be higher than $O(kl)$ using reasonable structures, where l is an upper bound on the alignment lengths. Because we have $O(k)$ sequences to add, we end up with $O(k^2 l)$ for the joining phase. The total time complexity is then $O(kn^2 + k^2 l)$. If we want to know the resulting score, this costs an extra $O(k^2 l)$, but the asymptotic complexity remains the same.

How should we select the center sequence? One way is to just try them all and then pick the best resulting score. Another way is to compute all $O(k^2)$ optimal pairwise alignments and select as the center the string that maximizes

$$\sum_{i \neq c} \text{sim}(s_i, s_c). \tag{3.11}$$

Example 3.1 Consider the following five DNA sequences. We begin by constructing a table with the pairwise similarities among the sequences (see Figure 3.15). The score system used is the same as in Section 3.2.1, where we explain the basic algorithm.

$s_1 =$ ATTGCCATT

$s_2 =$ ATGGCCATT

$s_3 =$ ATCCAATTTT

$s_4 =$ ATCTTCTT

$s_5 =$ ACTGACC

	s_1	s_2	s_3	s_4	s_5
s_1		7	−2	0	−3
s_2	7		−2	0	−4
s_3	−2	−2		0	−7
s_4	0	0	0		−3
s_5	−3	−4	−7	−3	

FIGURE 3.15

Pairwise scores for the sequences in Example 3.1.

From the table we see that the first sequence, s_1, is the one that maximizes expression (3.11). Our next step is then to choose optimal alignments between s_1 and all the other sequences. Suppose we choose the following optimal alignments:

```
ATTGCCATT
ATGGCCATT
```

```
ATTGCCATT--
ATC-CAATTTT
```

```
ATTGCCATT
ATCTTC-TT

ATTGCCATT
ACTGACC
```

In this example the only spaces introduced in s_1, the center sequence, and in all alignments were the two spaces at the end forced by sequence s_3. Therefore, s_1 will have just these two gaps in the final multiple alignment. The other sequences are aligned to s_1 as in the chosen alignments. The result is shown here:

```
ATTGCCATT--
ATGGCCATT--
ATC-CAATTTT
ATCTTC-TT--
ACTGACC----
```

3.4.3 TREE ALIGNMENTS

An alternative to comparing multiple sequences is discussed in this section. The motivation for this approach is that sometimes we have an evolutionary tree for the sequences involved (evolutionary trees are the subject of Chapter 6). In this case, we can compute the overall similarity based on pairwise alignments along tree edges. Contrast this with the SP measure, which takes into account *all* parwise similarities.

Suppose that we are given k sequences and a tree with exactly k leaves, with a one-to-one correspondence between leaves and sequences. If we assign sequences to the interior nodes of the tree, we can compute a *weight* for each edge, which is the similarity between the two sequences in the nodes incident to this edge. The sum of all these weights is the *score* of the tree with respect to this particular sequence assignment to interior nodes. Finding a sequence assignment that maximizes the score is what has been called the **tree alignment** problem. Star alignments can be viewed as particular cases of tree alignments in which the tree is a star.

Example 3.2 Consider the input shown in Figure 3.16. Assigning sequence CT to vertex x and sequence CG to vertex y, we have a score of 6. The scoring system used is $p(a,b) = 1$ if $a = b$ and 0 otherwise, and $p(a,-) = -1$.

The tree alignment problem is NP-hard. There exists an algorithm that finds an optimal solution, but it is exponential in the number of sequences. By using the ideas developed in Section 3.4.1 it is possible to improve space and time requirements in practice. Approximation algorithms with good performance guarantees have been designed for the case when edge weights are defined in terms of distance rather than similarity (see

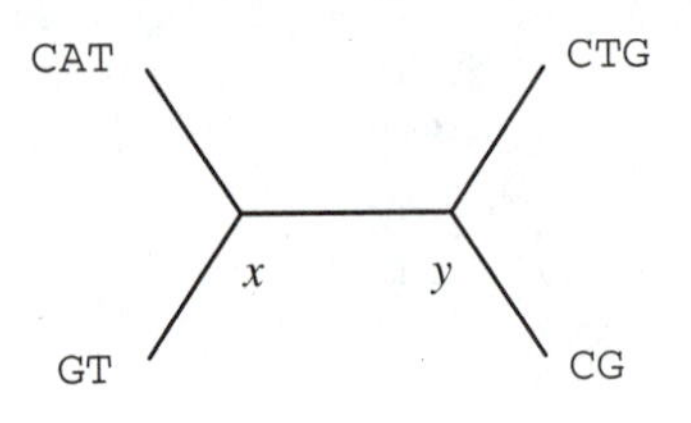

FIGURE 3.16

Input for a tree alignment problem.

Section 3.6.1). In this case we look for a sequence assignment that minimizes the distance sum. References for these results are given in the bibliographic notes.

3.5 DATABASE SEARCH

With the advent of fast and reliable technology for sequencing nucleic acids and proteins, centralized databases were created to store the large quantity of sequence data produced by labs all over the world. This created a need for efficient programs to be used in queries of these databases. In a typical application, one has a *query* sequence that must be compared to all those already in the database, in search of local similarities. This means hundreds of thousands of sequence comparisons.

The quadratic complexity of the methods we have seen so far for computing similarities and optimal alignments between two sequences makes them unsuitable for searching large databases. To speed the search, novel and faster methods have been developed. In general, these methods are based on heuristics and it is hard to establish their theoretical time and space complexity. Nevertheless, the programs based on them have become very important tools and these techniques deserve careful study.

In this section we concentrate on two of the most popular programs for database search. Neither uses pure dynamic programming, although one of them runs a variant of the dynamic programming method to refine alignments obtained by other methods.

Before we start describing these programs we make a little digression to explain the foundations of certain scoring matrices for amino acids, which are very important in database searches and in protein sequence comparison in general.

3.5.1 PAM MATRICES

When comparing protein sequences, simple scoring schemes, such as $+1$ for a match, 0 for a mismatch, and -1 for a space, are not enough. Amino acids, the residues that make up protein sequences, have biochemical properties that influence their relative replaceability in an evolutionary scenario. For instance, it is more likely that amino acids of similar sizes get substituted for one another than those of widely different sizes. Other properties such as the tendency to bind with water molecules also influence the probability of mutual substitution. Because protein comparisons are often made with evolutionary concerns in mind, it is important to use a scoring scheme that reflects these probabilities as much as possible.

The factors that influence the probability of mutual substitution are so numerous and varied that direct observation of actual substitution rates is often the best way of deriving similarity scores for pairs of residues. A standard procedure toward this goal is based on an important family of scoring matrices, the so-called PAM matrices, very popular among practitioners in the area. The acronym PAM stands for *Point Accepted Mutations,* or *Percent of Accepted Mutations,* in a reference to the fact that the basic 1-PAM matrix reflects an amount of evolution producing on average one mutation per hundred amino acids. In this section we briefly describe how PAMs are computed and what they mean.

Our description departs slightly from the original derivation of the matrices, but it highlights the role played by amino acid frequencies and mutabilities in coming up with the scores. In particular, the definition of mutability may be different in different sources.

Before anything else, the user has to choose an evolutionary distance at which to compare the sequences. These matrices are functions of this distance. For instance, a 250-PAM matrix is suitable for comparing sequences that are 250 units of evolution apart. In our derivation, we first construct the matrix corresponding to 1 PAM and then obtain the matrices for other distances from this one. Also, mutations are viewed at the amino acid level only, not at the DNA level.

For each evolutionary distance we have a *probability transition matrix* M and a *scores matrix* S. The scores matrix is obtained from the probability matrix, so we start our description with M. The necessary ingredients for building the 1-PAM matrix M are the following:

- A list of *accepted mutations*
- The *probabilities of occurrence* p_a for each amino acid a

An *accepted mutation* is a mutation that occurred and was positively selected by the environment; that is, it did not cause the demise of the particular organism where it occurred. One way to collect accepted mutations is to align two homologous proteins from different species, for example, the hemoglobin alpha chain in humans and in orangutans. Each position where the sequences differ will give us an accepted mutation. We consider these accepted mutations as undirected events; that is, given a pair a, b of aligned amino acids, we do not know for sure which one mutated into the other. Whichever was present in this same position in the ancestral sequence is the one that mutated into the other, but we do not know the ancestral sequence. It is even possible that the ancestral sequence contained a third amino acid that mutated into the two present ones, but this possibility is minimized taking very closely related sequences. It is important for the basic 1-PAM matrix that we consider immediate mutations, $a \to b$, not mediated ones like $a \to c \to b$.

The probabilities of occurrence can be estimated simply by computing the relative frequency of occurrence of amino acids over a large, sufficiently varied protein sequence set. These numbers satisfy

$$\sum_a p_a = 1.$$

From the list of accepted mutations we can compute the quantities f_{ab}, the number of times the mutation $a \leftrightarrow b$ was observed to occur. Recall that we are dealing with undirected mutations here, so $f_{ab} = f_{ba}$. We will also need the sums

$$f_a = \sum_{b \neq a} f_{ab},$$

which is the total number of mutations in which a was involved, and

$$f = \sum_a f_a, .$$

the total number of amino acid occurrences involved in mutations. The number f is also twice the total number of mutations.

The frequencies f_{ab} and the probabilities p_a are all that is needed to build a 1-PAM transition probability matrix M. This is a 20×20 matrix with M_{ab} being the probability of amino acid a changing into amino acid b. Note that a and b may be the same, in which case we have the probability of a remaining unchanged during this particular evolutionary interval. For PAM matrices, the computation of M_{aa} is done based on the *relative mutability* of amino acid a, defined as

$$m_a = \frac{f_a}{100 f p_a}. \tag{3.12}$$

The mutability of an amino acid is a measure of how much it changes. It is the probability that the given amino acid will change in the evolutionary period of interest. Hence, the probability of a remaining unchanged is the complementary probability

$$M_{aa} = 1 - m_a.$$

On the other hand, the probability of a changing into b can be computed as the product of the conditional probability that a will change into b, given that a changed, times the probability of a changing. We estimate the conditional probability as the ratio between the $a \leftrightarrow b$ mutations and the total number of mutations involving a. Therefore,

$$\begin{aligned} M_{ab} &= \Pr(a \to b) \\ &= \Pr(a \to b \mid a \text{ changed}) \Pr(a \text{ changed}) \\ &= \frac{f_{ab}}{f_a} m_a. \end{aligned}$$

Notice that we are computing these probabilities using a simplified model of protein evolution. For instance, an amino acid is supposed to mutate independently of its past history, which may not be true given the nature of the genetic code. Also, we are ignoring the influence that other amino acids in the same sequence may have on the mutation of a given residue. The independence from past history in particular leads to a Markov-type model of evolution, which has good mathematical properties, some of which will be mentioned in the sequel.

It is easy to verify that M has the following properties.

$$\sum_b M_{ab} = 1 \tag{3.13}$$

$$\sum_a p_a M_{aa} = 0.99 \tag{3.14}$$

Equation (3.13) is merely saying that by adding up the probability of a staying the same and the probabilities of it changing into every other amino acid we get 1. Thus we are justified in calling these numbers "probabilities." Recall that these probabilities refer to a unit of evolutionary change. It is tempting to think of unit of evolution as unit of time, but it is generally accepted that different things change at different speeds, so time and amount of evolution are not directly proportional in a universal sense.

The unit of evolution used in this model is the amount of evolution that will change 1 in 100 amino acids on average. This will be referred to as 1 PAM evolutionary distance. The transition probability matrix has been normalized to reflect this fact as we can see from Equation (3.14). We achieve this normalization by using 100 in the denominator of Equation (3.12). Had we used another number, say 50, instead of 100 there, we would have obtained a matrix with exactly the same properties except that (3.14) would reflect a 1 in 50 average change, which is to say a 0.98 chance of no change. Thus, the unit of evolution would mutate one in 50 amino acids on average.

Once we have the basic matrix M we can derive transition probabilities for larger amounts of evolution. For instance, what is the probability that a will change into b in two PAM units of evolution? Well, in the first unit period a changes into any amino acid c, including itself, with probability M_{ac} and then c changes into b in the second period with probability M_{cb}. Adding this all up, we conclude that the final figure is nothing more than M^2_{ab}, that is, an entry in the square of M. In general, M^k is the transition probability matrix for a period of k units of evolution.

One interesting fact here (that can be proved) is that as k grows very large, say, on the order of a thousand, M^k converges to a matrix with identical rows. Each row will contain the relative frequency p_b in column b. That is, no matter what amino acid you pick to start with, after this long period of evolution the resulting amino acid will be b with probability p_b.

We are now ready to define the scoring matrices. The entries in these matrices are related to the ratio between two probabilities, namely, the probability that a pair is a mutation as opposed to being a random occurrence. This is called a *likelihood* or *odds* ratio.

Let us then compute this ratio for two amino acids a and b. Suppose that we paired a with b in a given alignment. Taking the point of view of a, the probability that b is there in the other sequence due to a mutation is M_{ab}. On the other hand, there is a chance of p_b for a random occurrence of b. The ratio is then

$$\frac{M_{ab}}{p_b}.$$

The reasoning is the same even if b equals a. The actual score is 10 times the logarithm of this ratio, because when we align several pairs the sum of the individual scores will then correspond to the product of the ratios. In practice, the actual values used are rounded to the nearest integer to speed up calculations. The logarithm is multiplied by 10 to reduce the discrepancy between the correct value and the integer approximation.

The foregoing discussion refers to 1 PAM. But we can use exactly the same scheme for an arbitrary evolutionary distance. The scoring matrix for k PAM distance is defined as follows:

$$score_k(a, b) = 10 \log_{10} \frac{M^k_{ab}}{p_b}.$$

It may not be clear at first sight, but this is actually a symmetric matrix. Scoring matrices should be symmetric to yield a comparison score that does not depend on which sequence is given first. It is easy to verify the symmetry for $k = 1$, since

$$\frac{M_{ab}}{p_b} = \frac{f_{ab} m_a}{f_a p_b} = \frac{f_{ab}}{100 f p_a p_b}.$$

For larger values of k, the symmetry can be shown to hold as well (see Exercise 15). This stems from the fact that the accepted mutations considered are undirected so that the same quantity $f_{ab} = f_{ba}$ is used in the computation of the probability of a changing into b and of b changing into a.

In spite of being generated from observed accepted mutations and overall amino acid frequencies alone, these scoring matrices end up reflecting several important chemical and physical properties of amino acids, such as their affinity to water molecules and their size. In fact, it is usually the case that residues with similar properties have high pairwise scores and that unrelated residues have lower scores. This should not be too surprising, because similar amino acids are more frequently interchanged in accepted mutations than are unrelated residues. Substitutions of similar residues tend to preserve most properties of the protein involved.

We said in the beginning that we need to fix the PAM distance in order to pick a scoring matrix and perform our comparisons. But sometimes we have two sequences and no information whatsoever on their true evolutionary distance. In this case, the recommended approach is to compare the sequences using two or three matrices that cover a wide range, for instance, 40 PAM, 120 PAM, and 250 PAM. In general, low PAM numbers are good for finding short, strong local similarities, while high PAM numbers detect long, weak ones.

3.5.2 BLAST

In this section we give an overview of the BLAST family of sequence similarity tools. The BLAST programs are among the most frequently used to search sequence databases worldwide. BLAST is an acronym for Basic Local Alignment Search Tool.

It is important to observe that here the term *database* refers simply to a usually large set of catalogued sequences. It does not imply any extra capabilities of fast access, data sharing, and so on, commonly found in standard database management systems. For us, therefore, this "database" is merely a collection of sequences, although sequence information is copiously complemented with additional information such as the origin of the data, bibliographic references, sequence function (if known), and others.

BLAST returns a list of *high-scoring segment pairs* between the query sequence and sequences in the database. Before explaining what this means and how BLAST obtains its results, we will briefly introduce some terminology on segment pairs to keep this discussion consistent with the original paper describing BLAST.

A *segment* is a substring of a sequence. Given two sequences, a *segment pair* between them is a pair of segments of the same length, one from each sequence. Because the substrings in a segment pair have the same length, we can form a gapless alignment with them. This alignment can be scored using a matrix of substitution scores. No gap-penalty functions are needed, as there are no gaps. The score thus obtained is by definition the score of the segment pair. An example of a segment pair scored with the PAM120 substitution matrix is given in Figure 3.17. Segment pairs are basically gapless local alignments.

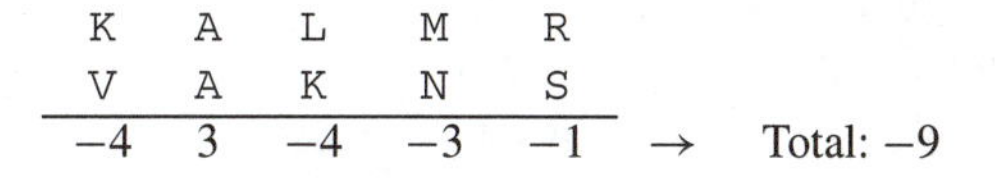

FIGURE 3.17

Segment pair and its score under PAM120.

We are now in possession of all the information necessary to describe precisely what BLAST does. Given a query sequence, BLAST returns all segment pairs between the query and a database sequence with scores above a certain threshold S. The parameter S can be set by the user, although a default value is provided in most servers that run the program. We have mentioned that a characteristic of reported alignments is the absence of gaps. This is actually a major reason why BLAST is so fast, since looking for good alignments with gaps is a good deal more time consuming.

A *maximum segment pair* (MSP) between two sequences is a segment pair of maximum score. This score is a measure of sequence similarity and can be computed precisely by dynamic programming. However, BLAST estimates this number much faster than any dynamic programming method. The program also returns *locally* maximal segment pairs, that is, those that cannot be improved further by extending or shortening them.

How BLAST Works

The BLAST approach to computing high-scoring segment pairs is as follows. It finds certain "seeds," which are very short segment pairs between the query and a database sequence. These seeds are then extended in both directions, without including gaps, until the maximum possible score for extensions of this particular seed is reached. Not all extensions are looked at. The program has a criterion to stop extensions when the score falls below a carefully computed limit. There is a very small chance of the right extension not being found due to this time optimization, but in practice this tradeoff is highly acceptable.

We may think of BLAST as a three-step algorithmic procedure, undertaking the following tasks.

1. Compile list of high-scoring strings (or *words,* in BLAST jargon).

2. Search for hits — each hit gives a seed.

3. Extend seeds.

The particular algorithmic steps depend on the type of sequences compared: DNA or protein. We detail each case next.

For protein sequences, the list of high-scoring words consists of all words with w characters (called w-mers) that score at least T with some w-mer of the query, using a PAM matrix to compute scores. Here w and T are program parameters. Note that this list may not contain all query w-mers! If a query w-mer consists of very common amino acids, it may be left out because even its score with itself may fall below T. However, there is an option to force inclusion of all query w-mers. The recommended value of w, the seed size, is 4 for protein searches.

Two approaches are tried for scanning the database in search for hits in the list constructed in the previous step. One of them is to arrange the list words into a hash table. Then, for each database word of size w, it is easy to get their index in the table and compare it to the words there, which will be a small fraction of all list words.

The second method uses a *deterministic finite automaton* to search for hits. This device has states and transitions and operates like a machine. It begins in a fixed initial state, and for each character in the database a transition is made to another state. Depending on the state and on the transition, a word from the list is recognized. The automaton is built only once, using the list of high-scoring words as input, and is a compact way of storing all these words. The search is fast, as it requires only a transition per character.

The final extension is straightforward. As we mentioned, to save time the algorithm stops when the score falls a certain distance below the best one obtained to that point for shortest extensions. This is done in both directions, and the high-scoring segment pair originated from this seed is kept. There is a small probability of missing important extensions, but it is negligible.

For DNA searches, the initial list contains only the query w-mers. Because scoring of DNA sequences is easier, this is enough for all practical purposes. The scanning strategy is radically different from the protein case. Taking advantage of the fact that the alphabet size is 4, the database is first compressed so that each nucleotide is represented using 2 bits. Four nucleotides fit in a byte. Apart from the space saved, the search can now be made much faster, because a byte is compared each time. There is an extra filtering step that removes from the initial list very common words from the database, to avoid a large number of spurious hits.

Extension is done in a way similar to the case of proteins. In both cases, the extension is based on a well-founded statistical theory that gives exact distribution of gapless local maximum score for random sequences, and permits a very accurate computation of the probability that the segment pair found could be possible due to chance alone. The smaller this probability, the more significant is the match.

We briefly sketch now the main points of this statistical theory. The distribution of the MSP score (defined above) for random sequences s and t of lengths m and n, respectively, can be accurately approximated as described next. This approximation gets better as m and n increase.

Given a matrix of replacement costs s_{ij} for the pairs of characters in the alphabet, and the probability p_i of occurrence of each individual character in the sequences, we first compute a value λ, solving the equation

$$\sum_{i,j} p_i p_j e^{\lambda s_{ij}} = 1.$$

The parameter λ is the unique positive solution to this equation and can be obtained by Newton's method. Once λ is known, the expected number of distinct segment pairs between s and t with score above S is

$$Kmne^{-\lambda S},$$

where K is a calculable constant. Actually, the distribution of the number of segment pairs scoring above S is a Poisson distribution with mean given by the previous formula. From this, it is easy to derive expressions for useful quantities like the average score, intervals where the score will fall 90% of the time, and so on.

3.5.3 FAST

FAST is another family of programs for sequence database search. The first program to be made generally available, FASTP, is designed to search a database of sequences looking for similarities to a given query sequence. The basic algorithm used by FASTP is to compare each database string in turn to the query and report those found significantly similar to it, along with alignments and other relevant information. The speed of FASTP is therefore mainly due to its ability to compare two sequences very quickly. Accordingly, our focus in this section is on how FASTP performs this basic step.

Let s and t denote the two sequences being compared, and let us assume for the moment that they are protein sequences. Their lengths are denoted by $m = |s|$ and $n = |t|$. The comparison starts by determining k-tuples common to both sequences, with $k = 1$ or 2. The value of k is a parameter called *ktup* in the program. In addition, the *offset* of a common k-tuple is important in the algorithm. The offset is a value between $-n + 1$ and $m - 1$ that determines a relative displacement of one sequence relative to the other. Specifically, if the common k-tuple starts at positions $s[i]$ and $t[j]$, we say that the offset is $i - j$.

The following data structures are needed: (1) a lookup table, and (2) a vector indexed by offsets and initialized with zeros. Sequence s is scanned and a table listing all positions of a given k-tuple in s is produced (see Figure 3.18). Then sequence t is scanned and each k-tuple in it is looked up in the table. For all common occurrences the vector entry of the corresponding offset is incremented. Figure 3.18 exhibits the final contents of the offset vector and the lookup table for sequences s = `HARFYAAQIVL` and t = `VDMAAQIA`. Notice that offset $+2$ has the highest entry value, meaning that many matches were found at this offset. This method is known as the *diagonal* method, because an offset can be viewed as a diagonal in a dynamic programming matrix. One possible use of the highest offset is to run a dynamic programming algorithm around the diagonal corresponding to this offset.

But FASTP does a more detailed analysis of the common k-tuples and joins two or more such k-tuples when they are in the same diagonal and not very far apart. The exact criteria are heuristic. These combined k-tuples form what is called a *region*. A region can be thought of as a segment pair, in the terminology of BLAST, or else as a gapless local alignment. Regions are given a certain score depending on the matches and mismatches contained in them. The important thing to remember is that regions have no gaps.

The next step consists in rescoring the five best regions received from the previous

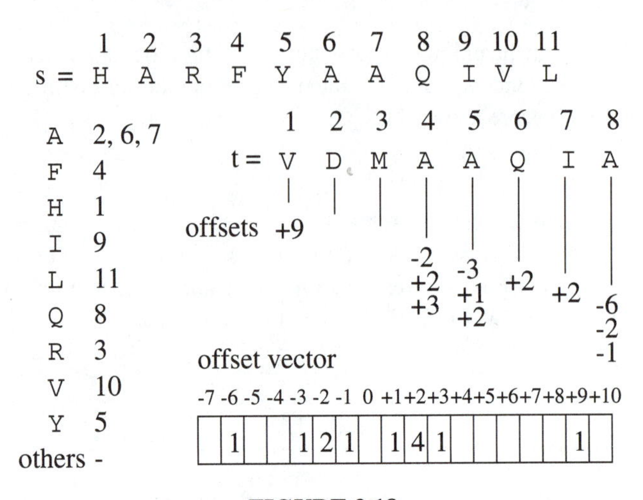

FIGURE 3.18

Lookup table and offset vector for sequence comparison by FASTP. The value of ktup is 1.

phase using a PAM matrix, usually PAM120 or PAM250. The best of these new scores is a first measure of the similarity between s and t and is termed the *initial score*. An initial score is computed for every database sequence with respect to the query sequence. These values are reported in histogram form along with the mean score. The initial score is also used to rank all database sequences. For the highest ranking such sequences, an optimized score is computed running a dynamic programming algorithm restricted to a band around the initial alignment — the one that produced the initial score. This procedure is much like the method exposed in Section 3.3.4. In practice, when sequences are truly related, the optimized score is usually significantly higher than the initial score. This observation often helps distinguish between good alignments occurring by chance and true relationships.

The value of *ktup* affects the performance of the algorithm in terms of its sensitivity and selectivity. *Sensitivity* is the ability of a search tool to recognize distantly related sequences. *Selectivity* is the ability of the tool to discard false positives — matches between unrelated sequences. In general, sensitivity and selectivity are opposite goals. When using FASTP and other programs in the family, low *ktup* increases sensitivity and high *ktup* favors selectivity.

Apart from the search tools, the FAST family includes a program that is useful in assessing the statistical significance of a score. This tool works by scrambling one of the sequences, say, t, maintaining the amino acid composition but changing their order, and running a full dynamic programming algorithm between the scrambled version of t and the original s. This is repeated many times, so that the average score and its standard deviation can be computed. From these, a z-value is determined by the formula

$$z = \frac{\text{score} - \text{average score}}{\text{standard deviation of scores}}.$$

However, since the statistical distribution of similarity scores for random sequences is not a normal distribution, these z-values have limited usefulness.

Improvement to the FASTP program led to FASTA. One added feature is the ability to process DNA as well as protein sequences. In this respect, FASTA can be seen as a combination of FASTP with FASTN, a program designed specifically to work with nucleotide sequences. Another program in the family — TFASTA — is devoted to comparing a protein query sequence to a DNA database, doing the necessary translations as it goes.

Another addition is related to the computation of initial scores. FASTA takes an extra step after the best regions have been selected and tries to join nearby regions, even if they do not belong to the same diagonal. With this, initial scores improve significantly for related sequences and get closer to the improved scores ("optimized scores" in FASTA parlance). Furthermore, the ten best regions are kept in FASTA as opposed to five best in FASTP. Other programs using the same techniques are also part of the package. LFASTA is a tool for local similarity, in the sense that it reports more than one good alignment between a pair of sequences.

The statistical significance tool was also improved. Its most outstanding addition is the possibility of doing local shuffling of sequences. It was observed that very often a biologically unimpressive alignment got a high z-value because shuffling destroyed an uneven distribution of residues in the sequence. Local shuffling mitigates this problem. Shuffling is done in blocks of 10 to 20 residues, thus yielding shuffled sequences that are random yet have the same local composition of the original one. Then, if high scores were due to biased local composition, this will affect the mean score of the locally shuffled sequences. Other improvements in this tool include flexibility in choosing the scoring matrix and calculation of more scores for each shuffled sequence.

3.6 OTHER ISSUES

We now study some miscellaneous topics related to sequence comparison. The first has to do with the notion of *distance* between sequences and what its relationship is to similarity. In the second topic we discuss rules for the various choices we have when comparing sequences. Finally we briefly discuss the topics of string matching and exact sequence comparison.

⋆ 3.6.1 SIMILARITY AND DISTANCE

Similarity and distance are two approaches to comparing character strings. In the similarity approach, we are interested in the best alignment between two strings, and the score of such an alignment gives a measure of how much the strings are alike. In the distance approach, we assign costs to elementary edit operations and seek the less expensive composition of these that transforms one string into the other. Thus, the distance is a measure of how much the strings differ. In either case we are looking for a numeric value that measures the degree by which the sequences are alike or are different.

Up to now we have concentrated on computing similarity. This section introduces the concept of distance and relates it to similarity. We show that in many cases these two measures are related by a simple formula, so that we can easily obtain one measure from the other.

We must stress that our study of distances is restricted to global comparison only. The distance approach is not suitable for local comparison. This limitation is one of the reasons we based this chapter on the similarity approach. Table 3.2 summarizes the main differences between similarity and distance. Some of the terms may not be clear to the reader now, but they are explained in what follows.

TABLE 3.2

Summary of properties of similarity and distance.

	Similarity	*Distance*
Triangle inequality?	no	yes
Local comparison?	yes	no
$p(a, a) \neq p(b, b)$ possible?	yes	no

We start by defining precisely our notions of similarity and distance. Let s, t be two sequences over an alphabet Σ. This alphabet will usually be the DNA or amino acid alphabet, but the results here hold for more general cases. Recall that we are looking for a number that measures how much the strings are similar or are different. As before, we denote similarity by $\text{sim}(s, t)$, distance by $\text{dist}(s, t)$.

Similarity

A similarity measure is always based on alignments. Let us recall and refine the precise definition of alignment. An *alignment* between s and t is a pair of sequences (s', t') obtained from s and t, respectively, by insertion of space characters in them. The alignment $\alpha = (s', t')$ must satisfy:

1. $|s'| = |t'|$.
2. Removal of all spaces from s' gives s.
3. Removal of all spaces from t' gives t.
4. For each i, either $s'[i]$ or $t'[i]$ is not a space.

An alignment creates a connection between symbols $s'[i]$ and $t'[i]$ that occupy the same position i in each sequence. Symbols $s'[i]$ and $t'[i]$ are said to be aligned under α.

The similarity is the highest score of any alignment. We assume an additive scoring system. Let us also review and refine what we mean by a scoring system. A scoring system is composed of a pair (p, g) whose members are a function $p : \Sigma \times \Sigma \mapsto \mathbf{R}$, used to score pairs of aligned characters, and a space penalty g used to penalize spaces. Usually, $g < 0$, but we do not require this. With such a scoring system we are able to assign a numerical value, or *score*, to each possible alignment. We add $p(a, b)$ each time a is matched with b in α, and add g every time a character a is paired with a space symbol. The total sum is the score of α, denoted $score(\alpha)$. The *similarity* between two sequences s and t according to this scoring system is

$$\text{sim}(s,t) = \max_{\alpha \in \mathcal{A}(s,t)} score(\alpha),$$

where $\mathcal{A}(s,t)$ is the set of all alignments between s and t. This scoring system is called *additive* because if we cut any alignment in two consecutive blocks, the score of the entire alignment is the sum of the scores of the blocks.

Distance

A *distance* on a set E is a function $d : E \times E \mapsto \mathbf{R}$ such that

1. $d(x,x) = 0$ for all $x \in E$ and $d(x,y) > 0$ for $x \neq y$
2. $d(x,y) = d(y,x)$ for all $x, y \in E$ (d is *symmetric*)
3. $d(x,y) \leq d(x,z) + d(y,z)$ for all x, y, and $z \in E$.

The third condition is known as the *triangle inequality*. This property is very useful in many contexts, and many algorithms rely on its validity. In the case of strings it is possible to define a distance on the set of all strings over Σ based on the amount of effort needed to transform one of them into the other.

By successive application of a number of admissible operations, any string can be transformed into any other. If we assign a cost to each admissible operation, we can define the distance between two strings as the minimum total cost needed to transform one into the other. Admissible operations are as follows.

1. Substitution of a character a by a character b
2. Insertion or deletion of an arbitrary character

To charge for these operations, we use a *cost measure* (c,h), where c is a function $c : \Sigma \times \Sigma \mapsto \mathbf{R}$ and h is a real value. The substitution of b for a costs $c(a,b)$. Insertion or deletion of a character costs h. The cost of a series σ of operations is just the sum of individual costs and is denoted by $cost(\sigma)$.

The *distance* between two sequences s and t according to a cost measure is

$$\text{dist}(s,t) = \min_{\sigma \in \mathcal{S}(s,t)} cost(\sigma),$$

where $\mathcal{S}(s,t)$ is the set of all series of operations transforming s into t.

Some restrictions on (c,h) are necessary to make this definition sound. First of all, we deal only with nonnegative cost values, otherwise the minimum would not make sense. We also assume that the cost function c is a symmetric function, otherwise the distance would not be necessarily symmetric. These minimal requirements guarantee that $\text{dist}(s,t)$ is symmetric and satisfies the triangle inequality. To make sure it is a distance, we also need the property that $c(a,b) > 0$ for $a \neq b$ and that $h > 0$. This is necessary to avoid cases where two sequences s and t are not equal but $\text{dist}(s,t) = 0$. We want the distance to be zero only when the sequences are identical.

We will also assume that c satisfies the triangle inequality, that is,

$$c(x, y) \leq c(x, z) + c(y, z) \tag{3.15}$$

for all x, y, and z in Σ. The reason for this last assumption is that even if we start with a pair (c, h) that does not satisfy it, we can always define a new pair (c', h) that does satisfy it and produces the very same distance function. For instance, if three characters x, y, and z are such that $c(x, y) > c(x, z) + c(z, y)$, then every time we need to replace x by y we will not do it directly but rather replace x by z and later z by y, producing the same effect at a lower cost. For all practical matters, it looks as though the effective cost of replacing x by y is $c(x, z) + c(z, y)$, not $c(x, y)$, as the former is the actual cost incurred for the substitution. By using (3.15) we avoid such situations.

Example 3.3 Taking

$$c(x, y) = \begin{cases} 0 & \text{if } x = y, \\ 1 & \text{if } x \neq y, \end{cases}$$

and $h = 1$, we have the so-called *edit distance* between two sequences, also known as the Levenshtein distance.

Computing the Distance

Given a cost measure (c, h) and a constant M, we can define a scoring system (p, g) as follows:

$$p(a, b) = M - c(a, b), \tag{3.16}$$

$$g = -h + \frac{M}{2}. \tag{3.17}$$

If we have an alignment α between two sequences s and t, it is possible to define a series σ of operations such that

$$score(\alpha) + cost(\sigma) = \frac{M}{2}(m + n), \tag{3.18}$$

where $m = |s|$ and $n = |t|$. To get σ, it suffices to divide the alignment α in columns as we did previously to compute the score. The columns correspond to admissible operations in a natural way. Character matches correspond to substitutions. Spaces correspond to insertions or deletions. The operations can be applied in any order because they act on disjoint regions of the alignment and do not interfere with one another.

Let us call σ the series of these operation done left to right. We shall now compute the score of α and the cost of σ. Suppose there are exactly l character matches in α, with the ith match formed by a_i in s and b_i in t. Suppose further that there are exactly r spaces in α. Granted that, we have the following expression for the score of α.

$$score(\alpha) = \sum_{i=1}^{l} p(a_i, b_i) + rg.$$

On the other hand, the cost of σ is

$$cost(\sigma) = \sum_{i=1}^{l} c(a_i, b_i) + rh.$$

Memberwise addition of these equations and relations (3.16) and (3.17) give

$$score(\alpha) + cost(\sigma) = lM + r\frac{M}{2}. \tag{3.19}$$

Observe that the values of l and r are not independent. They must satisfy a relation involving also the total number of characters in the two sequences. Indeed, each match used two characters, while a space uses just one character. Therefore, the total number of characters $m + n$ must be

$$m + n = 2l + r. \tag{3.20}$$

With this, we can see that (3.19) can be rewritten as

$$score(\alpha) + cost(\sigma) = \frac{M}{2}(m + n),$$

which is exactly Equation (3.18).

This is true for any alignment. In particular, if α is an optimal alignment, we have

$$\text{sim}(s, t) + cost(\sigma) = \frac{M}{2}(m + n),$$

where σ represent the series of operations derived from α. Since the distance between s and t is the minimum cost of any series of operations, we end up with

$$\text{sim}(s, t) + \text{dist}(s, t) \leq \frac{M}{2}(m + n).$$

We now summarize the above result in a theorem for future reference.

THEOREM 3.2 Given a cost measure (c, h) and the corresponding scoring system as above for a certain M, for every pair of strings s and t we have

$$\text{dist}(s, t) + \text{sim}(s, t) \leq \frac{M}{2}(|s| + |t|).$$

Theorem 3.2 gives us an upper bound for the value of $\text{dist}(s, t)$. We now want to show that this is also a lower bound and that the distance can in fact be obtained from the similarity in a straightforward fashion.

THEOREM 3.3 Under the conditions of Theorem 3.2, we have

$$\text{sim}(s, t) + \text{dist}(s, t) \geq \frac{M}{2}(|s| + |t|).$$

Proof. We follow the steps of Theorem 3.2. First of all, given a series σ of admissible operations, we build an alignment α with the property that

$$score(\alpha) + cost(\sigma) \geq \frac{M}{2}(m + n). \tag{3.21}$$

Notice that we cannot have equality here in general, as we did in the previous proof.

This case is a little bit different, because a series of operations can go around and around and the corresponding alignment will show only the net effect. For instance, if σ at some point inserts a character a somewhere and then immediately removes it, the net effect is null and we are charged for two operations just the same. The foregoing equation has to account for all series of operations, be them efficient or not, and that is why we have an inequality instead of an equality there.

Let us prove relation (3.21) by induction on the number of operations in σ. If $|\sigma| = 0$, no operation was performed, so $s = t$. The alignment (s, s) has score $Mm = Mn = M(m+n)/2$, and the relation holds in this case.

If σ has at least one operation, consider its last operation. We can write $\sigma = \sigma' u$, where u is this last operation, so that

$$s \xrightarrow{\sigma'} t' \xrightarrow{u} t,$$

that is, t' is the string obtained from s by applying all but the last operation, and t is obtained from t' by applying u.

Our induction hypothesis is that we have an alignment α' between s and t' such that

$$score(\alpha') + cost(\sigma') \geq \frac{M}{2}(m + n'),$$

where $n' = |t'|$. Because $cost(\sigma) = cost(\sigma') + cost(u)$, all we need is to find an alignment α between s and t such that

$$score(\alpha) \geq score(\alpha') + \frac{M}{2}(n - n') - cost(u). \tag{3.22}$$

Memberwise addition of these two inequalities will give us the desired result. There are three cases to consider, according to the type of operation that u is: substitution, insertion, deletion.

Case 1: u is a substitution.

In this case $cost(u) = c(a, b)$, $n = n'$, and t differs from t in just one position, as shown here:

s	$\ldots x \ldots$
t'	$\ldots a \ldots$
t	$\ldots b \ldots$

We can build α simply replacing a by b in α', which in the figure corresponds to ignoring the middle line. If the symbol x in s is an ordinary character, then

$$\begin{aligned} score(\alpha) &= score(\alpha') - p(a, x) + p(b, x) \\ &= score(\alpha') - M + c(a, x) + M - c(b, x) \\ &= score(\alpha') + c(a, x) - c(b, x) \\ &\geq score(\alpha') - c(a, b), \end{aligned}$$

where the last step is a consequence of the triangle inequality for c. Thus, we have what we need in this case.

If x is a space, the scores of α and α' are equal, hence

$$score(\alpha) = score(\alpha') \geq score(\alpha') - c(a, b)$$

because $c(a, b) \geq 0$ by hypothesis. This concludes the case where u is a substitution.

Case 2: u is an insertion.

In this case $cost(u) = h$ and $n' = n - 1$. The alignment α' has to be "opened" in a certain position to allow the insertion of a character, as shown here:

$$\begin{array}{lc|c} s & \dots & \dots \\ t' & \dots & \dots \end{array} \qquad \begin{array}{lc|c|c} s & \dots & - & \dots \\ t & \dots & \mathrm{a} & \dots \end{array}$$

An extra space is introduced in s. With this, the difference in score between α and α' is due to this extra space in α:

$$\begin{aligned} score(\alpha) &= score(\alpha') + g \\ &= score(\alpha') - h + \frac{M}{2}. \end{aligned}$$

This proves the validity of relation (3.22) in the case where u is an insertion.

Case 3: u is a deletion.

Now $cost(u) = h$ again but $n' = n + 1$. We have this situation:

$$\begin{array}{ll} s & \dots x \dots \\ t' & \dots a \dots \\ t & \dots - \dots \end{array}$$

We distinguish two subcases, according to x being a space or not. If x is a character, then

$$\begin{aligned} score(\alpha) &= score(\alpha') - p(a, x) + g \\ &= score(\alpha') - M + c(a, x) - h + \frac{M}{2} \\ &= score(\alpha') - h - \frac{M}{2} + c(a, x) \\ &\geq score(\alpha') - h - \frac{M}{2}, \end{aligned}$$

as desired, because $c(a, x) \geq 0$. If x is a space, we have

$$\begin{aligned} score(\alpha) &= score(\alpha') - g \\ &= score(\alpha') - h + \frac{M}{2} \\ &\geq score(\alpha') - h - \frac{M}{2}, \end{aligned}$$

since $h > 0$.

This completes the inductive argument and shows that relation (3.21) holds. In particular, it holds for a minimum cost σ and therefore

$$score(\alpha) + \mathrm{dist}(s, t) \geq \frac{M}{2}(m + n),$$

or taking into account that $\text{sim}(s, t)$ is the largest possible alignment score between s and t,

$$\text{sim}(\alpha) + \text{dist}(s, t) \geq \frac{M}{2}(m + n),$$

proving the theorem. ■

Combining Theorems 3.2 and 3.3, we can write

$$\text{sim}(\alpha) + \text{dist}(s, t) = \frac{M}{2}(m + n), \tag{3.23}$$

which shows us how to compute the distance given the similarity. Thus, distance computations can be reduced to similarity computations. To compute a distance, all we need to do is select a suitable M, define scoring parameters p and g as in (3.16) and (3.17), and apply one of the algorithms for global comparison we have seen so far. The resulting similarity is converted to distance by the above formula.

For instance, in the case of the edit distance, we may choose $M = 0$ and run a similarity algorithm with match $= 0$, mismatch $= -1$, and space $= -1$. Or we may take $M = 2$ and have match $= 2$, mismatch $= 1$ and space $= 0$. Both scoring systems yield the same optimal alignments, although with different scores. But after applying formula (3.23) the distance is the same.

Before we close, a comment on the constant M is in order. It may seem suspicious that any value of M will do for Theorem 3.2. If M is a large, positive number, the scoring system may result in negative values of the space penalties $w'(k)$ for many, and possibly all, values of k. This contradicts our intuition, given that spaces should be penalized instead of rewarded. The same goes for a negative, but very large in absolute value, M. The function p this time would be negative, again contradicting our intuition. The reason for this apparent anomaly lies in the fact that we consider global comparisons only. When we change the value of M, all alignments increase or decrease on their score in a uniform manner, so that the optimal ones always remain the same. For local alignments, Equation (3.20) is not valid, and by varying M we can give preference to longer or shorter local alignments.

3.6.2 PARAMETER CHOICE IN SEQUENCE COMPARISON

In this section we present considerations concerning the choice of parameters in a scoring system and the choice of algorithm given the particular sequence comparison that must be made.

Many issues must be taken into consideration when choosing the scoring system for a particular application. This includes, in its simplest form, the scores for a match (M), for a mismatch (m), and for a space ($g < 0$).

In any scoring system it is important to assure that a match is worth more than a mismatch, so that we encourage alignments of identical characters. Another rule that is normally used is to assure that a mismatch is preferred over a pair of spaces. For instance, the following leftmost alignment should score higher than the rightmost one:

```
A  -A
C  C-.
```

The rules above translate at once into inequalities involving the scoring parameters:

$$2g < m < M.$$

Notice that if we multiply all weights by a positive constant, the optimal alignments remain the same. This property can be used to transform all weights to integers, which are generally processed with much greater speed than floating point numbers in the majority of modern computers.

Consider now the transpositions. If we have, for instance, sequences `AT` and `TA`, there are essentially two alignments that compete for the best score:

```
AT  -AT
TA  TA-.
```

(There is still a third alignment with the same score as the second one above aligning the `T`'s.) The corresponding scores are $2m$ and $M + 2g$, respectively. Thus, if m is equal to the arithmetic mean between M and $2g$, the two preceding alignments are equivalent in terms of score. To give preference to one of them, we must choose m closer to one of the extremes of the interval $[2g, M]$.

In the same vein, it is possible to imagine other instances of pairs of short sequences and postulate which is the most desirable alignment in each case. This gives us more inequalities involving m, M, and g. For instance, when comparing `ATCG` to `TCGA`, we may prefer the first of the two following alignments:

```
ATCG-  ATCG
-TCGA  TCGA.
```

This will be reflected in the score if $4m < 3M + 2g$. The values used in Section 3.2.1 were chosen based on such criteria, among other reasons.

In practice, scoring systems more sophisticated than the simple M, m, g method are often needed. The subadditive space penalty functions mentioned in Section 3.3.3 are usually preferred. Among these, affine functions are very popular because of the quadratic running time algorithm, as opposed to cubic for general functions. Another important property, especially in comparisons of protein sequences, is the ability to distinguish among the various matchings of amino acids. A match involving amino acids with similar chemical or physical characteristics, such as size, charge, hydrophobicity, and so on, receives more points than a matching between not so similar ones. In this context, scoring systems based on identity only are in general insufficient. For protein comparison, the use of PAM matrices is commonplace.

Nevertheless, the simple identity/nonidentity method is versatile enough to include as particular cases several well-known problems in sequence comparison. One of these problems is to find the *longest common subsequence* (LCS) of a pair of sequences. This is equivalent to using $M = 1, m = g = 0$ in the basic algorithm. Thus, we can solve this problem in $O(mn)$ time and $O(\min(m, n))$ space for sequences of sizes m and n. This problem has received a great deal of attention, and several faster algorithms have been described for particular cases.

We now discuss the choice of algorithms. The decision whether to charge for end spaces and the choice of local or global methods depends heavily on the kind of application we are interested in and the results sought. If we want to compare sequences that are approximately the same length and relatively alike, a global comparison charging for all spaces is likely to be more appropriate. For instance, here we could include the case of two tRNAs of different organisms, or else two tRNAs of the same organism but carrying different amino acids.

If, on the other hand, one of the sequences is short and the other much longer, it is more advisable to charge for end spaces in the shorter sequence only. This will allow us to find all approximate occurrences of the short sequence in the long one. Such a search is useful when trying to locate relatively well-conserved structures in recently sequenced DNA.

Local comparison should be used when we have two relatively long sequences that may contain regions of high similarity. A typical case is protein sequences with similar functionality from reasonably well-separated organisms in terms of evolution. Because the proteins perform similar functions, it is probable that some high-similarity regions (active sites, motifs, functionally equivalent structures, etc.) exist, separated by unrelated regions that accumulated mutations and that do not have much influence on the protein's functionality.

If sequence comparison is done to test the hypothesis of common origin, care must be used when interpreting results. In general, optimal alignments are the most unlikely to have occurred by chance in some probabilistic sense. However, it is always advisable to compare the score obtained to what would be expected on average from completely unrelated sequences with the same characteristics as the two sequences compared. If the optimal score is well above average, this is a good indication that the similarity between the sequences is not due to chance. Even then, this result per se does not imply homology or any kind of evidence of common origin. Further experiments, based on the information the alignment gives, are in general carried out to strengthen or refute the hypothesis of common ancestry. On the other hand, if the similarity is nearly equal to what is expected by chance, it is likely that the sequences are unrelated. However, in biology there are no rules without exceptions, and cases of homologous proteins with no traces of similarity at the sequence level are known. Other pieces of evidence, such as three-dimensional structure, were used in these cases to ascertain homology.

3.6.3 STRING MATCHING AND EXACT SEQUENCE COMPARISON

Two other important problems that have relevance in computational molecular biology are the *string matching problem* and *exact sequence comparison*. In string matching we are given a string s, $|s| = n$, and a string t, $|t| = m$, and we want to find all occurrences of t in s. In other words, is t a substring of s? If it is, what are all the positions in s where we can find t? This is a classic computer science problem, and can be solved efficiently. In fact, there are algorithms that can solve this problem in time $O(n+m)$, which is much faster than the quadratic complexity of the basic algorithm for two-sequence comparison. We will not describe such algorithms here; they can be found on any good textbook on algorithms, such as those referenced in the bibliographic notes of Chapter 2. We note that efficient algorithms have also been developed for the *approximate* string matching problem, where we allow a certain number of errors when looking for a match. Such algorithms are important, of course, in molecular biology. References are given in the bibliographic notes.

We have devoted most of this chapter to sequence comparisons where we allow errors to be present in the sequences. This is certainly the case in the majority of applications in real-life sequence comparison, but in some situations we need *exact comparisons*. Many problems of this nature can be solved by a very useful and versatile structure known as *suffix tree*. A sample of such problems follows.

- We will see in Chapter 4 that repeated substrings of a sequence are one of the major problems when doing large-scale DNA sequencing. Given a DNA sequence, a natural problem then is to find its longest repeated substring.
- When using the polymerase chain reaction (described in Section 1.5.2), we need to find a short piece of DNA called a *primer* that does not occur in the DNA we want to multiply. This problem can be formalized as follows. Given two strings, A (long) and B (short), for each position i in B, find the shortest string from B starting at position i that is nowhere in A.
- We also saw in Section 1.5.2 that the restriction sites for restriction enzymes are even-length palindromes. Another problem is to find all maximal palindromes of a given DNA sequence.

We now describe what a suffix tree is. It basically contains all suffixes of a string s, factoring out common prefixes as much as possible in the tree structure. Formally, a suffix tree for $s = s_1 s_2 \ldots s_n$ is a rooted tree T with $n + 1$ leaves with the following properties:

- Edges of T are directed away from the root, and each edge is labeled by a substring from s.
- All edges coming out of a given vertex have different labels, and all such labels have different prefixes (not counting the empty prefix).
- To each leaf there corresponds a suffix from s, and this suffix is obtained by concatenating all labels on all edges on the path from the root to the leaf.

Figure 3.19 shows an example of a suffix tree. In order to avoid problems with the empty suffix, we usually append to s an extra character \$ not occurring anywhere else in the string. Suffix trees can be built using a nice algorithm that we will not describe but that runs in time $O(n)$, assuming that the alphabet size is a constant. Once a suffix tree is built, most problems using it can be solved in time $O(n)$. That is the case of the problems just described. The primer problem can be solved in time $O(|B|)$. We leave as exercises to the reader the task of figuring out how to apply suffix trees to solve such problems.

A related structure called *suffix array* can also be used to solve exact comparison problems. This structure consists basically of all suffixes of s lexicographically sorted.

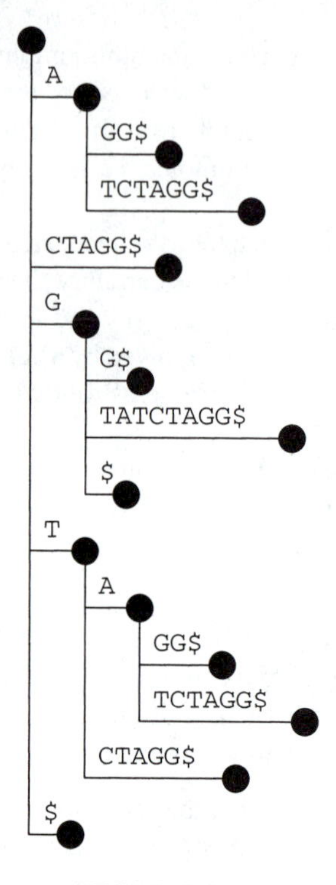

FIGURE 3.19

An example of a suffix tree for string GTATCTAGG. *A dollar sign marks the end of the string.*

SUMMARY

We began our study by looking at the comparison of just two sequences. An algorithm design technique called dynamic programming can be successfully applied in this case, leading to precise and efficient algorithms. We then studied several extensions to the basic dynamic programming algorithms. Some of these extensions allow us to save on the amount of time or space needed for the computation in certain cases. Others are aimed at solving more general forms of the problem necessary in some biological applications.

The following is a brief reminder of the various types of comparison and their uses:

- ***Global:*** compares sequences in their entirety
- ***Local:*** finds similar pieces in sequences
- ***Semiglobal:*** finds suffix-prefix or containment similarities

All these algorithms have quadratic complexity in both time and space. However, space-saving versions exist for all of them. These versions consume linear space, but may double the running time.

The following table summarizes the characteristics of the algorithms according to the type of gap penalty function used.

Type of function	*Expression*	*Complexity*	*Space used*
linear	$w(k) = ak$	$O(n^2)$	one array
affine	$w(k) = ak + b$	$O(n^2)$	three arrays
general	any $w(k)$	$O(n^3)$	three arrays

The multiple-sequence case was studied next. A dynamic programming approach here led to an inefficient algorithm, so other techniques had to be considered; we described a number of these alternative techniques. Another situation in which dynamic programming is not fast enough is when we want to search a database for a given sequence or pattern. We studied this problem in Section 3.5, examining heuristics used in software systems specifically designed for database search.

Most of this chapter is based on the notion of similarity. However, this is not the only approach to sequence comparison. The notion of distance, which attempts to measure by how much two sequences differ, is also important. In Section 3.6.1 we studied the relationship between these two concepts. We finished the chapter with a brief discussion on the several choices of comparison style and parameters one faces when entertaining a sequence comparison task.

EXERCISES

1. Find the correspondence between the algorithms listed below and the problems cited in Section 3.1.

   semiglobal comparison
   filter and then apply semiglobal comparison
   K-Band
   database search
   local comparison

2. Score alignment (3.1) of page 49 according to the following system: $p(a, b) = 1$ if $a = b$, $p(a, b) = 0$ if $a \neq b$, and $g = -1$.

3. Find all optimal alignments between AAAG and ACG, with the scoring system used in Section 3.2.1.

4. The downmost and upmost alignments (see end of Section 3.2.1) often comprise all optimal alignments between the two sequences compared. However, this is not always true. Find an example when this does not occur.

5. Consider the scoring system of Exercise 2 and the one used in Section 3.2.1. Find two sequences for which the optimal alignments are different for the two scoring systems.

6. Modify algorithm *Align* of Figure 3.3 so that it finds an optimal semiglobal alignment between s and t not charging for gaps in any extremity of s or t. Assume that array a

has been filled in the proper way for this style of comparison.

7. Find the best local alignment between ATACTACGGAGGG and GAACGTAGGCGTAT. Use your favorite scoring system.

8. Given two sequences, which value is larger: their local similarity or their global similarity? Why? How does their semiglobal similarity compare with the other two values?

★9. Develop linear space versions for local and semiglobal similarity, including the construction of an optimal alignment.

10. Decide whether the following statement is true or false and justify your answer. Two optimal alignments between s and t will have the same number of blocks (see Section 3.3.2 for a definition of alignment blocks).

11. Recall the concepts from Section 3.3.4. Find a pair of sequences such that the best score for $k = 0$ is the same as the best for $k = 1$, but strictly lower than the best for $k = 2$. Use $M = 1, m = g = -1$ in this problem.

12. Modify algorithm *KBand* (Figure 3.9) so that it handles sequences of unequal length. What is the resulting time complexity?

13. Find all optimal multiple alignments between ATC, CAGC, and CGC using the sum of pairs measure with $p(a, b) = 1$ if a and b are both letters and $a = b$; $p(a, b) = 0$ if a and b are both letters and $a \neq b$; and $p(a, -) = -1$.

14. How much space does algorithm *Multiple-Sequence Alignment* of Figure 3.14 need apart from the pool of relevant cells?

15. Prove that the k-PAM scores matrix is symmetric for every k.

16. We saw in the text that a 1-PAM matrix changes on average 1% of amino acids. Can we say that a 2-PAM matrix changes on average 2%?

17. Query sequence GAATTCCAATAGA against GenBank using BLAST. Which database sequences did you hit?

18. A gap penalty function w is *concave* when $\Delta w(k) \geq \Delta w(k+1)$. Prove that a concave function for which $w(0) = 0$ is subadditive. (See page 64 for a definition of $\Delta w(k)$.)

★19. True or false: A subadditive, nondecreasing gap penalty function w is necessarily concave.

20. Consider an alignment of length l between sequences s and t. Divide this alignment in two parts, one going from column 1 to column k, and the other going from $k + 1$ to l. Which properties of the scoring system guarantee that the score of whole alignment is the sum of the scores of its parts, for any k?

21. Is the assignment presented in Example 3.2 optimal?

22. According to the relationship between similarity and distance explored in Section 3.6.1, give one possible scoring system corresponding to edit distance.

◇23. Given two random DNA sequences of lengths m and n, respectively, find the distribution of the global similarity between them. Does it approximate any known distribution as n and m grow?

24. Design an algorithm to find the *Shortest Common Supersequence* between two sequences. *Hint:* This can be done with a simple adaptation of the basic sequence comparison algorithm.

25. Show how to use suffix trees to solve the problems mentioned in Section 3.6.3.

BIBLIOGRAPHIC NOTES

Sankoff and Kruskal's book [168] is a classic reference in sequence comparison. The work by Waterman [196] is a good review on optimal alignments, similarity, distance, and related algorithms. Pearson and Miller's review [156] concentrates on dynamic programming methods and is rather complete. Heijne's book [191] is another useful source on sequence comparison.

The well-known paper by Needleman and Wunsch [146] is generally considered the first important contribution in sequence comparison from the point of view of biologists, although S. Ulam had already considered distances on sequence spaces in the 1950s. The algorithm described by Needleman and Wunsch has a fixed penalty for a gap, independent of its length. In that paper no complexity analysis was given. Later analysis showed that the algorithm as presented ran in cubic time. Now we know how to perform the same computation in quadratic time, using the algorithm for affine space penalty functions, which includes constant functions. In any case, the name "Needleman and Wunsch algorithm" is often used to designate any kind of global alignment algorithm based on dynamic programming.

A similar phenomenon occurred with local alignments. The seminal paper of Smith and Waterman [175] was very influential, lending its name to designate almost any local alignment dynamic programming algorithm. The paper itself is very short (three pages only) and provides a formula for filling in the array, but no complexity analysis. The example shown in that paper uses an affine space penalty function. Later, Huang and coworkers developed quadratic time, linear space versions for this algorithm [100]. Waterman [197] described versions that can efficiently retrieve near optimal alignments.

It is not clear when the basic algorithm presented in Section 3.2.1 first appeared. It was probably rediscovered many times in different contexts [172]. The useful space-saving trick is due to Hirschberg [94]. It was later extended to most of the important versions of dynamic programming sequence comparison algorithms [100, 145].

Much attention has been devoted in the literature to concave space penalty functions [140], partly because they model well the idea of charging more for separate spaces, just as the subadditive functions do. Concavity and subadditivity are in general incomparable, although concavity plus the additional condition $2w(1) \geq w(2)$ implies subadditivity (see Exercise 18).

Our section on comparison of similar sequences is based mainly on a paper by Fickett [63]. Ukkonen [188], Myers [142], and Chang and Lawler [34] have developed more efficient algorithms for scoring schemes with certain special properties.

PAM matrices were introduced by Margareth Dayhoff and coworkers [45]. A good introduction is given in [70]. Altschul [10] gives a sound analysis on which PAM distances should be chosen for particular comparisons. This depends heavily on considerations of statistical significance covered by Karlin and Altschul [109] and Karlin, Dembo, and Kawabata [110]. See also Altschul and others [11] for a discussion of many issues in searching sequence databases. Alternatives to PAM matrices have been proposed by Henikoff and Henikoff [93]. These are known as the BLOSUM matrices.

Researchers have also studied ways of computing optimal alignments simultaneously for many score choices. Such methods are reviewed by Waterman [198].

The FAST implementation is described by Lipman and Pearson [127], Pearson and Lipman [155], and Pearson [153]. A thorough comparison between FASTA and pure dynamic programming was done by Pearson [154]. The BLAST program is described by Altschul et al.[12]. The idea of using common windows that can be found very quickly so that larger alignments can be built is a recurring theme. Among the various applications, we cite the work by Chao, Zhang, Ostell, and Miller [36] on local comparison between extremely long sequences. See also Chao and Miller [35] and Joseph, Meidanis, and Tiwari [104].

A number of papers discuss the relationship between distance and similarity. The standard transformation used in Section 3.6.1 was already considered by Smith, Waterman, and Fitch [176]. The review chapter by Waterman [197] devotes a section to this topic. An algorithm for alignments that allow space–space matches is given in [134].

The projections method for saving time in multiple alignment is due to Carrillo and Lipman [32]. Altschul and Lipman [13] generalized it to star alignments. Star alignments were also considered by Gusfield [83], who proved performance guarantees with respect to a sum-of-pairs distance measure. Gupta, Kececioglu, and Schäffer [81] describe their implementation of Carrillo and Lipman's algorithm, known as the Multiple Sequence Alignment (or MSA) program. Kececioglu [113] formalized the notion of "a multiple alignment that is as close as possible to a set of pairwise alignments," calling it the *maximum weight trace problem*. He proved the problem NP-hard and presented a branch-and-bound algorithm. See also [137] for an alternative, heuristic method for sequence alignment based on the notion of trace. The NP-hardness of the multiple alignment with SP measure problem and the tree alignment problem were shown by Wang and Jiang [193]. The exponential-time algorithm for tree alignment is due to Sankoff [166]. Approximation algorithms for distance-based tree alignment were described by Wang and Gusfield [192], who improved on an earlier result by Jiang, Lawler, and Wang [102]. Some heuristics for multiple alignment must solve the problem of aligning alignments. Two references on this topic are Miller [139] and Gotoh [77]. For a survey on multiple-sequence comparison methods see Chan, Wong, and Chiu [33].

Because certified methods for finding the best alignments are in practice sometimes prohibitively slow, the use of parallel computers is often regarded as a way of speeding up the process. One such attempt is described by Jones et al. [103].

Suffix trees and many other algorithms for strings are described in the book by Crochemore and Rytter [39] and in the book by Stephen [180]. Suffix arrays appeared in a paper by Manber and Myers [130]. The application of suffix trees to the primer selection problem and to the palindrome problem cited in Section 3.6.3 were given in a talk by Dan Gusfield. Gusfield is preparing a new book that will describe in depth many string algorithms and their applications in computational biology [85].

CHAPTER

4

FRAGMENT ASSEMBLY OF DNA

In Chapter 1 we saw the biological aspects of DNA sequencing. In this chapter we discuss the computational task involved in sequencing, which is called fragment assembly. The motivation for this problem comes from the fact that with current technology it is impossible to sequence directly contiguous stretches of more than a few hundred bases. On the other hand, there is technology to cut random pieces of a long DNA molecule and to produce enough copies of the pieces to sequence. Thus, a typical approach to sequencing long DNA molecules is to sample and then sequence fragments from them. However, this leaves us with the problem of assembling the pieces, which is the problem we study in this chapter. We present formal models for the problem and algorithms for its solution.

4.1 BIOLOGICAL BACKGROUND

To **sequence** a DNA molecule is to obtain the string of bases that it contains. In large-scale DNA sequencing we have a long target DNA molecule (thousands of bp) that we want to sequence. We may think of this problem as a puzzle in which we are given a double row of cards facing the table, as in Figure 4.1. We do not know which letter from the set {A, C, G, T} is written on each card, but we do know that cards in the same position of opposite strands form a complementary pair. Our goal is to obtain the letters using certain *hints*, which are (approximate) substrings of the rows. The long sequence to reconstruct is called the **target**.

In the biological problem, we know the length of the target sequence approximately, within 10% or so. It is impossible to sequence the whole molecule directly. However, we may instead get a piece of the molecule starting at a random position in one of the strands and sequence it in the canonical ($5' \rightarrow 3'$) direction for a certain length. Each such sequence is called a **fragment**. It corresponds to a substring of one of the strands of the

```
5′  ···□□□□□□□···  3′
3′  ···□□□□□□□···  5′
```

FIGURE 4.1

Unknown DNA to be sequenced.

target molecule, but we do not know which strand or its position relative to the beginning of the strand in addition it may contain errors. By using the *shotgun method* (described in Section 1.5.2), we obtain a large number of fragments and then we try to reconstruct the target molecule's sequence based on fragment overlap. Depending on experimental factors, fragment length can be as low as 200 or as high as 700. Typical problems involve target sequences 30,000 to 100,000 base-pairs long, and total number of fragments is in the range 500 to 2000.

The problem is then to deduce the whole sequence of the target DNA molecule. Because we have a collection of fragments to put together, this task is known as **fragment assembly**. We note that it suffices to determine one of the strands of the original molecule, since the other can be readily obtained because of the complementary pair rule.

In the remainder of this section we give additional details of a biological nature that are important in the design of fragment assembly algorithms.

4.1.1 THE IDEAL CASE

The best way of studying the issues involved is by looking at an example. Be aware, however, that real instances are much larger than the examples we present. Suppose the input is composed of the four sequences

```
ACCGT
CGTGC
TTAC
TACCGT
```

and we know that the answer has approximately 10 bases. One possible way to assemble this set is

```
--ACCGT--
----CGTGC
TTAC-----
-TACCGT--
---------
TTACCGTGC
```

Notice that we aligned the input set, ignoring spaces at the extremities. We try to align in the same column bases that are equal. The only guidance to assembly, apart from the approximate size of the target, are the *overlaps* between fragments. By overlap here we mean the fact that sometimes the end part of a fragment is similar to the beginning of another, as with the first and second sequences above. By positioning fragments so that they align well with each other we get a **layout**, which can be seen as a multiple alignment of the fragments.

The sequence below the line is the **consensus sequence**, or simply *consensus,* and is the answer to our problem. The consensus is obtained by taking a majority vote among all bases in each column. In this example, every column is unanimous, so computing the consensus is straightforward. This answer has nine bases, which is close to the given target length of 10, and contains each fragment as an exact substring. However, in practice fragments are seldom exact substrings of the consensus, as we will see.

4.1.2 COMPLICATIONS

As mentioned above, real problem instances are very large. Apart from this fact, several other complications exist that make the problem much harder than the small example we saw. The main factors that add to the complexity of the problem are errors, unknown orientation, repeated regions, and lack of coverage. We describe each factor in the sequel.

Errors

The simplest errors are called *base call* errors and comprise base substitutions, insertions, and deletions in the fragments. Examples of each kind are given in Figures 4.2, 4.3, and 4.4, respectively. Transpositions are also common but we can treat them as compositions of an insertion and a deletion or two substitutions.

```
Input:                 Answer:

ACCGT                  --ACCGT--
CGTGC                  ----CGTGC
TTAC                   TTAC-----
TGCCGT                 -TGCCGT--
                       ---------
                       TTACCGTGC
```

FIGURE 4.2

In this instance there was a substitution error in the second position of the last fragment, where A *was replaced by* G*. The consensus is still correct because of majority voting.*

Base call errors occurs in practice at rates varying from 1 to 5 errors every 100 characters. Their distribution along the sequence is not uniform, as they tend to concentrate towards the 3′ end of the fragment. As we can see from the examples, it is still possible to reconstruct the correct consensus even in the presence of errors, but the computer program must be prepared to deal with this possibility and this usually means algorithms that require more time and space. For instance, it is possible to find the best alignments between two sequences in linear time if there are no errors, whereas we saw in Chapter 3 that quadratic algorithms are needed to account for gaps.

Apart from erroneous base calls, two other types of errors can affect assembly. One of them is the artifact of *chimeric* fragments, and the other is *contamination* by host or vector DNA. We explain each type in what follows.

Input:

```
ACCGT
CAGTGC
TTAC
TACCGT
```

Answer:

```
--ACC-GT--
----CAGTGC
TTAC------
-TACC-GT--
----------
TTACC-GTGC
```

FIGURE 4.3

In this instance there was an insertion error in the second position of the second fragment. Base A *appeared where there should be none. The consensus is still correct because of spaces introduced in the multiple alignment and majority voting. Notice that the space in the consensus will be discarded when reporting the answer.*

Input:

```
ACCGT
CGTGC
TTAC
TACGT
```

Answer:

```
--ACCGT--
----CGTGC
TTAC-----
-TAC-GT--
---------
TTACCGTGC
```

FIGURE 4.4

In this instance there was a deletion in the third (or fourth) base in the last fragment. The consensus is still correct because of spaces in the alignment and majority voting.

Chimeric fragments, or *chimeras*, arise when two regular fragments from distinct parts of the target molecule join end-to-end to form a fragment that is *not* a contiguous part of the target. An example is shown in Figure 4.5. These misleading fragments must be recognized as such and removed from the fragment set in a preprocessing stage.

Sometimes fragments or parts of fragments that do not have anything to do with the target molecule are present in the input set. This is due to contamination from host or vector DNA. As we saw in Section 1.5.2, the process of replicating a fragment consists of inserting it into the genome of a vector, which is an organism that will reproduce and carry along copies of our fragment. In the end, the fragment must be purified from the vector DNA, and here is where contamination occurs — if this purification is not complete. If the vector is a virus, then the infected cell — generally a bacterial cell — can also contribute some genetic material to the fragment.

Contamination is a rather common phenomenon in sequencing experiments. Witness to this is the significant quantity of vector DNA that is present in the community databases. Scientists sometimes fail to screen against the vector sequence prior to assembly and submit a contaminated consensus to the database.

As with chimeras, the remedy for this problem is to screen the data before starting assembly. The complete sequences of vectors commonly used in DNA sequencing are well known, and it is not difficult to screen all fragments against these known sequences

```
Input:                    Answer:

    ACCGT                   --ACCGT--
    CGTGC                   ----CGTGC
    TTAC                    TTAC-----
    TACCGT                  -TACCGT--
    TTATGC                  ---------
                            TTACCGTGC

                            TTA---TGC
```

FIGURE 4.5

The last fragment in this input set is a chimera. The only way to deal with chimeras is to recognize and remove them from the input set before starting assembly proper.

to see whether a substantial part of them is present in any fragment. Chimeric fragment detection leads to interesting algorithmic problems, but we will not detail them any further in this book. Pointers to relevant references are given in the bibliographic notes.

Unknown Orientation

Fragments can come from any of the DNA strands and we generally do not know to which strand a particular fragment belongs to. We do know, however, that whatever the strand the sequence read goes from 5′ to 3′. Because of the complementarity and opposite orientation of strands, the fact that a fragment is a substring of one strand is equivalent to the fact that its reverse complement is a substring of the other. As a result, we can think of the input fragments as being all approximate substrings of the consensus sought either as given or in reverse complement. Figure 4.6 shows an assembly problem involving fragments in both orientations, initially unknown, but with no errors. In practice, we have to deal with both errors and unknown orientation at the same time.

Because the orientations are unknown, in principle we should try all possible com-

```
Input:                Answer:

    CACGT               →    CACGT--------
    ACGT                →    -ACGT--------
    ACTACG              ←    --CGTAGT-----
    GTACT               ←    -----AGTAC---
    ACTGA               →    --------ACTGA
    CTGA                →    ---------CTGA
                             -------------
                             CACGTAGTACTGA
```

FIGURE 4.6

Fragment assembly with unknown orientation. Initially we do not know the orientation of fragments. Each one can be used either in direct or reverse orientation. In the solution, we indicate by an arrow the chosen orientation: → means fragment as is, ← means its reverse complement.

binations, which are 2^n for a set with n fragments. Of course, this method is unacceptable and is not the way it is done in an assembly program, but it does hint at the complexity introduced by the orientation issue.

Repeated Regions

Repeated regions or **repeats** are sequences that appear two or more times in the target molecule. Figure 4.7 shows an example. Short repeats, that is, repeats that can be entirely covered by one fragment, do not pose difficulties. The worst problems are caused by longer repeats. Also, the copies of a repeat do not have to be identical to upset assembly. If the level of similarity between two copies of a repeat is high enough, the differences can be mistaken for base call errors. Remember that the assembler must be prepared to deal with errors, so there is usually some degree of tolerance in overlap detection.

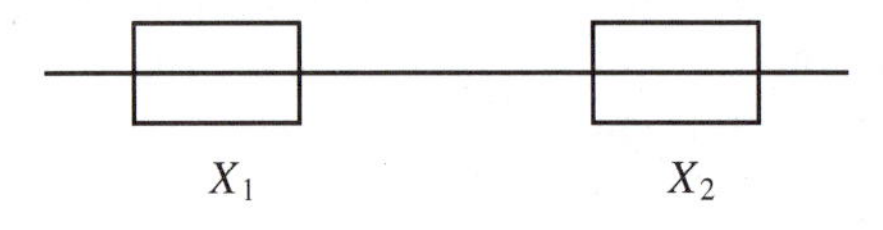

FIGURE 4.7

Repeated regions. The blocks marked X_1 and X_2 are approximately the same sequence.

The kinds of problems that repeats cause are twofold. First, if a fragment is totally contained in a repeat, we may have several places to put it in the final alignment, as it may fit reasonably well in the several repeat copies. One could argue that it does not matter where we put it, since in any copy of the same repeat the consensus will be approximately the same. But the point is that, when the copies are not exactly equal, we may weaken the consensus by placing a fragment in the wrong copy.

Second, repeats can be positioned in such a way as to render assembly inherently ambiguous; that is, two or more layouts are compatible with the input fragments and approximate target length at equivalent levels of fitness. Two such cases are shown in Figures 4.8 and 4.9, respectively. The first one features three copies of the same repeat, and the second has two interleaving copies of different repeats. The common feature between these two cases is the presence of two different regions flanked by the same repeats. In the first example, both B and C are flanked by X and X. In the second, both B and D are flanked by X and Y.

So far we discussed **direct repeats**, namely, repeated copies in the same strand. However, **inverted repeats**, which are repeated regions in opposite strands, can also occur and are potentially more dangerous. As few as two copies of a long, inverted repeat are enough to make the instance ambiguous. An example is given in Figure 4.10.

Lack of Coverage

Another problem is lack of adequate coverage. We define the **coverage** at position i of the target as the number of fragments that cover this position. This concept is well

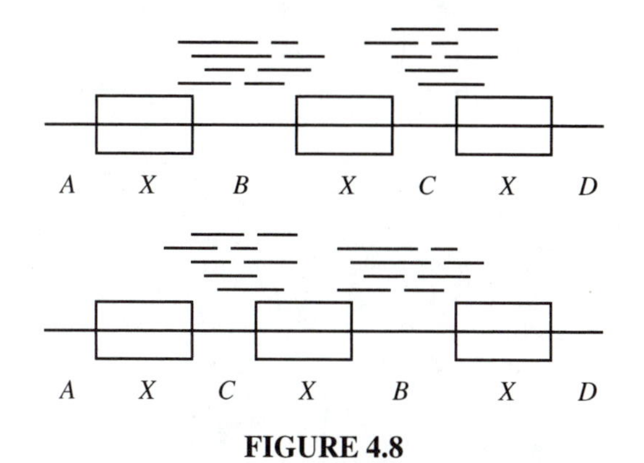

FIGURE 4.8

Target sequence leading to ambiguous assembly because of repeats of the form XXX.

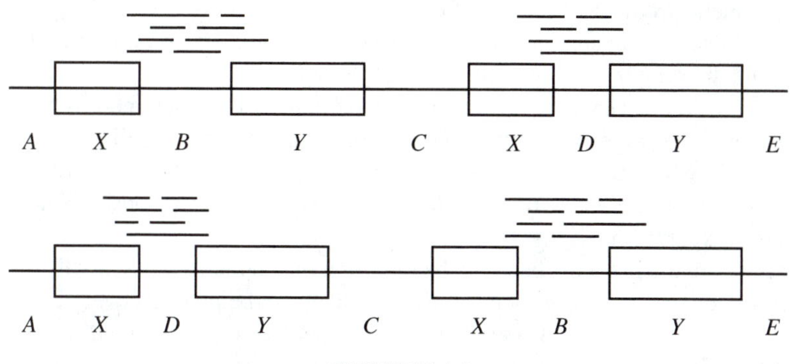

FIGURE 4.9

Target sequence leading to ambiguous assembly because of repeats of the form $XYXY$.

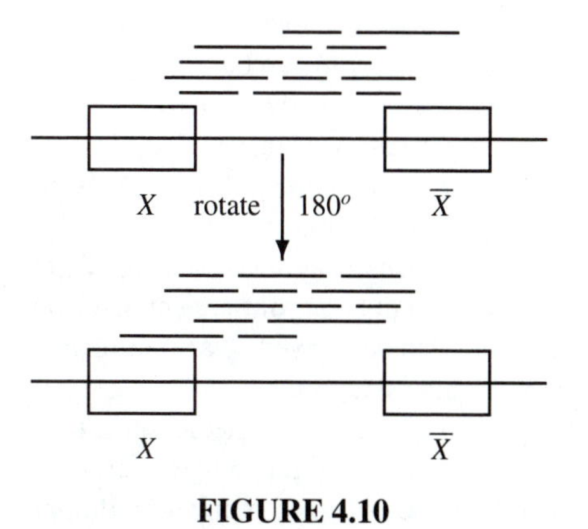

FIGURE 4.10

Target sequence with inverted repeat. The region marked $\overline{X}$ is the reverse complement of the region marked X.

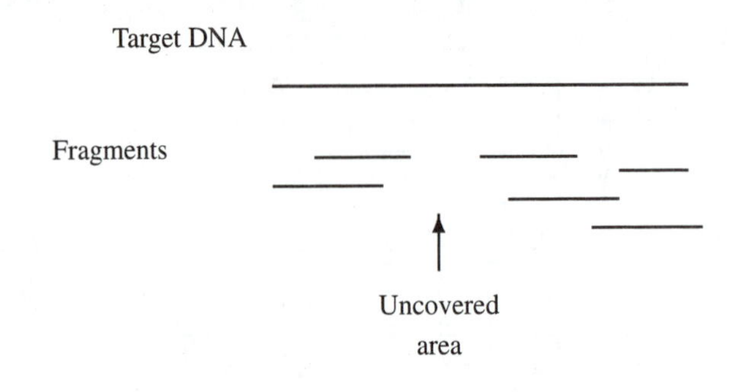

FIGURE 4.11
Example of insufficient coverage.

defined, but it is impossible to compute it because we do not know the actual positions of the fragments in the target. Even after assembly, all we have is our best guess about these positions. We can compute the *mean coverage,* though, by adding up all fragment lengths and dividing by the approximate target length.

If the coverage is equal to zero for one or more positions, then there is not enough information in the fragment set to reconstruct the target completely. Figure 4.11 shows an example in which there is an uncovered area. In such cases, the best we can hope to achieve is a layout for every one of the contiguously covered regions, called **contigs**. There are two contigs in Figure 4.11.

Lack of coverage occurs because the sampling of fragments is essentially a random process; it is therefore possible that some parts end up not being well covered or not covered at all. In general, we want not only one, but several different fragments covering any given point. The more fragments we have, the safer is our assessment of the consensus based on voting. It is also desirable to have fragments from both strands covering a given region, for it has been observed in practice that certain kinds of errors occur consistently in one strand only.

Insufficient coverage can usually be dealt with by sampling more fragments, but we must be careful with this approach. If everything is well covered except for a small portion of the target, sampling at random can be a very inefficient way of covering the gaps. An approach, called *directed sequencing* or *walking* (see Section 4.1.3), can be used in this case. To avoid the formation of these little gaps, some researchers advocate sampling at very high coverage rates, collecting fragments whose combined length is enough to cover the target molecule 8 times or more.

It is important to know how many fragments we need to generate in order to achieve a given coverage. The answer cannot be absolute, of course, since sampling is a random process, but reasonable bounds can be derived. We mention here an important result that is valid under certain simplifying assumptions.

Let T denote the length of the target molecule. Assume that all fragments have about the same length l and that we can safely recognize overlaps of at least t bases. If we sample n fragments at random, the expected number p of *apparent contigs* is given by the formula

$$p = ne^{-n(l-t)/T}. \tag{4.1}$$

The term "apparent contig" refers to the fact that we are assuming that we are able to recognize overlaps only if they have size at least t. Thus some of the contigs we have may in fact overlap by less that t bases, so they are really parts of a longer true contig but "appear" to be two separated contigs to us.

Notice that p approaches zero as the number n of fragments grows, which seems strange because we know there will always be at least one apparent contig. This happens because formula (4.1) is an approximation. It can be used also in the context of DNA mapping, where the number of contigs is much larger, so a difference of one or two is negligible. The formula is useful in the sense that it provides a reliable ballpark in a wide variety of situations.

A formula for the fraction of the target molecule covered by the fragments is also available. With the same notation as above, the fraction covered by exactly k fragments is given by

$$r_k = \frac{e^{-c}c^k}{k!}, \tag{4.2}$$

where $c = nl/T$ is the mean coverage.

4.1.3 ALTERNATIVE METHODS FOR DNA SEQUENCING

We mention here a few supplementary and alternative approaches to DNA sequencing. We start with *directed sequencing*, a method that can be used to cover small remaining gaps in a shotgun project. In direct sequencing a special primer is derived from the sequence near the end of a contig, so that fragments spanning the region including the end of this contig and the continuation of it in the target are generated. These new fragments are then sequenced and give the sequence adjacent to the contig, thereby augmenting it. Continuing in this fashion, we can cover the gap to the next contig. The problem with this approach is that it is expensive to build special primers. Also, the next step can be accomplished only after the current one, so the process is essentially sequential rather than parallel (but can be done for all gaps in parallel).

Another technique that has become very popular is called *dual end sequencing*. In a shotgun experiment several copies of the target DNA molecule are broken randomly and short pieces are selected for cloning and sequencing. We recall from Section 1.5.2 that these pieces are called *inserts* because they are inserted into a vector for amplification. Inserts sizes range from 1 to 5 kbp, but only about 200 to 700 bases can be directly read from one extremity to yield a fragment. However, it is also possible to read the other extremity if we have a suitable primer. The two fragments thus obtained belong to opposite strands, and they should be separated in the final alignment by roughly the insert size minus the fragment size. This extra information is extremely useful in closing gaps, for instance. Sometimes the dual end is sequenced only if it is necessary to close a gap. Notice that dual end sequencing exploits the fact that inserts are usually larger than the portion read from them.

A radically different approach from shotgun methods has been proposed recently. Called *sequencing by hybridization* (SBH), it consists of assembling the target molecule based on many hybridization experiments with very short, fixed length sequences called *probes*. A hybridization simply checks whether a probe binds (by complementarity) to a DNA molecule. The idea is to design a *DNA chip* that performs all the necessary hybridizations simultaneously and delivers a list of all strings of length w, or w-mers, that exist in the target. With current technology it is possible to construct such chips for probes of length up to eight bases. Larger probe sizes seem still prohibitive at this point.

Important issues in SBH include the following. It is clear that not all target molecules can be reconstructed with a probe size as small as eight (see Exercise 10). One important problem then is to characterize the molecules that can actually be reconstructed with a given probe size. Another problem is that the hybridization experiments do not provide the number of times a given w-mer appears in the target, just whether an w-mer does appear. Clearly, it would help to have the extra information on how many copies do appear. Finally, problems with errors in the experiments and orientation need to be tackled. One proposed strategy is to reduce the shotgun data to hybridization data by simply using all the w-mers of the fragments instead of the fragments themselves and to think of them as generated by hybridization experiments. However, this strategy throws away information, because it is generally impossible to recover the fragments from the w-mers, while it is possible to generate the w-mers from the fragments. Algorithms for the SBH problem are not covered in this book, but references are given in the bibliographic notes at the end of this chapter.

MODELS

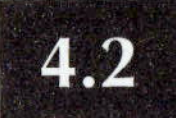

4.2

We will now examine formalisms for fragment assembly. We present in this section three models for the problem: shortest common superstring (SCS), RECONSTRUCTION, and MULTICONTIG. Each throws light on the diverse computational aspects of the problem, although none of them completely addresses the biological issues. All three assume that the fragment collection is free of contamination and chimeras.

4.2.1 SHORTEST COMMON SUPERSTRING

One of the first attempts to formalize fragment assembly was through a string problem in which we seek the shortest superstring of a collection of given strings. Accordingly, this is called the *Shortest Common Superstring* problem, or SCS. Although this model has serious shortcomings in representing the fragment assembly problem — it does not account for errors, for instance — the techniques used to tackle the resulting computational problem have application in other models as well. It is worthwhile, therefore, to study these techniques. The Shortest Common Superstring problem is defined as follows:

PROBLEM: SHORTEST COMMON SUPERSTRING (SCS)
INPUT: A collection $\mathcal{F}$ of strings.

OUTPUT: A shortest possible string S such that for every $f \in \mathcal{F}$, S is a superstring of f.

Example 4.1 Let $\mathcal{F} = \{\texttt{ACT}, \texttt{CTA}, \texttt{AGT}\}$. The sequence $S = \texttt{ACTAGT}$ is the shortest common superstring of $\mathcal{F}$. It obviously contains all fragments in $\mathcal{F}$ as substrings. To see that it is the shortest, notice that any string S' that has both $u = \texttt{ACT}$ and $v = \texttt{AGT}$ as substrings must have length at least 6. Moreover, with $|S'| = 6$ we have just the concatenations uv and vu. But CTA is a substring of $uv = S$ only.

In the context of fragment assembly, the collection $\mathcal{F}$ corresponds to the fragments, each one given by its sequence in the correct orientation, and S is the sequence of the target DNA molecule.

Notice that the computational problem specifies that S should be a perfect superstring of each fragment, not an approximate superstring, so it does not allow for experimental errors in the fragments. Furthermore, the orientation of each fragment must be known, which is seldom the case. Finally, even in a perfect assembly project in which all these factors could be somehow controlled, the shortest common superstring may not be the actual biological solution because of repeated sections in the target DNA sequence, as the following example shows.

Example 4.2 Suppose that the target molecule has two copies of an exact repeat and that fragments are sampled as shown in Figure 4.12. Notice that the repeat copies are long and contain many fragments. In this case, even if the fragments are exact substrings of the consensus, and even if we know their correct orientation, finding the shortest common superstring may not be what we want.

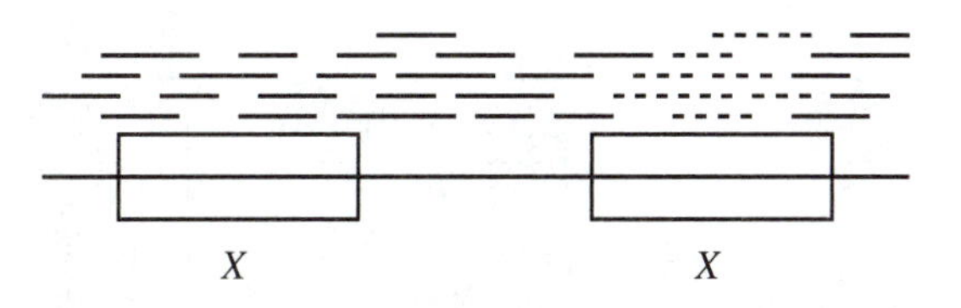

FIGURE 4.12

Target sequence with long repeat that contains many fragments. This example shows that even with no errors and known orientation, the SCS formulation fails in the presence of repeats.

Indeed, Figure 4.13 shows a different assembly, with a shorter consensus for the same fragment set. Observe that because the repeat copies are identical, a superstring may contain only one copy, which will absorb all fragments totally contained in any of the copies. The other copy can be shorter, as it must contain only fragments that cross the border between X and its flanking regions. The shortened version of X is denoted by X' in this figure.

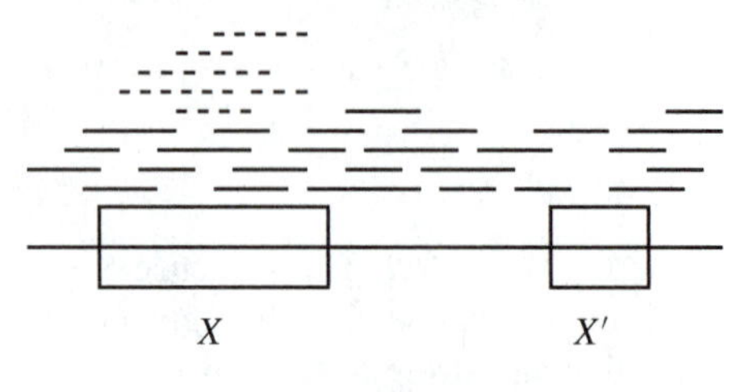

FIGURE 4.13

Alternative assembly for the fragments in the previous figure. This assembly leads to a shorter consensus, because all fragments totally contained in the rightmost copy (the dashed fragments) were moved to the leftmost copy, causing a decrease in length of the rightmost copy.

Notice that although shorter, this alternative assembly is poorer in terms of both coverage and linkage. The coverage is uneven, with many fragments spanning X and much fewer spanning X'. The linkage is poor because no fragment links the leftmost part of X' to its rightmost part. In fact, this consensus is the concatenation of two unrelated parts u and v, with u going from the beginning to about the middle of X', and v going from the middle of X' to the end. As far as shortest substrings are concerned, vu would be a perfectly equivalent consensus.

The Shortest Common Superstring problem is NP-hard, but approximation algorithms exist. However, given all the shortcomings of this model with respect to the real biological problem, such algorithms are primarily of theoretical interest.

4.2.2 RECONSTRUCTION

This model takes into account both errors and unknown orientation. To deal with errors, we need a few preliminary definitions. Recall from Chapter 3 that we can configure the basic dynamic programming sequence comparison algorithm to suit many different needs. Here we will use distance rather than similarity and a version that charges for gaps in the extremities of the second sequence only. The scoring system is edit distance; that is, one unit of distance is charged for every insertion, deletion, or substitution, except for deletion in the extremities of the second sequence, which are free of charge. The distance thus obtained will be called *substring edit distance*, to distinguish it from the classical edit distance that charges for end deletions in both strings. We denote it by d_s and define it formally by the expression

$$d_s(a, b) = \min_{s \in S(b)} d(a, s),$$

where $S(b)$ denotes the set of all substrings of b and d is the classical edit distance. Notice that substring edit distance is asymmetric, that is, in general $d_s(a, b) \neq d_s(b, a)$.

```
-----GC-GATAG----
CAGTCGCTGATCGTACG
```

FIGURE 4.14

Optimal alignment for substring edit distance, which does not charge for end deletions in the second string.

Example 4.3 If a = GCGATAG and b = CAGTCGCTGATCGTACG, then the best alignment is as indicated in Figure 4.14 and the distance is $d_s(a, b) = 2$.

Let ϵ be a real number between 0 and 1. A string f is an *approximate substring* of S at error level ϵ when

$$d_s(f, S) \leq \epsilon |f|,$$

where $|f|$ is the length of f. This means that we are allowed on average ϵ errors for each base in f. For instance, if $\epsilon = 0.05$ we are allowed five errors per hundred bases. We are now ready for the definition of fragment assembly according to the RECONSTRUCTION model.

PROBLEM: RECONSTRUCTION
INPUT: A collection $\mathcal{F}$ of strings and an error tolerance ϵ between 0 and 1.
OUTPUT: A shortest possible string S such that for every $f \in \mathcal{F}$ we have

$$\min(d_s(f, S), d_s(\overline{f}, S)) \leq \epsilon |f|,$$

where $\overline{f}$ is the reverse complement of f.

The idea is to find a string S as short as possible such that either f or its reverse complement must be an approximate substring of S at error level ϵ. This formulation will assemble correctly all the examples in Section 4.1 except for the one involving a chimeric fragment. However, the general problem is still NP-hard. This is not surprising, though, as it contains the SCS as a particular case with $\epsilon = 0$ (see Exercise 18).

In summary, RECONSTRUCTION models errors and orientation but does not model repeats, lack of coverage, and size of target.

4.2.3 MULTICONTIG

The MULTICONTIG model adds a notion of good linkage to the answer. The previous models do not care about the internal linkage of the fragments in the layout — only the final answer matters. Thus, it is also necessary to accept answers formed by several contigs, and for this reason it is called "multicontig."

We define first an error-free version of the MULTICONTIG model. Given a collection $\mathcal{F}$ of fragments, we consider a multiple alignment, or layout. This layout must be such that every column contains only one kind of base. This is where the error-free hypothesis comes in. Also, to contemplate orientation, we require that either the fragment or its reverse complement be in the alignment, but not both.

Let us call this layout $\mathcal{L}$. We number the columns from 1 to the length $|\mathcal{L}|$ of the layout. According to this numbering, each fragment f has a left endpoint $l(f)$ and a right endpoint $r(f)$, so that $|f| = r(f) - l(f) + 1$. We say that fragments f and g *overlap* in this layout if the integer intervals $[l(f)..r(f)]$ and $[l(g)..r(g)]$ intersect. The nonempty intersection $[l(f)..r(f)] \cap [l(g)..r(g)]$ is called the *overlap* between f and g. The *size* of this overlap is just the size of the intersection. Because we are using integer intervals, all sets are finite.

Among all the overlaps between fragments, we are interested in the most important ones, the ones that provide linkage. We say that an overlap $[x..y]$ is a *nonlink* if there is a fragment in $\mathcal{F}$ that properly contains the overlap on both sides, that is, if the fragment contains interval $[(x-1)..(y+1)]$. If no fragment has this property, the overlap is a *link*. The *weakest link* in a layout is the smallest size of any link in it. Finally, we say that a layout is a *t-contig* if its weakest link is at least as large as t. If it is possible to construct a t-contig from the fragments of a collection $\mathcal{F}$, we say that $\mathcal{F}$ *admits* a t-contig.

We can now formalize the fragment assembly problem according to the MULTICONTIG model. Given a collection of fragments $\mathcal{F}$ and an integer t, we want to partition $\mathcal{F}$ in the minimum number of subcollections C_i, $1 \leq i \leq k$, such that every C_i admits a t-contig.

Example 4.4 Let $\mathcal{F} = \{\text{GTAC, TAATG, TGTAA}\}$. We want to partition $\mathcal{F}$ in the minimum number of t-contigs. If $t = 3$, this minimum is two, as follows.

```
--TAATG          GTAC
TGTAA--
```

There is no way to make one single contig with all three fragments, so two contigs is the minimum possible in this case. Notice that GTAC is its own reverse complement.

If $t = 2$, the same solution is a partition in 2-contigs, since every 3-contig is a 2-contig. However, another solution exists now:

```
TAATG---          GTAC
---TGTAA
```

Here again it is impossible to assemble $\mathcal{F}$ into one t-contig.

Finally, if $t = 1$, we see that a solution with one t-contig exists:

```
TGTAA-----
--TAATG---
------GTAC
```

A version contemplating errors can be defined as follows. The layout is not required to be error-free, but with each alignment we must associate a *consensus* sequence S of the same length, possibly containing space characters (-). The numbers $l(f)$ and $r(f)$ are still defined as the leftmost and rightmost columns of f in the alignment, but now we may have $|f| \neq r(f) - l(f) + 1$. The *image* of an aligned fragment f in the consensus is $S[l(f)..r(f)]$. Given an error tolerance degree ϵ, we say that S is an *ϵ-consensus* for this contig when the edit distance between each aligned fragment f and its image in the consensus is at most $\epsilon|f|$. We are now ready for the formal definition.

PROBLEM: MULTICONTIG
INPUT: A collection $\mathcal{F}$ of strings, an integer $t \geq 0$, and an error tolerance ϵ between 0 and 1.
OUTPUT: A partition of $\mathcal{F}$ in the minimum number of subcollections C_i, $1 \leq i \leq k$, such that every C_i admits a t-contig with an ϵ-consensus.

Notice that in the RECONSTRUCTION model a fragment is required to be an approximate substring of S, but the particular place in S where it is aligned is not important. In contrast, in the MULTICONTIG formulation we need to specify explicitly where each fragment should go.

The MULTICONTIG formalization of fragment assembly is NP-hard, even in the simplest case of no errors and known orientation. It contains as a special case the problem of finding a Hamiltonian path in a restricted class of graphs.

This formulation models errors, orientation, and lack of coverage but has no provision to use information on the approximate size of the target molecule. In addition, it partially models repeats, in the sense that it can satisfactorily solve some instances with repeats, although not all of them. For instance, it can correctly reconstruct the instance given in Figure 4.12.

⋆ ALGORITHMS 4.3

In this section we present two algorithms for the case of fragments with no errors and known orientation. One of them is known as the *greedy* algorithm, and many practical systems are based on the same idea, with additions that contemplate errors and unknown orientation. The other algorithm is based on the MULTICONTIG model and is useful when we can obtain an acyclic overlap graph by discarding edges with small weight. It should be mentioned that both algorithms are of little practical value in themselves, because of the restrictive hypothesis. Nevertheless, the ideas behind them can be useful in designing algorithms that deal with real instances of the problem.

4.3.1 REPRESENTING OVERLAPS

Common superstrings correspond to *paths* in a certain graph structure based on the collection $\mathcal{F}$. We can translate properties of these superstrings to properties of the paths. Because researchers have studied paths in graphs for a long time, many people feel more comfortable dealing with graphs, and relating a new problem to them is often a good idea.

The **overlap multigraph** $\mathcal{OM}(\mathcal{F})$ of a collection $\mathcal{F}$ is the directed, weighted multigraph defined as follows. The set V of nodes of this structure is just $\mathcal{F}$ itself. A directed edge from $a \in \mathcal{F}$ to a different fragment $b \in \mathcal{F}$ with weight $t \geq 0$ exists if the suffix of a with t characters is a prefix of b. In symbols, this is equivalent to

$$\mathit{suffix}(a, t) = \mathit{prefix}(b, t),$$

or

$$\kappa^{|a|-t} a = b \kappa^{|b|-t},$$

or

$$(a\kappa^t)b = a(\kappa^t b),$$

where κ is the killer agent defined in Section 2.1. Notice that we must have $|a| \geq t$ and $|b| \geq t$ for this edge to exist. Observe also that there may be many edges from a to b, with different values of t. This is why the structure is called a multigraph instead of simply a graph. Note that we disallow self-loops, that is, edges going from a node to itself. On the other hand, edges of zero weight are allowed.

A word must be said about edge weights. These weights represent overlaps between fragments, and we use the overlaps to join fragments into longer strings, sharing the overlapping part. This joining means that we believe that the two fragments come from the same region in the target DNA. Now this belief has to be based on solid evidence. In our case, this evidence is the size of the overlap. A short overlap provides weak evidence, whereas a long overlap gives strong evidence. When dealing with hundreds of fragments, each one a few hundred characters long, it is foolish to assume that two fragments share ends based on an overlap of, say, three characters. For this reason, sometimes we impose a minimum acceptable amount of overlap, and throw away all edges with weight below this threshold. But, in principle, we will keep all the edges in the multigraph, including the zero weight edges. There are $n(n-1)$ such edges, because any two strings share a common part of zero characters.

4.3.2 PATHS ORIGINATING SUPERSTRINGS

The overlap multigraph is important because directed paths in it give rise to a multiple alignment of the sequences that belong to this path. A consensus sequence can then be derived from this alignment, providing a common superstring of the involved sequences. Let us investigate this process more closely.

Given a directed path P in $\mathcal{OM}(\mathcal{F})$, we can construct a multiple alignment with the sequences in P as follows. Each edge $e = (f, g)$ in the path has a certain weight t, which means that the last t bases of the tail f of e are a prefix of the head g of e. Thus, f and g can be aligned, with g starting t positions before f's end.

An example is given in Figure 4.15. Only edges with strictly positive weight were drawn. The collection $\mathcal{F}$ is $\{a, b, c, d\}$, where

$$a = \mathtt{TACGA},$$
$$b = \mathtt{ACCC},$$
$$c = \mathtt{CTAAAG},$$
$$d = \mathtt{GACA}.$$

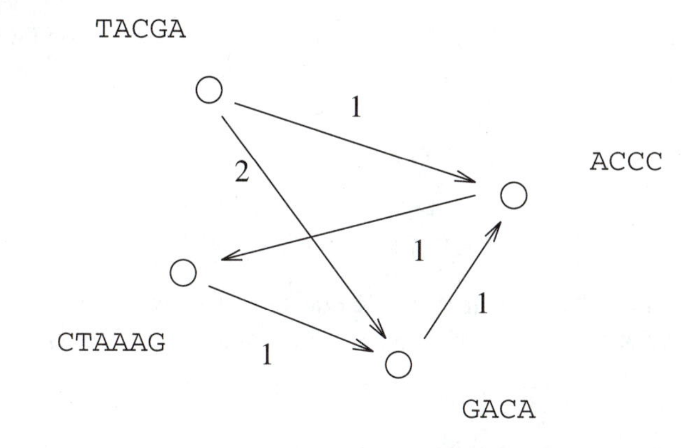

FIGURE 4.15

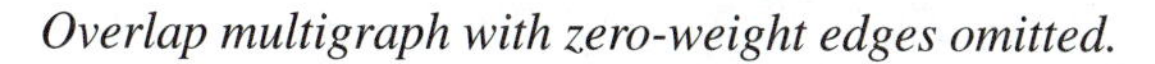

Overlap multigraph with zero-weight edges omitted.

Usually, we indicate paths by a list of the form vertex, edge, vertex, edge, ..., vertex, with each edge connecting its two neighboring vertices in the list. However, in this example there are no parallel edges, so, for simplicity, we will indicate the paths by just a list of the vertices involved. The path $P_1 = dbc$ leads to the following alignment:

```
GACA--------
---ACCC-----
------CTAAAG
```

whereas path $P_2 = abcd$ originates the following alignment.

```
TACGA-----------
----ACCC--------
-------CTAAAG---
------------GACA
```

In general, let P be a path in $\mathcal{OM}(\mathcal{F})$, and let A be the set of fragments involved in P. Paths by definition cannot repeat vertices, so we have exactly $|A| - 1$ edges in P. The common superstring derived from P as above will be called $S(P)$.

The relationship between the total length of A, the path's weight, and the superstring length is given by the following equation.

$$\|A\| = w(P) + |S(P)|, \tag{4.3}$$

where $\|A\| = \sum_{a \in A} |a|$ is the sum of lengths of all sequences in A, and $w(P)$ is the weight of P. Is this really true? Let us check it for some simple cases.

Start with a path with just one fragment and no edges. Then $A = \{f\}$, $\|A\| = |f|$, and $w(P) = 0$, since P has no edges. Furthermore, $S(P)$ is simply f. It is easy to see that Equation (4.3) holds in this case. Let us try something larger. Suppose that $P = f_1 e_1 f_2 e_2 \cdots f_l$ and that the formula holds for P. What happens if we consider an additional edge? We have $P' = P e_{l+1} f_{l+1}$ and

$$w(P') = w(P) + w(e_{l+1}) = w(P) + k,$$

if $w(e_{l+1}) = k$.

On the other hand, $S(P') = S(P)\mathit{suffix}(f_{l+1}, |f_{l+1}| - k)$, because the first k characters of f_{l+1} are already present in $S(P)$. We need to concatenate the remaining suffix of f_{l+1} only. Then,

$$\begin{aligned}|S(P')| &= |S(P)| + |f_{l+1}| - k \\ &= \|A\| - w(P) + |f_{l+1}| - k \\ &= \|A'\| - w(P')\end{aligned}$$

and the formula holds. We have just proved Equation (4.3) by induction.

Another way of seeing it is by using the killer character. If $P = f_1 e_1 f_2 e_2 \ldots f_l$,

$$S(P) = f_1 \kappa^{w(e_1)} f_2 \kappa^{w(e_2)} \cdots \kappa^{w(e_{l-1})} f_l.$$

The right-hand side is not ambiguous in this case because

$$(f_i \kappa^{w(e_i)}) f_{i+1} = f_i (\kappa^{w(e_i)} f_{i+1}),$$

by definition of edge weight.

Computing the length,

$$\begin{aligned}|S(P)| &= |f_1| - w(e_1) + |f_2| - w(e_2) + \cdots - w(e_{l-1}) + |f_l| \\ &= \sum_{i=1}^{l} |f_i| - \sum_{i=1}^{l-1} w(e_i) \\ &= \|A\| - w(P).\end{aligned}$$

So, every path originates a common superstring of the fragments involved. This is particularly important when we have a path that goes through every vertex. Such paths are called *Hamiltonian paths* (see Section 2.2). If we get a Hamiltonian path, we have a common superstring of $A = \mathcal{F}$. In this case Equation 4.3 becomes

$$|S(P)| = \|\mathcal{F}\| - w(P) \tag{4.4}$$

and, since $\|\mathcal{F}\|$ is constant, that is, independent of the particular Hamiltonian path we take, we see that minimizing $|S(P)|$ is equivalent to maximizing $w(P)$.

4.3.3 SHORTEST SUPERSTRINGS AS PATHS

We have seen that every path corresponds to a superstring. Is the converse true? Not always, as the following example shows. For the fragment set that originated the graph in Figure 4.15, the superstring

GTATACGACCCAAACTAAAGACAGGG

does not correspond to any path, and it easy to see why. Superstrings may contain unnecessary characters not present in any fragment.

But shortest superstrings cannot waste any characters. So, do shortest superstrings always correspond to paths? The answer is affirmative, and we will prove this result in the sequel. We first observe that it is true for certain collections that are *substring-free,* defined below, and then generalize the result to any collection.

A collection $\mathcal{F}$ is said to be substring-free if there are no two distinct strings a and b in $\mathcal{F}$ such that a is a substring of b. The advantage of dealing with substring-free collections is that a partial converse of the path-to-superstring construction is easier to prove, as illustrated by the following result.

THEOREM 4.1 Let $\mathcal{F}$ be a substring-free collection. Then for every common superstring S of $\mathcal{F}$ there is a Hamiltonian path P in $\mathcal{OM}(\mathcal{F})$ such that $S(P)$ is a subsequence of S.

Proof. The string S is a common superstring of $\mathcal{F}$, so for each $f \in \mathcal{F}$ there is an interval $[l(f)..r(f)]$ of S such that $S[l(f)..r(f)] = f$. For some fragments there may be many positions in which to anchor them. Never mind. Take one of them and fix it for the rest of this proof.

No interval of the form $[l(f)..r(f)]$ is contained in another such interval, because $\mathcal{F}$ is substring-free. These intervals have the following important property: All left endpoints are distinct and all right endpoints are distinct, although there may be left endpoints equal to right endpoints. Moreover, if we sort the fragments in increasing order of left endpoint, the right endpoints end up in increasing order as well. Indeed, if we had two strings f and g in $\mathcal{F}$ with $l(f) \leq l(g)$ and $r(f) \geq r(g)$, then the interval of g would be contained in the interval of f, which is impossible. So, we have the intervals sorted by increasing initial *and* final endpoints. Let $(f_i)_{i=1,\ldots,m}$ be this ordering. The fragments in this order make up the claimed path.

Consider the relationship between f_i and f_{i+1}, for each i such that $1 \leq i \leq m-1$. If $l(f_{i+1}) \leq r(f_i)+1$, then there is an edge in $\mathcal{OM}(\mathcal{F})$ with weight $r(f_i)+1-l(f_{i+1})$. If $l(f_{i+1}) > r(f_i)+1$, then take the edge of weight zero from f_i to f_{i+1}. The characters $S[j]$ for $r(f_i)+1 \leq j \leq l(f_{i+1})-1$ will be discarded, because they do not belong to any interval and are therefore superfluous as far as making a common superstring is concerned. These characters are in S but not in $S(P)$, which is therefore a subsequence of S. ■

COROLLARY 4.1 Let $\mathcal{F}$ be a substring-free collection. If S is a shortest common superstring of $\mathcal{F}$, there is a Hamiltonian path P such that $S = S(P)$.

Proof. By the previous theorem, there is such a path with $S(P)$ a subsequence of S. But $S(P)$ is also a common superstring, because P is Hamiltonian. Because S is a *shortest* common superstring, we have $|S(P)| \geq |S|$ and therefore $S = S(P)$. ■

We now generalize this result to any collection. First, a few definitions are in order. A collection of strings $\mathcal{F}$ *dominates* another collection $\mathcal{G}$ when every $g \in \mathcal{G}$ is a substring of some $f \in \mathcal{F}$. For instance, if $\mathcal{G} \subseteq \mathcal{F}$, then $\mathcal{F}$ dominates $\mathcal{G}$. Two collections $\mathcal{F}$ and $\mathcal{G}$ are said to be *equivalent*, denoted $\mathcal{F} \equiv \mathcal{G}$, when $\mathcal{F}$ dominates $\mathcal{G}$ and $\mathcal{G}$ also dominates $\mathcal{F}$. Equivalent collections have the same superstrings, so this notion is important for us.

LEMMA 4.1 Two equivalent substring-free collections are identical.

Proof. If $\mathcal{F} \equiv \mathcal{G}$ and $\mathcal{F}$ is substring-free, then $\mathcal{F} \subseteq \mathcal{G}$. To see that, consider $f \in \mathcal{F}$. Since $\mathcal{F}$ is dominated by $\mathcal{G}$, there must be $g \in \mathcal{G}$ such that f is a substring of g. But $\mathcal{F}$

also dominates $\mathcal{G}$, so there is a superstring h of g in $\mathcal{F}$. By transitivity, f is a substring of h. But $\mathcal{F}$ is substring-free, so $f = h$, and, since g is "in between" them, $f = g = h$. It follows that $f = g \in \mathcal{G}$. Since f is arbitrary, we conclude that $\mathcal{F} \subseteq \mathcal{G}$. The same argument with $\mathcal{F}$ and $\mathcal{G}$ interchanged shows that $\mathcal{G} \subseteq \mathcal{F}$. Hence, $\mathcal{F} = \mathcal{G}$. ■

The next theorem helps us understand how to obtain a substring-free collection from an arbitrary collection of fragments.

THEOREM 4.2 Let $\mathcal{F}$ be a collection of strings. Then there is a unique substring-free collection $\mathcal{G}$ equivalent to $\mathcal{F}$.

Proof. Uniqueness is guaranteed by the previous lemma. Now let us turn to the existence. We want to show that for every $\mathcal{F}$ there is a substring-free collection $\mathcal{G}$ equivalent to $\mathcal{F}$. We do that by induction on the number of strings in $\mathcal{F}$.

For $|\mathcal{F}| = 0$, $\mathcal{F}$ itself is substring-free, so we are done. For $|\mathcal{F}| \geq 1$, $\mathcal{F}$ may or may not be substring-free. If it is, we are again done. If it is not, then there are a and b in $\mathcal{F}$ such that a is a substring of b. Remove a from $\mathcal{F}$, obtaining a smaller collection $\mathcal{F}' = \mathcal{F} - \{a\}$. By the induction hypothesis, $\mathcal{F}'$ has an equivalent substring-free collection $\mathcal{G}$. But it is easy to verify that $\mathcal{F} \equiv \mathcal{F}'$. Obviously $\mathcal{F}$ dominates $\mathcal{F}'$, which is a subset of $\mathcal{F}$. On the other hand, the only string that is in $\mathcal{F}$ but not in $\mathcal{F}'$ is a, which is covered by $b \in \mathcal{F}'$. Hence, $\mathcal{G}$ is also equivalent to $\mathcal{F}$ and we are done. ■

This result tells us that if we are looking for common superstrings, we might as well restrict ourselves to substring-free collections, because given any collection there is a substring-free one equivalent to it, that is, having the same common superstrings. Incidentally, the proof of Theorem 4.2 also gives us a way of obtaining the unique substring-free collection equivalent to a given $\mathcal{F}$: Just remove from $\mathcal{F}$ all strings that are substrings of other elements of $\mathcal{F}$.

4.3.4 THE GREEDY ALGORITHM

We now know that looking for shortest common superstrings is the same as looking for Hamiltonian paths of maximum weight in a directed multigraph. Moreover, because our goal is to maximize the weight, we can simplify the multigraph and consider only the heaviest edge between every pair of nodes, discarding other, parallel edges of smaller weight. Any path that does not use the heaviest edge between a pair of nodes can be improved, so it is not a maximum weight path. Let us call this new graph the *overlap graph* of $\mathcal{F}$, denoted by $\mathcal{OG}(\mathcal{F})$. In an actual implementation we would not construct these edges one by one. Using suffix trees (see Section 3.6.3), we can build the edges simultaneously, saving both space and time in the computer (see the bibliographic notes). The same structures can help in deciding which fragments are substrings of other fragments and may therefore be left out of the graph.

The following algorithm is a "greedy" attempt at computing the heaviest path. The basic idea employed in it is to continuously add the heaviest available edge, which is one that does not upset the construction of a Hamiltonian path given the previously chosen edges. Because the graph is complete, that is, there are edges between every pair of nodes, this process stops only when a path containing all vertices is formed.

In a Hamiltonian path, or in any path for that matter, we cannot have two edges leaving from the same node, or two edges leading to the same node. In addition, we have to prevent the formation of cycles. So these are the three conditions we have to test before accepting an edge in our Hamiltonian path. Edges are processed in nonincreasing order by weight, and the procedure ends when we have exactly $n - 1$ edges, or, equivalently, when the accepted edges induce a connected subgraph.

Figure 4.16 shows the algorithm. For each node, we keep its current indegree and outdegree with respect to the accepted edges. This information is used to check whether a previously chosen edge has the same tail or head as the currently examined edge. In addition, we keep the disjoint sets that form the connected components of the graph induced by the accepted edges. It is easy to see that a new edge will form a cycle if and only if its tail and head are in the same component. Therefore, this structure is needed to check for cycles. The disjoint sets are maintained by a disjoint-set forest data structure (mentioned in Section 2.3).

We described the greedy algorithm in terms of graphs, but it can be implemented using the fragments directly. The algorithm corresponds to the following procedure being continuously applied to the collection of fragments, until only one fragment remains. Take the pair (f, g) of fragments with largest overlap k, remove the two fragments from $\mathcal{F}$, and add $f\kappa^k g$ to $\mathcal{F}$. We assume that the collection $\mathcal{F}$ is substring-free.

This greedy strategy does not always produce the best result, as shown by the following example.

```
Algorithm Greedy
    input: weighted directed graph OG(F) with n vertices
    output: Hamiltonian path in OG(F)
    // Initialize
    for i ← 1 to n do
        in[i] ← 0 // how many selected edges enter i
        out[i] ← 0 // how many selected edges exit i
        MakeSet(i)
    // Process
    Sort edges by weight, heaviest first
    for each edge (f, g) in this order do
        // test edge for acceptance
        if in[g] = 0 and out[f] = 0 and FindSet(f) ≠ FindSet(g)
            select (f, g)
            in[g] ← 1
            out[f] ← 1
            Union(FindSet(f), FindSet(g))
        if there is only one component
            break
    return selected edges
```

FIGURE 4.16

Greedy algorithm to find Hamiltonian path.

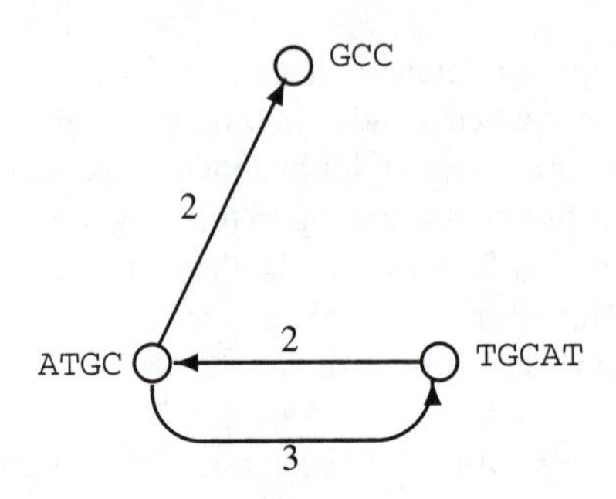

FIGURE 4.17

A graph where the greedy algorithm fails.

Example 4.5 Suppose we have

$$\mathcal{F} = \{\texttt{GCC}, \texttt{ATGC}, \texttt{TGCAT}\}.$$

The overlap graph looks like Figure 4.17, where we omitted the zero edges to avoid cluttering the drawing.

In this graph, the edge of weight 3 is the first one to be examined and is obviously accepted because the first edge is always accepted. However, it invalidates the two edges of weight 2, so the algorithm is forced to select an edge of weight zero to complete the Hamiltonian path. The two rejected edges form a path of total weight 4, which is the heaviest path in this example.

This example shows that the greedy algorithm will not always return the shortest superstring. Is there an algorithm that works in all cases? Apparently not. We are trying to solve the SCS problem through the Hamiltonian path (HP) problem; but as we have seen in Section 2.3, the HP problem is NP-complete. Perhaps we could have better luck trying another approach to solve the SCS problem. Unfortunately this is probably not the case either, since, as already mentioned, it can be shown that the SCS problem is NP-hard.

4.3.5 ACYCLIC SUBGRAPHS

The hardness results we mentioned in the previous section apply to an arbitrary collection of fragments. In this section we consider the problem of assembling fragments without errors and known orientation assuming that the fragments have been obtained from a "good sampling" of the target DNA. As shown in the sequel, this assumption enables us to derive algorithms that deliver the correct solution and are particularly efficient.

What do we mean by a "good sampling"? Basically, we want the fragments to cover the entire target molecule, and the collection as a whole to exhibit enough linkage to guarantee a safe assembly. Recall from Section 4.2.3 the definition of t-contig. Our definitions here are based on similar concepts.

Suppose that S is a string over the alphabet {A, C, G, T}. Recall from Section 2.1

that an *interval* of S is an integer interval $[i..j]$ such that $1 \leq i \leq j+1 \leq |S|+1$. A *sampling* of S is a collection $\mathcal{A}$ of intervals of S. The sampling $\mathcal{A}$ *covers* S if for any i such that $1 \leq i \leq |S|$ we have at least one interval $[j..k] \in \mathcal{A}$ with $i \in [j..k]$.

We say that two intervals α and β are *linked at level t* if $|\alpha \cap \beta| \geq t$. The entire sampling $\mathcal{A}$ is said to be *connected at level t* if for every two intervals α and β in $\mathcal{A}$ there is a series of intervals α_i for $0 \leq i \leq l$ such that $\alpha = \alpha_0$, $\beta = \alpha_l$ and α_i is linked to α_{i+1} at level t for $0 \leq i \leq l-1$. Note that we are using parameter t to measure how *strong* a link is. A sampling $\mathcal{A}$ will be considered good if it is connected at level t.

Our goal is to study fragment collections coming from connected at level t samplings that cover a certain string S. The value of t is around 10 in typical, real instances, but in our study we will not fix it. We do assume that t is a nonnegative integer, though.

The fragment collection *generated* by a sampling $\mathcal{A}$ is

$$S[\mathcal{A}] = \{S[\alpha] \mid \alpha \in \mathcal{A}\}.$$

We say that a sampling $\mathcal{A}$ is *subinterval-free* if there are no two intervals $[i..j]$ and $[k..l]$ in $\mathcal{A}$ with $[i..j] \subseteq [k..l]$. The following observation about subinterval-free samplings will be coming into play again and again. In a subinterval free sampling, no two intervals can have the same left endpoint. Otherwise, whichever has the larger right endpoint would contain the other. An analogous argument holds for right endpoints.

We consider also modified forms of our overlap graphs. Given a collection $\mathcal{F}$ and a nonnegative integer t define $\mathcal{OM}(\mathcal{F}, t)$ as the graph obtained from $\mathcal{OM}(\mathcal{F})$ by keeping only the edges of weight at least t. A similar construction defines $\mathcal{OG}(\mathcal{F}, t)$. This reduces the overall number of edges, keeping only the stronger ones.

We will use the symbol $\alpha \overset{w}{\to} \beta$ to mean that $|\alpha| \geq w$, $|\beta| \geq w$, and $l(\beta) + w - 1 = r(\alpha)$. Similarly, $f \overset{w}{\to} g$ means that there is an edge from f to g in $\mathcal{OM}(\mathcal{F})$ of weight w.

Suppose now that we got such a collection $\mathcal{F}$ and would like to recover S. How can we do that? What are the good algorithms for it? It turns out that the answer depends strongly on the repeat structure of S. If S does not have any repeats of size t or more, then it is easy to reconstruct it. But if S has such repeats, the problem gets more complex, and in some cases S cannot be recovered without ambiguity.

One thing that is independent of the repeat structure of S is the existence of a Hamiltonian path in the graph $\mathcal{OM}(\mathcal{F}, t)$. We state and prove this basic result here, before embarking on an analysis of the repeat structure of S. First, an auxiliary lemma.

LEMMA 4.2 Let S be a string over {A, C, G, T} and $\mathcal{A}$ be a subinterval-free, connected at level t sampling of S, for some $t \geq 0$. If α is an interval in $\mathcal{A}$ such that there is another interval β in $\mathcal{A}$ with $l(\alpha) < l(\beta)$, then β can be chosen so that

$$r(\alpha) + 1 - l(\beta) \geq t.$$

Proof. We know $\mathcal{A}$ is connected at level t, therefore there is a sequence $(\alpha_i)_{0 \leq i \leq l}$ of intervals in $\mathcal{A}$ such that $\alpha = \alpha_0$, $\beta = \alpha_l$, and $|\alpha_i \cap \alpha_{i+1}| \geq t$ for $0 \leq i \leq l-1$.

Let i be the smallest index such that $l(\alpha) < l(\alpha_i)$. Because $\beta = \alpha_l$ satisfies this property, we know that $i \leq l$. On the other hand, $i > 0$ because $\alpha = \alpha_0$ does not satisfy the property. Because i is the smallest index with the property, we have $l(\alpha_{i-1}) < l(\alpha) < l(\alpha_i)$. It follows that $r(\alpha_{i-1}) < r(\alpha) < r(\alpha_i)$, since $\mathcal{A}$ is subinterval-free. The picture looks like Figure 4.18.

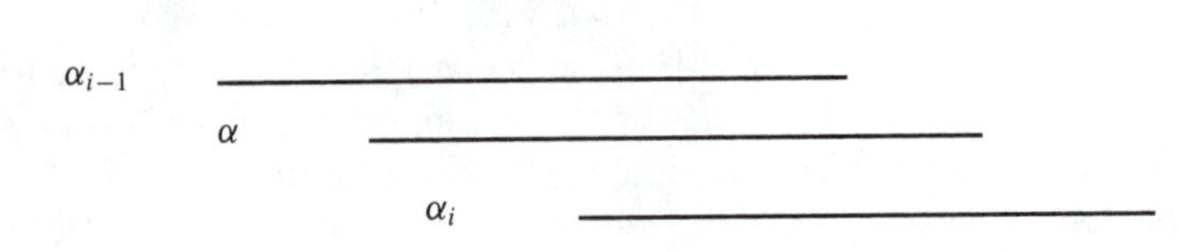

FIGURE 4.18

Three mutually overlapping intervals.

We also know that $|\alpha_{i-1} \cap \alpha_i| \geq t$. But $|\alpha \cap \alpha_i| \geq |\alpha_{i-1} \cap \alpha_i| \geq t$ (look at Figure 4.18), so α_i is an interval in $\mathcal{A}$ such that

$$r(\alpha) + 1 - l(\alpha_i) \geq t.$$

■

Now, the Hamiltonian path existence result.

THEOREM 4.3 Let S be a string over {A, C, G, T} and $\mathcal{A}$ be a subinterval-free, connected at level t sampling of S, for some $t \geq 0$. Then the multigraph $\mathcal{OM}(\mathcal{F}, t)$, where $\mathcal{F}$ is the fragment collection generated by $\mathcal{A}$, admits a Hamiltonian path P. Furthermore, if $\mathcal{A}$ covers S, then P can be chosen so that $S(P) = S$.

Proof. Let $(\alpha_i)_{1 \leq i \leq n}$ be all the intervals in $\mathcal{A}$ sorted by left endpoint. Because $\mathcal{A}$ is subinterval-free, this sequence is also sorted by right endpoint. Let $f_i = S[\alpha_i]$ for $1 \leq i \leq n$. We claim that $f_i \stackrel{k_i}{\rightarrow} f_{i+1}$ for some $k_i \geq t$, so that the f_i in the above order form a Hamiltonian path.

Let α_i be any interval except the last one. By the previous lemma, there is another interval $\beta \in \mathcal{A}$ such that $l(\alpha_i) < l(\beta)$ and $r(\alpha_i) + 1 - l(\beta) \geq t$. The sequence $(\alpha_i)_{1 \leq i \leq n}$ is sorted by left endpoint, so $l(\alpha_i) < l(\alpha_{i+1}) \leq l(\beta)$. (The intervals α_{i+1} and β may be the same.) Hence,

$$k_i = |\alpha_i \cap \alpha_{i+1}| = |[l(\alpha_{i+1})..r(\alpha_i)]| = r(\alpha_i) - l(\alpha_{i+1}) + 1 \geq r(\alpha_i) + 1 - l(\beta) \geq t.$$

This shows that an edge $f_i \rightarrow f_{i+1}$ really exists in $\mathcal{OM}(\mathcal{F}, t)$ and concludes the proof of the first part.

The second part, which says that $S(P) = S$, is a consequence of the following fact. Every time we have $\alpha \stackrel{w}{\rightarrow} \beta$, we also have $S[\alpha] \stackrel{w}{\rightarrow} S[\beta]$ and $S[\alpha]\kappa^w S[\beta] = S[\alpha \cup \beta]$. ■

The presence of repeated regions, or repeated elements, in the target string S is related to the existence of cycles in the overlap graph. In this and following sections we make this relationship more explicit.

A repeated element, or just a *repeat* in S, occurs when two different intervals of S give rise to the same substring. In other words, there are two intervals $\alpha \neq \beta$ such that $S[\alpha] = S[\beta]$. The *size* of this repeat is $|S[\alpha]|$, which is the same as $|S[\beta]|$. The repeat is *self-overlapping* when $\alpha \cap \beta \neq \emptyset$, otherwise we have *disjoint occurrences* of the repeat.

Our next result shows that cycles in an overlap graph are necessarily due to repeats in S. The converse is not necessarily true; that is, we may have repeats but still an acyclic overlap graph. Again, a couple of intermediate results are needed before we can prove an important fact.

LEMMA 4.3 If $\alpha \xrightarrow{w} \beta$ then $l(\alpha) \leq l(\beta)$ and $r(\alpha) \leq r(\beta)$.

Proof. From $|\alpha| \geq w$ we have $l(\alpha)+w-1 \leq r(\alpha)$. From $\alpha \xrightarrow{w} \beta$ we get $l(\beta)+w-1 = r(\alpha)$. Therefore $l(\alpha) + w - 1 \leq l(\beta) + w - 1$, that is, $l(\alpha) \leq l(\beta)$. Similarly, $|\beta| \geq w$ implies $l(\beta) + w - 1 \leq r(\beta)$, or $r(\alpha) \leq r(\beta)$, in view of previous equations. ■

We need a few more definitions before proceeding. A *false positive* at level t is a pair of intervals α and β such that there is a $w \geq t$ with $S[\alpha] \xrightarrow{w} S[\beta]$ but $\alpha \xrightarrow{w} \beta$ does not hold. The name refers to the fact that we assumed that $S[\alpha]$ and $S[\beta]$ came from intersecting regions in the target molecule because their sequences overlap, but this is not the case. They *look* overlapping but they are not, so they constitute a false positive.

LEMMA 4.4 The existence of a false positive at level t implies the existence of a repeat of size at least t.

Proof. Because $S[\alpha] \xrightarrow{w} S[\beta]$, we have $|\alpha| \geq w$ and $|\beta| \geq w$. If $l(\beta) = r(\alpha) - w + 1$, then we would have $\alpha \xrightarrow{w} \beta$, since $\alpha \neq \beta$ by hypothesis. Hence, $l(\beta) \neq r(\alpha) - w + 1$.

The relation $S[\alpha] \xrightarrow{w} S[\beta]$ also implies that $S[\alpha'] = S[\beta']$, where α' is the last w elements of α and β' is the first w elements of β. Because $l(\beta) \neq r(\alpha) - w + 1$, α' and β' are different intervals giving rise to the same substring; that is, we have a repeat of size $w \geq t$. ■

THEOREM 4.4 Let $\mathcal{F}$ be a collection generated by a sampling $\mathcal{A}$ of S. If $\mathcal{OG}(\mathcal{F}, t)$ has a directed cycle, then there is a repeat in S of size at least t.

Proof. Let $f_1 \rightarrow f_2 \rightarrow \cdots \rightarrow f_l \rightarrow f_1$ be a directed cycle in $\mathcal{OG}(\mathcal{F}, t)$. Consider intervals $\alpha_1, \ldots, \alpha_l$ such that $S[\alpha_i] = f_i$, for $1 \leq i \leq l$. We will show that at least one of the pairs (α_i, α_{i+1}), including (α_l, α_1), is a false positive, which implies the existence of a repeat in S by Lemma 4.4.

Suppose they are all true positives. Then $\alpha_i \rightarrow \alpha_{i+1}$ for $1 \leq i < l$ and $\alpha_l \rightarrow \alpha_1$. It follows that $l(\alpha_1) \leq l(\alpha_2) \leq \cdots \leq l(\alpha_l) \leq l(\alpha_1)$. So, all the $l(\alpha_i)$ are equal. An analogous observation holds for the $r(\alpha_i)$. But then $\alpha_1 = \alpha_2 = \cdots = \alpha_l$, which is impossible, since the f_i are distinct by hypothesis. Hence, there is at least one false positive and the result is proved. ■

All these considerations lead to a satisfactory solution for fragment assembly when the target string S does not have repeats.

THEOREM 4.5 Let S be a string over {A, C, G, T}, $\mathcal{A}$ a subinterval-free, connected at level t sampling of S that covers S, and $\mathcal{F} = S[\mathcal{A}]$. If S has no repeats of size t or larger, then the graph $\mathcal{OG}(\mathcal{F}, t)$ has a unique Hamiltonian path P and $S(P) = S$.

Proof. We have seen that $\mathcal{OG}(\mathcal{F}, t)$ is acyclic (Theorem 4.4) and that $\mathcal{OM}(\mathcal{F}, t)$ has a Hamiltonian path P with $S(P) = S$.

To begin with, notice that $\mathcal{OM}(\mathcal{F}, t) = \mathcal{OG}(\mathcal{F}, t)$ if S has no repeats of size t or larger. This follows from the absence of false positives. It is clear that the vertex sets of $\mathcal{OM}(\mathcal{F}, t)$ and $\mathcal{OG}(\mathcal{F}, t)$ are the same. It is also clear that every edge of $\mathcal{OG}(\mathcal{F}, t)$ is also in $\mathcal{OM}(\mathcal{F}, t)$. What we need is to convince ourselves that all edges in $\mathcal{OM}(\mathcal{F}, t)$ are in $\mathcal{OG}(\mathcal{F}, t)$.

Let $f \stackrel{w}{\to} g$ be an edge in $\mathcal{OM}(\mathcal{F}, t)$. Then $w \geq t$. If this edge is not in $\mathcal{OG}(\mathcal{F}, t)$, this is due to the existence of another edge $f \stackrel{x}{\to} g$ with $x > w$. Let α and β be intervals in $\mathcal{A}$ with $S[\alpha] = f$ and $S[\beta] = g$. Since there are no false positives, we must have at the same time $\alpha \stackrel{w}{\to} \beta$ and $\alpha \stackrel{x}{\to} \beta$, which is impossible. The weight in this edge is uniquely determined by $w = r(\alpha) - l(\beta) + 1$.

So, $\mathcal{OG}(\mathcal{F}, t)$ is an acyclic graph with a Hamiltonian path. It turns out that every time an acyclic graph has a Hamiltonian path this path is unique. To see why this is true, consider the nodes in the graph that have indegree zero (the *sources*). Because the graph is acyclic, it must have at least one source. But, having a Hamiltonian path, it can have at most one source: the starting point of this path. All other vertices have at least one incoming edge. We conclude that our graph must have exactly one source, and it must be the starting point of the Hamiltonian path.

Now, if we remove this source from the graph, the remaining graph is still acyclic, and has a Hamiltonian path, because the vertex we removed is an extremity of the old Hamiltonian path. Proceeding in this way, we see that the Hamiltonian path is unique, and we know how to construct it: just keep removing sources from the graph. ■

The algorithm just described is well known in computer science (see Exercise 12 of Chapter 2). It is called *topological sorting* and, in its more general context, is used to find an ordering of nodes that is consistent with an acyclic set of edges in the following sense: For all edges $f \to g$ the node f appears before g in the ordering. If the graph has a Hamiltonian path besides being acyclic, then this ordering is unique, as we just saw.

We end this section with an example where *Greedy* does not work, but there is an acyclic subgraph for a suitable threshold t, so that topological sorting will find the correct path. When we say that *Greedy* does not work, we mean that it does not do a good job regarding assembly, although it does do a good job finding a shortest common superstring. In this example the fragments come from a target sequence that does have repeats longer than t. This goes to show that the absence of long repeats is not necessary to yield an acyclic graph. It is merely sufficient, in mathematical jargon.

Example 4.6 Consider the following target string S and fragments w, z, u, x, and y:

$S =$ AGTATTGGCAATCGATGCAAACCTTTTGGCAATCACT

$w =$ AGTATTGGCAATC

$z =$ AATCGATG

$u =$ ATGCAAACCT

$x =$ CCTTTTGG

$y =$ TTGGCAATCACT

Choosing a threshold value of $t = 3$, we get the overlap graph depicted in Figure 4.19. A greedy assembly will result in a shortest common superstring, formed by concatenation of the contig $w\kappa^9 y$ with the contig $z\kappa^3 u\kappa^3 x$ in any order, as shown in Figures 4.20 and 4.21. This solution has length 36, one character shorter than the target sequence, which is also a superstring. However, the solution given by the Hamiltonian path has better linkage.

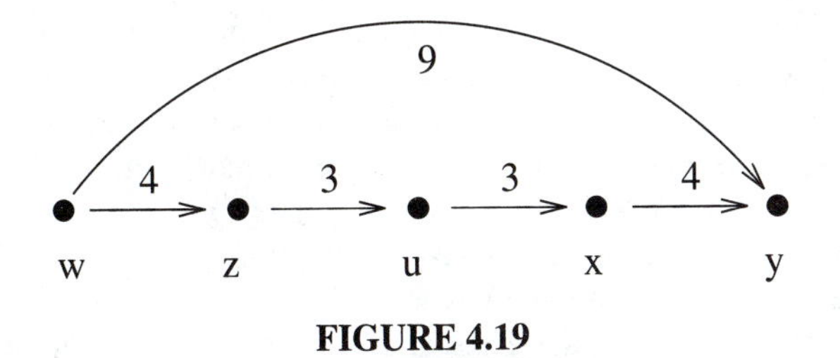

FIGURE 4.19

Overlap graph for a given set of five fragments.

```
AGTATTGGCAATC---AATCGATG------------
---------------------ATGCAAACCT-----
----TTGGCAATCACT------------CCTTTTGG
------------------------------------
AGTATTGGCAATCACTAATCGATGCAAACCTTTTGG
```

FIGURE 4.20

A bad solution for an assembly problem, with a multiple alignment whose consensus is a shortest common superstring. This solution has length 36 and is generated by the Greedy algorithm. However, its weakest link is zero.

```
AGTATTGGCAATC--------CCTTTTGG--------
---------AATCGATG--------TTGGCAATCACT
--------------ATGCAAACCT-------------
-------------------------------------
AGTATTGGCAATCGATGCAAACCTTTTGGCAATCACT
```

FIGURE 4.21

Solution according to the unique Hamiltonian path. This solution has length 37, but exhibits better linkage. Its weakest link is 3.

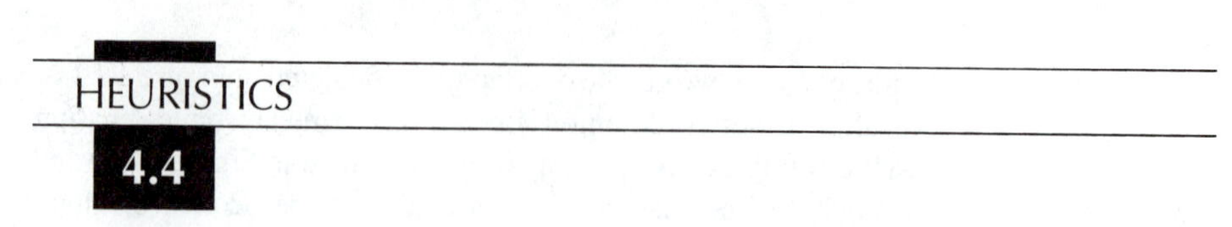

HEURISTICS 4.4

We have seen that none of the formalisms proposed for fragment assembly are entirely adequate, so in practice we must often resort to heuristic methods. We review some of these in this section, starting by pointing out relevant properties of a good assembly layout.

Fragment assembly can be viewed as a multiple alignment problem with some additional features. The most striking such feature is perhaps the fact that each fragment can participate with either the direct or the reverse-complemented sequence. A second special feature is that the sequences themselves are usually much shorter than the alignment itself. This leads us to charge internal gaps and external gaps in the fragments differently, with internal gaps being much more penalized. If we charge both kinds of gaps the same or nearly the same penalty, then there would be no incentive to keep fragment characters together, potentially causing them to scatter along the entire length of the alignment. Moreover, the fact that fragments are short compared to the alignment forces us to look at other criteria, besides the score, to assess the quality of an alignment. These extra criteria include coverage and linkage. The latter measures how strongly the sequences are linked in terms of their overlap, as we have seen in Sections 4.2.3 and 4.3.5. We now discuss these criteria in more detail.

Scoring: In each column of the multiple alignment we have a certain coverage (defined in Section 4.1.2). The ideal situation is when all sequences participating in a column have the same character there. In general, uniformity is good and variability is bad in a column.

This suggests the use of a concept called *entropy* to measure the quality of a column. Entropy is a quantity that is defined on a group of relative frequencies, and it is low when one of these frequencies stands out from the others, and high when they are all more or less equal. In our context, the lower the entropy, the better. The frequencies we use are the frequencies of each character, where a character here can be an ordinary base `A`, `C`, `G`, `T`, or an internal space.

Suppose that we have n sequences covering a certain column, and relative frequencies $p_{\mathtt{A}}$, $p_{\mathtt{C}}$, $p_{\mathtt{G}}$, $p_{\mathtt{T}}$, and p_{space} for each character. The entropy of this distribution is defined as

$$E = -\sum_c p_c \log p_c.$$

If $p_c = 1$ for a certain character c and all other frequencies are zero, then we have (assume $0 \log 0 = 0$)

$$E = 0.$$

If, on the other hand, all frequencies are equal, we have $p_c = 1/5$ for each c and then

$$E = -5 \times \frac{1}{5} \log \frac{1}{5} = \log 5,$$

which is the maximum value possible for five frequencies.

Recall from Section 3.4 that we used the sum-of-pairs scoring scheme when we discussed multiple alignments. This scheme combines in a single number both the coverage and the uniformity of a column. The larger the coverage, the larger the sum-of-pairs value, provided the uniformity is high. Also, for constant coverage, the sum-of-pairs value increases with uniformity. In contrast, the entropy is independent of coverage. It measures uniformity alone. Depending on the situation, a scientist may want to use sum-of-pairs, a combined measure, or the coverage and entropy separately.

Coverage: In Section 4.1.2 we defined the coverage of a set of fragments with respect to the target from which they were obtained. Here we define an analogous concept for a layout. A fragment covers a column i if it participates in this column either with a character or with an internal space. Let $l(f)$ and $r(f)$ be the leftmost position and the rightmost position, respectively, of a character of f (or of its reversed complement) in the layout. The previous definition means that fragment f covers column i if $l(f) \le i \le r(f)$.

Having defined what the coverage is in each column, we may speak of *minimum*, *maximum*, and *mean* coverage of a layout. The meanings are as expected. If the coverage ever reaches value zero for a column i, then we do not really have a connected layout. Indeed, no fragment image spans column i, so the parts of the layout to the left and to the right of i are in some sense independent. For instance, the right part could be placed to the left of the other, giving an alternative layout as good as the original one.

This means that if we have more columns with zero coverage, any permutation of the intervening regions is acceptable. Each one of these regions corresponds to what we have called a *contig*. Coverage is also important to give confidence to the consensus sequence. If there are several fragments covering a column, we have more information on the particular base call for this column. It is also important to have enough fragments from both strands in the layout.

Linkage: Apart from the score and the coverage, the way individual fragments are linked in the layout is another determinant of layout quality. For example, consider the layout of Figure 4.22. The coverage is high, but there are no real links between the fragments. Each block is composed of many almost exact copies of a certain sequence, but the blocks themselves do not seem related at all. Fragment images should have overlapping ends to show some evidence of linkage.

```
------ACTTTT------
TCCGAG------ACGGAC
------ACTTTT------
TCCGAG------ACGGAC
------ACTTTT------
TCCGAG------ACGGAC
------------------
TCCGAGACTTTTACGGAG
```

FIGURE 4.22

Good coverage but bad linkage.

Assembly in Practice

A good assembly program must balance these requirement and find all solutions that satisfy them reasonably well. This is no easy task, and, coupled with the fact that no perfect formalisms exist, it makes algorithm design very difficult. As a result, practical implementations often divide the whole problem in three phases, each one using results produced by the previous phases. A typical division encompasses the following phases:

- Finding overlaps
- Building a layout
- Computing the consensus

Such partitioning makes the problem more manageable, because each phase can be treated more or less independently as a different problem. However, it becomes more difficult to understand the relationship between the initial input and the final output. Often, the methods used in these phases are heuristic, and no clear guarantees can be given with respect to solution quality. Nevertheless, many existing implementations perform well in practice and it is worthwhile to know some of the techniques used.

4.4.1 FINDING OVERLAPS

The first step in any assembly program is fragment overlap detection. We need to try all pairs of fragments and their reverse complements to determine whether there is a prefix of a fragment that matches well with the suffix of another. In general, we will need all such pairs. Also, we are interested in fragments entirely contained in other fragments. All these comparisons are made within an error threshold.

We mention here a way of comparing two fragments with the goal of finding such overlaps. In a real problem we could apply this method to all pairs, including reverse complements, but this is seldom done, because it would take too long due to its being based on a quadratic time algorithm. However, it is one of the most reliable methods of assessing overlap.

The comparison method consists in applying the dynamic programming algorithm described in Section 3.2.3, which is a variant of the basic algorithm. This variant does not charge for gaps in any extremity. Figure 4.23 presents the results of using such an algorithm for two fragments. The score system used in this comparison is 1 for matches, -1 for mismatches, and -2 for gaps. Notice that the algorithm does not charge for the gaps after the first sequence or before the second one, so it was able to find a prefix–suffix similarity. The same algorithm would find an approximate containment.

4.4.2 ORDERING FRAGMENTS

In this section we briefly discuss one aspect of layout construction in a real assembly problem, namely, the issue of finding a good ordering of the fragments in a contig. Given such an ordering, and knowing that each fragment overlaps with the next one, a layout

```
AGGAGAAGAATTCACCGCTAT----------
----------TTCCCCT-TATTCAATTCTAA
```

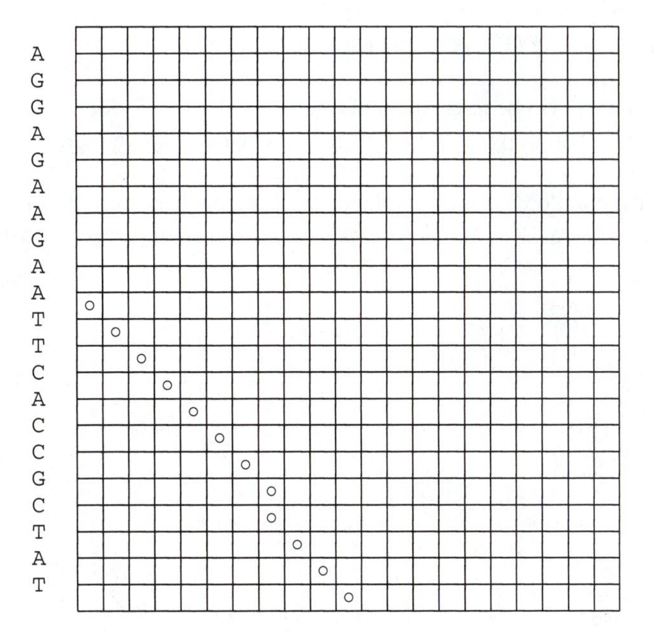

FIGURE 4.23

An optimal semiglobal alignment and its corresponding path in the dynamic programming matrix. The algorithm is configured to find prefix–suffix or containment alignments, which are appropriate for fragment assembly.

can be built using the methods of the next section. Unfortunately, there is no algorithm that is simple and general enough, even resorting to heuristics. Therefore, rather than presenting an algorithm, we point out important considerations that should be taken into account when trying to construct good fragment orderings.

In a real instance we have to deal with reverse complements, so it is necessary to extend the input collection $\mathcal{F}$ with the reverse complements of the fragments. Define

$$\overline{\mathcal{F}} = \{\overline{f} \mid f \in \mathcal{F}\}$$

and

$$\mathcal{DF} = \mathcal{F} \cup \overline{\mathcal{F}}.$$

In a real problem we have to worry about errors as well, and this means considering approximate matching. We will use the notation $f \to g$ in a slightly different sense from previously: It now indicates that the end of fragment f is approximately the same as the beginning of fragment g in some sense. We also assume that whenever $f \to g$ we also have $\overline{g} \to \overline{f}$, because it is reasonable to expect that whatever criterion is used to assess the similarity between two sequences, the same criterion would apply to their reverse complements. This can be shown as follows: If $f \to g$, we have that $f = uv$, $g = wx$, and v and w are similar. This implies that $\overline{g} = \overline{x}\,\overline{w}$, $\overline{f} = \overline{v}\,\overline{u}$, with $\overline{w}$ and $\overline{v}$ similar, which is equivalent to $\overline{g} \to \overline{f}$. We will use the notation $f \subseteq g$ meaning that f is an approximate substring of g. An argument analogous to the one just presented shows that $f \subseteq g \Rightarrow \overline{f} \subseteq \overline{g}$.

The graph we will be interested in is then $\mathcal{OG}(\mathcal{DF})$, with edges indicating approximate overlap. In addition it is important to know which fragments are approximate substrings of others.

Finding a good ordering of overlapping fragments is essentially equivalent to finding directed paths in $\mathcal{OG}(\mathcal{DF})$. These paths originate contigs. Notice that for any path

$$f_1 \to f_2 \to \cdots \to f_k$$

in this graph there is a corresponding "complementary" path

$$\overline{f_k} \to \overline{f_{k-1}} \to \cdots \to \overline{f_1},$$

which gives rise to a contig with a consensus that is the reverse complement of the first path. This means that both strands of the target DNA are being constructed simultaneously. Ideally we would like to get a pair of complementary paths such that every noncontained fragment belongs to exactly one path in the pair. Contained fragments can be positioned afterward so their presence in the paths is not essential. Actually, forcing their presence may prevent the formation of a path (see Exercise 14). In rare occasions contained fragments may provide an important link, so we consider them optional fragments when building a path.

Two obstacles may prevent the ideal goal of a pair of paths covering all noncontained fragments: lack of coverage and repeats. A disconnected overlap graph is usually an indication of lack of coverage. Repeats cause the appearance of cycles in the overlap graph, if coverage is good, or may show up as edges $f \to g$ and $f \to h$ with g and h unrelated (g and h are unrelated if $g \not\to h$, $h \not\to g$, $g \not\subseteq h$, and $h \not\subseteq g$). This happens when f is a fragment near the end of a repeat X and g and h are fragments leading from X to distinct flanking regions in the target. A cycle involving both f and $\overline{f}$ for a certain fragment f indicates the presence of an inverted repeat.

Another indication of repeated regions is unusually high coverage (see Figure 4.13). All fragments from all copies of a given repeat are piled together, causing coverage to increase to about the average value times the number of repeat copies (assuming an approximately uniform distribution of fragments along the target).

In conclusion, there are four issues to keep in mind when building paths: (1) Every path has a corresponding complementary path; (2) it is not necessary to include contained fragments; (3) cycles usually indicate the presence of repeats; and (4) unbalanced coverage may be related to repeats as well.

4.4.3 ALIGNMENT AND CONSENSUS

We briefly discuss in this section the problem of building a layout from a path in an overlap graph. This is straightforward if the overlaps are exact, but approximate overlaps require extra care. We present here two techniques related to alignment construction. The first one helps in building a good layout from a path in the presence of errors. The second one focuses on locally improving an already constructed layout. To introduce the first technique, we use the following example.

Example 4.7 Suppose we have a path $f \to g \to h$ with

$$f = \texttt{CATAGTC}$$
$$g = \texttt{TAACTAT}$$
$$h = \texttt{AGACTATCC}$$

The two optimal semiglobal alignments between f and g are

```
CATAGTC---            CATAGTC---
--TAA-CTAT            --TA-ACTAT
```

We do not have any reason to prefer one of them over the other, since their score is the same. However, when sequence h enters the scene, it becomes clear that the second optimal alignment (where the second A of g is matched to a T in f) is better:

```
CATAGTC-----
--TA-ACTAT--
---AGACTATCC
------------
CATAGACTATCC
```

The consensus S satisfies $d_s(f, S) = 1$, $d_s(g, S) = 1$, and $d_s(h, S) = 0$. Had we aligned A with G, the voting in column 6 would have been ambiguous, with T, A, and - all tied with one occurrence each. In the consensus any choice other than A in this column would have led to a worse result, in the sense that the sum $d_s(f, S) + d_s(g, S) + d_s(h, S)$ would be greater than with the first solution.

To avoid such problems, we use a structure that represents alignments distinguishing between bases that are actually matched to some other base, and bases that are "loose," as is the case of the second base A of sequence g in the above example. We can implement this idea as follows. Each sequence is represented by a linked list of bases. When an alignment is formed, matched bases are unified and become the same node in the overall structure. If unsure about matching a particular base, we can leave it unmatched, as we did in Figure 4.24 for the sequences f and g of Example 4.7. Then, as new sequences are added, some of their bases will be unified according to suitable alignments. The structure can be seen as an acyclic graph, and a traversal in topological order yields a layout. Applying this technique to Example 4.7, we get the best possible layout. When h is added, an arrow from G to the A node just below it will be added, as well as additional bases that belong to h alone.

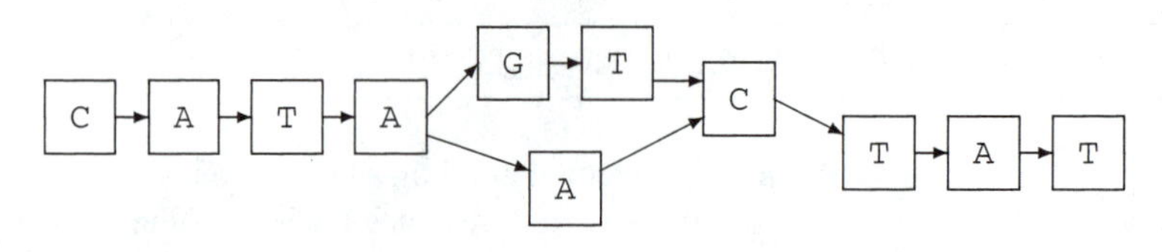

FIGURE 4.24

Linked structure for approximate overlaps.

```
ACT-GG          ACT-GG
ACTTGG          ACTTGG
AC-TGG          ACT-GG
ACT-GG          ACT-GG
AC-TGG          ACT-GG
------          ------
ACTTGG          ACT-GG
```

FIGURE 4.25

Two layouts for the same sequences. The layout to the right is better, but an assembly program might produce the left one due to ambiguities in pairwise alignments.

Our second technique concerns improving a given layout. One common problem in practice is illustrated by the following example.

Example 4.8 Figure 4.25 shows two layouts for the same set of sequences. In the left-hand layout, the consensus ends up with two Ts by majority voting, but four of the five sequences have just one T. The layout to the right is better, for its consensus agrees with the majority of the sequences.

One idea to solve such problems is to perform a multiple alignment locally. We select a short range of consecutive columns that need doctoring, extract from each fragment its portion contained in these columns, and align the portions. For moderate values of the local coverage and width of the range the multiple alignment algorithm will run in a reasonable amount of time, despite its exponential complexity. A sum-of-pairs scoring scheme seems adequate here. In Example 4.8, the right-hand layout maximizes the sum-of-pairs score, assuming $p(a, b) = 1$ if $a = b$, $p(a, b) = 0$ if $a \neq b$, $p(a, -) = -1$, and $p(-, -) = 0$.

SUMMARY

In this chapter we studied the fragment assembly problem, which appears in large-scale DNA sequencing. Basically, it consists of assembling in the correct order and orientation a collection of fragments coming from a long, unknown DNA sequence.

We began by reviewing the biological setting of this problem, and how the biological factors influence the input. Then we presented attempts that have been made to capture the problem into a mathematical framework. The proposed models fail to address all the issues involved. In particular, it is hard to deal with repeated regions and to take into account information about the size of the target sequence. Also, all models assume that the collection of fragments is free of chimeras and that there is no vector contamination. Algorithms based on the models were also presented.

Fragment assembly requires a particular kind of multiple alignment, also called layout, in which we have to worry about fragment orientation, coverage, and linkage. We do not charge for end spaces when scoring this alignment.

Because of the difficulties in finding a precise formulation for the problem, heuristic methods are often used in practice. The problem is usually broken into three phases: finding overlaps between fragments, building a layout, and computing the consensus. Programs based on this approach perform well in real instances without repeats. We presented an overview of such heuristics.

New techniques, such as sequencing by hybridization, may radically change the way large-scale sequencing is done. If this happens, new algorithms will be required.

EXERCISES

1. Suppose we have the following fragments

   f_1 : `ATCCGTTGAAGCCGCGGGC`
   f_2 : `TTAACTCGAGG`
   f_3 : `TTAAGTACTGCCCG`
   f_4 : `ATCTGTGTCGGG`
   f_5 : `CGACTCCCGACACA`
   f_6 : `CACAGATCCGTTGAAGCCGCGGG`
   f_7 : `CTCGAGTTAAGTA`
   f_8 : `CGCGGGCAGTACTT`

   and we know that the total length of the target molecule is about 55 base pairs. Assemble these fragments and obtain a consensus sequence. Be prepared to deal with errors. You may also have to use the reverse complement of some of the fragments.

2. What are the minimum, maximum, and average coverage for the layout on page 106?

3. What is the smallest value of ϵ such that the answer in Figure 4.2 is valid under the RECONSTRUCTION model?

4. Repeat Exercise 3, this time for the layouts in Figures 4.3 and 4.4.

5. In Figure 4.8, suppose there is a fragment composed by the end of B, a copy of X, and the beginning of D. How would this fragment affect the assembly?

6. Construct the overlap graph for $\mathcal{F} = \{\texttt{AAA}, \texttt{TTA}, \texttt{ATA}\}$. Find a shortest common superstring for this collection.

7. Find sequences that give rise to the following overlap graph, where only edges with positive weights are shown. The particular weights you come up with are not important, as long as they are strictly positive.

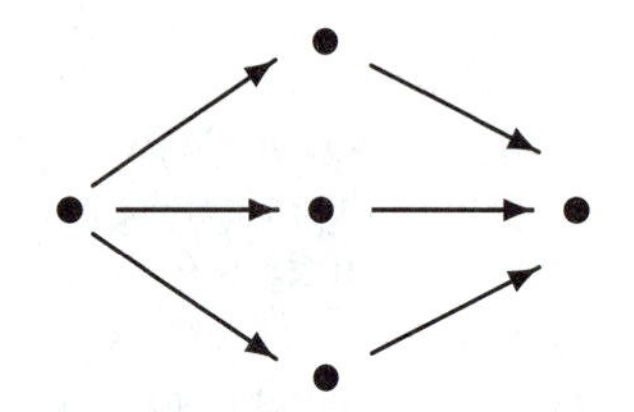

8. Describe an algorithm to find the largest exact overlap between two strings of lengths m and n respectively that runs in $O(m+n)$ time.

★ 9. Devise a way of locating approximate repeats in a DNA sequence.

10. Give an example of a DNA sequence that cannot be reconstructed unambiguously from the set of its 8-mers.

11. Show that the SCS formulation fails for the examples in Section 4.1. What is the shortest common superstring in each case?

12. Let $\mathcal{F} = \{\texttt{ATC}, \texttt{TCG}, \texttt{AACG}\}$. Find the best layout for this collection according to the RECONSTRUCTION model with $\epsilon = 0.1$ and $\epsilon = 0.25$. Be sure to consider reverse complements.

13. Let $\mathcal{F} = \{\texttt{TCCCTACTT}, \texttt{AATCCGGTT}, \texttt{GACATCGGT}\}$. Find the best set of contigs for this collection according to the MULTICONTIG model with $\epsilon = 0.3$ and $t = 5$.

14. Give an example of a collection $\mathcal{F}$ such that the shortest superstring of $\mathcal{F}$ does not correspond to a path in $\mathcal{OG}(\mathcal{F})$.

15. For every $m > 0$, find a substring-free collection $\mathcal{F}_m$ such that the greedy algorithm returns a superstring at least m characters longer than the shortest possible for $\mathcal{F}_m$.

16. Let A, B, C, D, E, X, Y, and Z be blocks in a target DNA sequence. Suppose that these blocks are larger than any fragment that can be sampled from the target molecule. Fill in the blanks below with letters X, Y, or Z so that this sequence becomes inherently ambiguous to assemble.

$$\text{target} = A \quad X \quad B \quad \ldots \quad C \quad Y \quad D \quad \ldots \quad E \quad Z$$

17. How would you use BLAST or other database search program to screen fragments for vector contamination?

18. Find a polynomial time reduction of SCS to RECONSTRUCTION.

★ 19. Suppose that a sequence has an overlapping repeat S. Show that S is either of the form XXX or of the form $X(YX)^k$ for some nonempty blocks X and Y and some $k \geq 2$.

◇ 20. Design an algorithm that receives a collection of DNA sequences and constructs a common superstring of length at most twice the minimum possible.

BIBLIOGRAPHIC NOTES

The shortest common superstring problem has been intensively studied. Turner [187] described several algorithms, including the greedy method, and showed a performance guarantee with respect to the sum of the overlaps. Li [126] gave the first performance guarantee for the actual length of the superstring, providing an $O(n \log n)$ upper bound, where n is the optimal length. Then Blum, Jiang, Li, Tromp, and Yannakakis [24] improved this to a $3n$ guarantee and also showed that the greedy algorithm yields superstrings at most $4n$ long. This prompted a series of papers using more and more refined techniques and achieving better and better ratios. In this line, Teng and Yao [185] proved a ratio of $2+8/9$; Czumaj, Gasienec, Piotrow, and Rytter [41] proved a ratio of $2+5/6$; Kosaraju, Park, and Stein [118] proved a ratio of $2+50/63$; and Armen and Stein proved ratios of $2+3/4$ [15] and $2+2/3$ [16]. Most of these algorithms run in time $O(|S|+n^3)$, where $|S|$ is the superstring size and n is the number of fragments (see, for instance, the work by Gusfield [84]). The graph $\mathcal{OG}(\mathcal{F})$ can be constructed in time $O(\|\mathcal{F}\|+n^2)$ with the help of suffix trees using a method due to Gusfield, Landau, and Schieber [86].

The RECONSTRUCTION model appeared in a paper by Peltola, Söderlund, Tarhio, and Ukkonen [157]. Kececioglu's thesis [114] and the paper by Kececioglu and Myers [115] derived from it are based on this formulation. The MULTICONTIG formalism was developed by the authors. Myers [144] proposed a method based on the distribution of fragment endpoints in the layout as a powerful formulation of fragment assembly, especially good in instances involving repeats. He also proposed an algorithm based on a special kind of graph different from those covered in this book. Figure 4.12 is based on an example that appeared in this paper.

Several groups have implemented programs to perform DNA fragment assembly. Early systems include [71, 138]. A pioneering effort toward integrating a sound mathematical formalism with actual implementation was done by Peltola, Söderlund, and Ukkonen [158]. Huang [98] described a successful fragment assembly program, later improved to deal with repeats, detect chimeric fragments, and automatically edit layouts [99]. Kececioglu and Myers [115] divided the entire process in three phases, providing careful formal models as well as exact and approximate algorithms for each phase. Staden [178, 46] maintains a sequence processing system with strong fragment assembly capabilities since the late 1970s. The GCG system [47], very popular in lab computers around the world, also provides assembly support. Design goals for fragment assembly systems were summarized by Myers [143]. A good general reference on automated DNA processing is the book edited by Adams, Fields, and Venter [1]. Details on the technique discussed in Section 4.4.3 can be found in an article by Meidanis and Setubal [137].

Engle and Burks [53, 54] described a set of tools for generating artificial instances for fragment assembly. Data generated in a real sequencing project was made available as a test instance for computerized assembly tools by Seto, Koop, and Hood [171]. Equations (4.1) and (4.2) describing the progress of a shotgun project appear in Waterman [199, Section 7.1.5]. Early work on this topic was done by Lander and Waterman [120].

Waterman [199, Section 7.3] presents an algorithm for SBH, giving also a pointer to a paper with more details.

Fragment assembly was one of the topics of an implementation workshop at the NSF Center for Discrete Mathematics and Theoretical Computer Science (DIMACS) in 1995. A volume with contributed papers from that workshop is forthcoming. See also

http://dimacs.rutgers.edu/SpecialYears/1994_1995/challenge.html

for more details on the workshop.

CHAPTER

5

PHYSICAL MAPPING OF DNA

As we saw in Chapter 1, a human chromosome is a DNA molecule with about 10^8 base pairs. The techniques we have studied so far for sequencing are restricted to pieces of DNA with up to tens of thousands of base pairs. This means that whenever we sequence such a piece we will be looking at an extremely small part of a chromosome. It is as though we were viewing Earth from the moon with a telescope that enabled us to see features that are a couple of inches apart (a DNA base) but with a field of vision no greater than a mile (a piece of DNA with 15,000 bases; see Figure 1.10). Such a telescope would not be a good instrument to study large-scale structures, such as mountain ranges, continents, and islands. To perceive those structures, we would need some other instrument. Analogously, molecular biologists use special techniques to deal with DNA molecules comparable in size to a chromosome. These techniques enable them to create *maps* of entire chromosomes or of significant fractions of chromosomes. In this chapter we examine computational techniques that can help biologists in the map-generation process.

5.1 BIOLOGICAL BACKGROUND

A *physical map* of a piece of DNA tells us the location of certain *markers* along the molecule. These markers are generally small but precisely defined sequences. Such a map helps molecular biologists further explore a genome. For example, suppose a certain stretch of DNA has been completely sequenced, giving us sequence S. If we know which chromosome S came from, and if we have a physical map of this chromosome, we could try to find one of the map's markers in S. If we succeed, we have located S in the chromosome. See an illustration in Figure 5.1.

So the question now is, How do we create such maps? The first task is to obtain

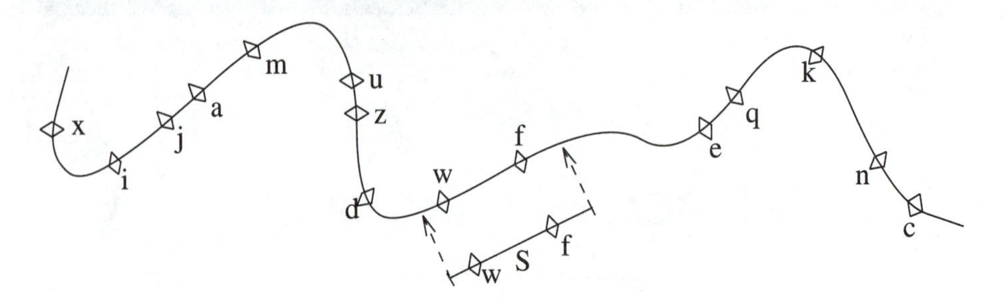

FIGURE 5.1

Sketch of a physical map. In this figure a very long DNA molecule (e.g., a chromosome) is represented by the curved solid line. Markers along the molecule are represented by the diamond symbols. Each marker is distinct, and this is shown by the letters associated with each. Segment S represents a DNA fragment that has been sequenced, containing, for example, a gene. Thanks to the map, S can be located on the chromosome.

several copies of the DNA molecule we want to map (known as the **target** DNA). Each copy must then be broken up into several fragments, using restriction enzymes. (DNA fragmentation by restriction enzymes is described in Section 1.5.2.) Mapping is done by carefully comparing the fragments obtained, in particular by observing overlaps. Note that in general each fragment is a DNA piece still too long to be sequenced, so we cannot determine overlap by sequencing and comparing fragments, as is done in fragment assembly (Chapter 4). Instead, we obtain overlap information by generating *fingerprints* of the fragments. A fingerprint should describe part of the information contained in a fragment in some unique way, just like our fingerprints uniquely describe some part of ourselves. Two popular ways of getting fingerprints are *restriction site analysis* and *hybridization*.

In restriction site mapping the aim is to locate the restriction sites of a given enzyme on the target DNA. The restriction sites are thus the markers previously mentioned. The technique used for location is based on fragment length measurement; a fragment's length is thus its fingerprint. In hybridization mapping we check whether certain small sequences bind to fragments; the subset of such small sequences that do bind to the fragment become its fingerprint. The fingerprint of a fragment can also be given by hybridization information *and* fragment lengths, but in this chapter we consider each technique separately for clarity of exposition. In any case, by comparing fingerprints we try to determine whether the fragments overlap and thus determine their relative order (which was lost during the break-up process).

Just as in other problems from molecular biology, possible lack of information and the presence of numerous experimental errors make the physical mapping problem especially hard. In particular, it may not be possible with a given collection of fragments to obtain one contiguous physical map. This may happen simply because the fragmentation process did not produce fragments covering certain sections of the target DNA. When this happens, the physical map pieces are called **contigs** (the same name used in fragment assembly). This is a feature common to all physical mapping processes. We now discuss each of the techniques mentioned above in more detail.

5.1.1 RESTRICTION SITE MAPPING

In this section we briefly present two possible techniques for mapping based on measuring lengths of fragments between restriction sites.

One way to obtain fragments of many different sizes is to apply not one, but *two* restriction enzymes to the target DNA. We will explain this technique by an example. Suppose we have two restriction enzymes, A and B, each recognizing a different sequence (restriction site). Applying A to one copy of the target DNA, we obtain fragments measuring 3, 6, 8, and 10 length units (for example, thousands of base pairs). Applying B to another copy of the target DNA, we obtain fragments measuring 4, 5, 7, and 11 units. Applying now *both* enzymes to a third copy of the target DNA we obtain fragments measuring 1, 2, 3, 3, 5, 6, and 7 units. The problem now is, How can we order these fragments in such a way that the order we determine is consistent with experimental results? This is known as the *double digest problem,* because the process of breaking a molecule by an enzyme is known as a digest. A solution for this particular example is shown in Figure 5.2.

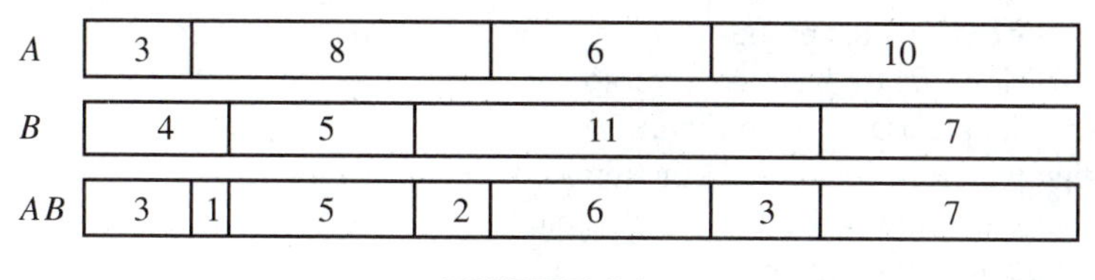

FIGURE 5.2

Solution to a double digest problem.

A variant of the double digest approach is *partial digest*. Here we subject the target DNA to one enzyme only but perform many experiments on copies of the DNA, varying the time during which the enzyme acts on each copy. By giving more or less time to the enzyme, more or less restriction sites will be recognized, thus yielding fragments of different lengths. Ideally the experiments should provide us with at least one fragment for every pair of restriction sites. We then try to pinpoint the location of the restriction sites by analyzing fragment lengths. For example, in Figure 5.2 enzyme A recognizes three restriction sites. A partial digest experiment based on this enzyme would result in the following fragment sizes: 3, 11, 17, and 27 (fragments between the left endpoint and all other sites); 8, 14, and 24 (fragments between the first restriction site and those to its right); 6 and 16 (fragments between the second restriction site and those to its right); and 10 (last fragment).

Several kinds of experimental errors may occur with digestion data. First, there is uncertainty in length measurement. This is done by gel electrophoresis (Section 1.5.2), where there could be an error of up to 5%. Second, if fragments are too small, it may not be possible to measure their lengths at all. Third, some fragments may be lost in the digestion process, leading to gaps in the DNA coverage. As can be expected, all these errors bring complications to the mapping process. We will see later on that even when there are no errors the associated computational problems seem very hard.

5.1.2 HYBRIDIZATION MAPPING

Recall that, in hybridization mapping, overlap information between fragments is based on partial information about each fragment's content. Before explaining what this partial information is, we note that in this context fragments are called **clones**, replicated using a technique called *cloning* (described in Section 1.5.2). After each copy of the target DNA has been broken up (in a different way), and after each resulting fragment has been cloned, we obtain a collection of many thousands of clones (a *clone library*), each clone being typically several thousands of base pairs long.

Partial information about a clone is obtained through *hybridization experiments*. In these experiments we try to verify whether a small sequence known as a **probe** binds (or *hybridizes*) to the clone. If the binding does occur, this means that the clone contains the sequence complementary to the probe sequence. We can do this experiment on a clone using several different probes. The clone's fingerprint will be the set of probes that successfully hybridized to the clone. Two clones that share part of their respective fingerprints are likely to have come from overlapping regions of the target DNA. For example, if we know that probes x, y, and z bound to clone A, whereas probes x, w, and z bound to clone B, we have good reason to believe that clones A and B overlap each other, for example as shown in Figure 5.3 (unless there are *repeats;* more on this later). Notice that with this kind of information we will not in general be able to tell the *location* of the probes along the target DNA, but only their relative order. We return to this important point further on.

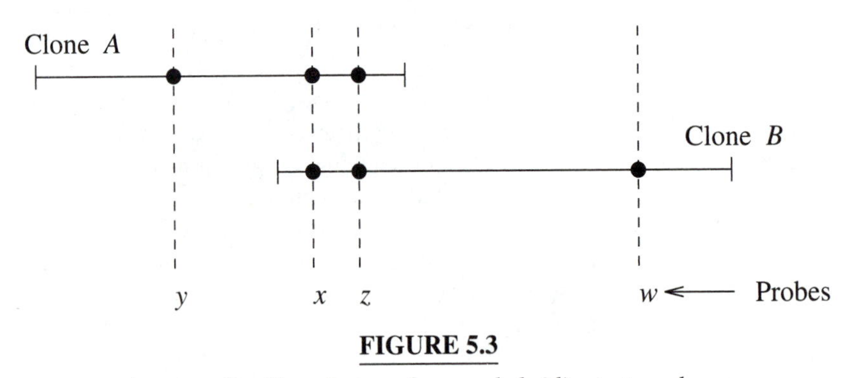

FIGURE 5.3

An example of how four probes can hybridize to two clones.

Many kinds of errors may happen in hybridization experiments. First, a probe may fail to bind where it should; this creates a **false negative**. Second, a probe may bind to a site where it should not; this creates a **false positive**. Sometimes there is human misreading of experimental results, also resulting in false negatives or false positives. Errors may have appeared even before the hybridization itself. During the cloning process, two separate pieces of the target DNA may join and be replicated as if they were one single clone. This is known as a **chimeric clone**, and from it false inference about relative probe order can be made. Chimerism occurs frequently and is thus one of the most severe problems in hybridization mapping. Some estimates say that in many clone libraries between 40% and 60% of all clones are in fact chimeric. Another type of error in the cloning process is the **deletion**, in which an internal piece of a clone is lost, thus again joining two fragments that are not contiguous in the target DNA.

Two other situations can cause problems to hybridization mapping. One happens when probes are not unique, meaning that there may be more than one site along the target DNA to which the probe can bind; these are the *repeats*. A technique called *Sequence Tagged Site* (STS) avoids this problem by generating probes that hybridize to unique sites along the target DNA. Another problematic situation is simply lack of data, as it may not be feasible to perform all hybridization experiments required. Pooling techniques have been developed to deal with this problem.

MODELS 5.2

Having looked at the various mapping strategies, we now review some of the corresponding mathematical models. Our aim is to explore the computational complexity of the mapping problems derived from these models. As we will see, mapping appears to be algorithmically hard, but it also brings up several interesting issues in problem solving that we discuss at the end of this section.

5.2.1 RESTRICTION SITE MODELS

Initially we will look at a model for the double digest problem, which we shorten to DDP. In this model we consider each fragment to be represented by its length. We will assume that the digestion process is perfect, meaning that lengths have no measurement errors, and all fragments are present. This is clearly unrealistic, but as we will see shortly even with these oversimplifications the resulting computational problem appears to be very hard.

Placing the experimental results in the model framework, we get a collection of lengths (positive integers) for each experiment performed. Digesting the target DNA by the first enzyme gives us collection $A = \{a_1, a_2, \ldots, a_n\}$. Some of these fragments may have the same length, which means that this collection is actually a multiset. Similarly, from the second enzyme we get $B = \{b_1, b_2, \ldots, b_m\}$. Finally, from the digestion with both A and B we get the multiset $O = \{o_1, o_2, \ldots, o_k\}$.

Given these multisets, what is it that we want to find? We want a permutation π_A of the elements from A and a permutation π_B of the elements from B, such that the following can be achieved. We plot the lengths from A on a line according to the order given by

π_A. Similarly, we plot the lengths from B following π_B, but we do it *on top* of the previous plot. Due to overlap, we may now have several new subintervals. It should now be possible to establish a one-to-one correspondence between each resulting subinterval and each element from O. If this can be done, we will say that π_A and π_B are solutions to the DDP.

Finding these permutations is an NP-complete problem, and here is the proof. First, it is not difficult to see that given a solution to the DDP it is easy to check whether it is a true solution. All we have to do is sort O and the plotted subintervals to determine whether we get identical results. To complete the proof, we note that, as stated, the DDP is simply a generalization of the *set partition problem,* a well-known NP-complete problem.

In the set partition problem we are given a set of integers $X = \{x_1, x_2, \ldots, x_l\}$ and want to know whether we can partition X into sets X_1 and X_2 such that the sum of all elements in X_1 is the same as those in X_2. This corresponds to a double digest problem in which $A = X$, $B = \{K/2, K/2\}$, and $O = A$, where $K = \sum_{x \in X} x_i$. In other words, the set partition problem is a DDP in which one of the enzymes produced only two fragments, and both of the same length.

Considering that we are always seeking the *true* solution to the DDP, this problem has one other facet that makes this search a great deal harder: The number of solutions may be exponential. To see this, consider the case where between two sites cut by enzyme A there are three sites b_1, b_2, b_3 cut only by enzyme B. In this case we will not be able to tell the order of fragments $[b_1, b_2]$ and $[b_2, b_3]$. In general, with a run of k restriction sites from one enzyme we will have $(k-1)!$ solutions. In fact, the situation is even worse than that because of the so-called *coincidences.* A coincidence happens when a restriction site for enzyme A is very close to a restriction site for enzyme B (they cannot exactly coincide; why?). When we have consecutive coincidences, the corresponding fragments for the A digestion and the B digestion will appear to have the same length (due to measurement inaccuracy). Therefore a run of n consecutive coincidences will also result in $n!$ possible solutions. Taking this argument still further, it is possible to show that the number of solutions increases exponentially with the target DNA length (see the bibliographic notes).

Thus we see that even with a very simple model double digestion poses serious computational problems. The NP-completeness of partial digest has not been proved, and in addition the problem seems to be more tractable for the following reason. In mapping by partial digest there are n restriction sites (including endpoints), and by experiment we obtain at least one fragment for every pair of sites, or $\binom{n}{2}$ fragments. When measured, these fragments yield a multiset of distances. The problem is to determine the location of the sites on the line such that the set of distances between these locations is the same as the lengths obtained by the digests. What makes the problem easier is that researchers have shown that the maximum number of solutions is small, thus making partial digest, from a combinatorial point of view, a good alternative to double digest.

5.2.2 INTERVAL GRAPH MODELS

We can have a first look at the computational complexity of hybridization mapping (or more generally, *fingerprint mapping*) using simple models based on *interval graphs* (see definition in Section 2.2). These models abstract away many of the complications of mapping; yet, simple as they are, they also give rise to problems that are NP-hard, as was the case with the double digest problem seen earlier.

In these models we create graphs whose vertices represent clones and whose edges represent overlap information between clones. If we had complete and correct information about clone overlapping, the resulting graph would be an interval graph, and this is a key observation in these models.

In the first model, we create two graphs. The first graph is $G_r = (V, E_r)$. If $(i, j) \in E_r$, this means that we know for sure that clones i and j overlap. The second graph is $G_t = (V, E_t)$, where E_t represents known *plus* unknown overlap information (thus $E_r \subseteq E_t$). G_t is not necessarily complete, because we may know for sure that certain pairs of clones do not overlap and hence the corresponding edge would be left out of E_t. The problem we want to solve is this: Does there exist a graph $G_s = (V, E_s)$ such that $E_r \subseteq E_s \subseteq E_t$ and such that G_s is an interval graph? Notice that it is not necessarily the case that G_r or G_t are interval graphs (in which case a solution would be trivial). See example in Figures 5.4, 5.5, and 5.6.

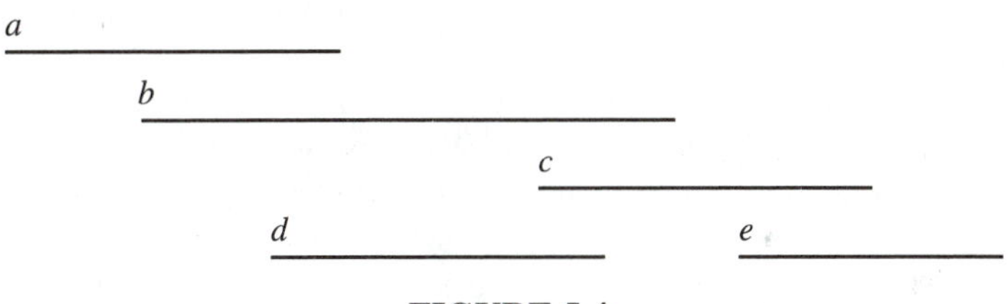

FIGURE 5.4

Example of clone overlapping patterns.

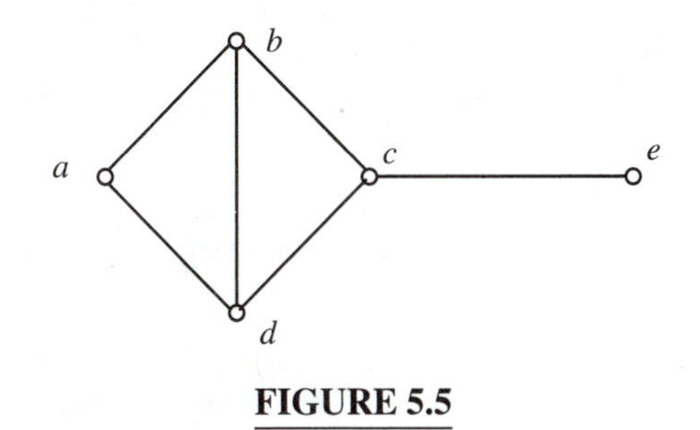

FIGURE 5.5

The interval graph corresponding to Figure 5.4.

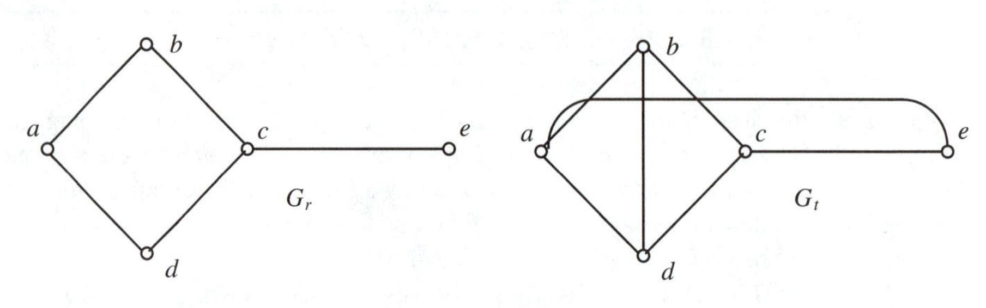

FIGURE 5.6

Possible graphs G_r and G_t for the example from Figure 5.4. None of them is an interval graph.

In the second model, we do not assume that the known overlap information is reliable. We construct a graph $G = (V, E)$ using that information anyway. The problem we want to solve is this: Does there exist graph $G' = (V, E')$ such that $E' \subseteq E$, G' is an interval graph and $|E'|$ is maximum? The requirement that $E' \subseteq E$ means that, to get a solution for this problem, we may have to discard some edges of G, which would be interpreted as false positives. The solution is the interpretation that has the minimum number of false positives.

In the third model, we use overlap information together with information about the source of each clone as follows. Because clones come from different copies of the same molecule, a reasonable idea is to label each clone with the identification of the molecule copy it came from. We assume we have k copies of the target DNA, and different restriction enzymes were used to break up each copy. We now build a graph $G = (V, E)$, with known overlap information between clones. Because clones are labeled, we can think that the corresponding vertices are *colored,* using k colors (one color for each molecule copy). The graph constructed will not have an edge between vertices of the same color, because they correspond to clones that came from the same molecule copy and hence cannot overlap. We thus say that the graph has a *valid* coloring. The problem we want to solve is this: Does there exist graph $G' = (V, E')$ such that $E \subseteq E'$, G' is an interval graph, and the coloring of G is valid for G'? In other words, can we add edges to G transforming it into an interval graph without violating the coloring?

We have just described three different models for hybridization mapping, and all are relatively simple. However, it turns out that all three problems described are NP-hard. We will not give the proofs here; they can be found in the references cited at the end of the chapter. We now proceed to describe more specialized models.

5.2.3 THE CONSECUTIVE ONES PROPERTY

The models described in the previous section apply to hybridization mapping, but in principle they can be used in any situation where we can obtain some kind of fingerprint for each fragment. We now present a model that uses as a clone fingerprint the set of probes that bind to it. In this model we make the following assumptions:

- The reverse complement of each probe's sequence occurs only once along the target DNA ("probes are unique").

- There are no errors.

- All "clones $\times$ probes" hybridization experiments have been done.

The reader might think that another NP-hard problem is coming up. For once, this is not the case. As we will see the resulting problem is polynomial-time solvable, although some of its generalizations are indeed NP-hard.

Assuming there are n clones and m probes, the experimental data will enable us to build an $n \times m$ binary matrix M, where entry M_{ij} tells us whether probe j hybridized to clone i ($M_{ij} = 1$) or not ($M_{ij} = 0$). Obtaining a physical map from this matrix becomes the problem of finding a permutation of the columns (probes) such that all 1s in each row (clone) are consecutive. A binary matrix for which such a permutation can be found is said to have the *consecutive 1s property* for rows, or C1P for short. Below we also use the terms C1P permutation and C1P problem, with the obvious meanings.

Verifying whether a matrix has this property and then finding a valid permutation is a well-known problem for which polynomial algorithms exist. We present one such algorithm in Section 5.3. However, the model is clearly too simple. If errors are present, even in the true permutation some rows may not have all 1s in consecutive positions. So why bother? The answer is that looking at the error-free case helps us to gain insights that are useful when designing algorithms for data with errors. In particular, we can use the following important assumption. Laboratory researchers try to minimize the number of errors in their experiments. If their experiments were *perfect,* the resulting hybridization matrix *would have* the C1P. This means that any algorithm that tries to find or approximate the true column permutation should give solutions that minimize the number of situations that must be explained by experimental error. Hence, such algorithms should be able to find a C1P permutation if one exists.

Note that even if a C1P permutation exists, we cannot claim that *it* is the true permutation. For one thing, there may be several such permutations for a given input matrix. How can we say which is the true one? The answer is that we cannot, in general. We can only say that all permutations that make all 1s consecutive are likely candidates for the true permutation. For this reason an algorithm for the C1P problem should give us not one possible solution, but all of them. In addition, given that errors do exist, it may even be the case, as noted above, that the true permutation is *not* a C1P permutation, even if there is one.

On the other hand, how can we modify the model to account for errors? There are many possibilities. Instead of requiring exactly one block of consecutive 1s per row, we may try to find a column permutation such that in each row there are at most k blocks of consecutive 1s, where k's value could be 2 or 3. Or we may try to minimize the total number of blocks of consecutive 1s in the matrix. By doing this, we may expect that most rows will have just one block of consecutive 1s, while an occasional row might have four or five. Unfortunately, as already hinted above, these generalizations yield NP-hard problems (see Exercise 16). NP-hardness comes up again if we relax the assumption that probes must be unique, even if no errors are present (see the bibliographic notes for references on these results).

5.2.4 ALGORITHMIC IMPLICATIONS

Before describing algorithmic techniques for obtaining physical maps, we would like to explore a bit the constraints forced upon us by the difficulty of the problem. These considerations give a new light on what it means to "solve" a problem using an algorithm, and should therefore be of interest outside physical mapping as well.

The first consideration is that what we are really trying to solve is a real-life problem, not an abstract mathematical problem. And the real-life problem is this: We want to know what the true ordering of the clones is in the target DNA. It is a combinatorial problem: There are finitely many possible orderings, and one of them is the one we want.

We will try to discover this true ordering by means of abstract models that give rise to optimization problems. These problems may be themselves very hard, notwithstanding the fact that they are abstractions of a yet harder problem. For example, the optimization versions of all problems we saw in Section 5.2.2 are NP-hard. In general we will not be able to find optimal solutions to these problems quickly; and even if we did solve one of them to optimality within a reasonable time, we will have no guarantee that the solution found corresponds to the true solution. Thus it looks like we are in a terrible predicament.

However, the situation is not as bleak as it seems. First, since we have no guarantee that any optimal solution is the true solution, good approximate solutions may serve our purposes just as well. Second, we must remember that algorithms in molecular biology should be used in an iterative process. The scientist obtains experimental data and feeds it to an algorithm. Based on the solution or solutions given by the algorithm, the scientist can make more experiments and obtain more data and thus use the algorithm again. At each step we may expect that "better" (or "closer to the truth") solutions will be found.

These observations suggest desirable features of a mapping algorithm, which we now list:

- It should work better with more data, assuming that the error rate stays the same.
- It should present a solution embedded in a rich framework of details, in particular showing how the solution was obtained, distinguishing "good" parts of the solution (groups of clones for which there was strong evidence for the ordering reported) from "not so good" parts. This greatly facilitates further experiments.
- If several candidate solutions meet the optimization criteria, all of them should be reported. If too many solutions are reported, the optimization criteria may be too weak (or the input data may contain too many errors). Conversely, if no solutions are reported, the optimization criteria may be too strong.

We would also like to have some guarantee that the true solution is among the set of approximate/optimal solutions found by the algorithm. In general this will not be possible. Instead, we try to obtain a few different optimization functions and respective algorithms whose optima share properties with the true solution. The hope is that the true solution will lie in the intersection of the solution sets found by each algorithm with high probability. We give one example of such a guarantee in Section 5.4.2.

Note that instead of using several algorithms as suggested, we may try to design an algorithm that can optimize multi-objective functions. This is an important research area.

5.3 AN ALGORITHM FOR THE C1P PROBLEM

In this section we present a simple algorithm that determines whether an $n \times m$ binary matrix M has the C1P for rows. The goal of the algorithm is to find a permutation of the columns such that in each row all 1s are consecutive. For simplicity we assume all rows are different (that is, no two clones have the same fingerprint) and that no row is all zeros (that is, every clone is hybridized by at least one probe).

Understanding the algorithm requires that we look first at the way rows can relate to each other. For this we need a definition.

DEFINITION 5.1 For each row i of M, let S_i be the set of columns k where $M_{i,k} = 1$.

With this definition we can now relate a row to another row. Given two rows i and j three situations can arise:

1. $S_i \cap S_j = \emptyset$.
2. $S_i \subseteq S_j$ or $S_j \subseteq S_i$.
3. $S_i \cap S_j \neq \emptyset$, and none of them is a subset of the other.

We can clearly deal separately with rows that fall into the first of these situations, since the permutation of columns corresponding to the elements of S_i will not interfere with those from S_j. In such a case we can place i and j into different *components* and deal with each component separately. Now, how about rows that have nonempty intersection? (We are abusing language a bit here and speaking of rows instead of their respective sets S.) Should they all be in the same component? Let us look at this case more carefully.

Let us initially lump together in a component all rows that have nonempty intersection. Suppose there is a row k in this component such that S_k is either a subset of or is disjoint from S_i for all $i \neq k$ in this component. It is clear that we can leave row k out (and any others with the same property) while dealing with this component. Row k will thus be in its own component, possibly sharing it with other rows. What exactly these other rows are will be made clear.

Based on the intuition just given, we give a formal way to determine when two rows belong to the same component. We build a graph G_C using matrix M. In G_C each vertex will be a row from M. There is an undirected edge from vertices i to j if $S_i \cap S_j \neq \emptyset$ and none of them is a subset of the other. The components we want will be the connected components of G_C. As an example, Figure 5.7 shows the graph G_C for the matrix in Table 5.1.

We can now sketch the algorithm: Separate rows into components according to the rule above, permute the columns of each component, and then somehow join components together. This leaves us with two subproblems: how to find the right permutation for one component, and how to join components together. Let us first work on the column permutation inside a component.

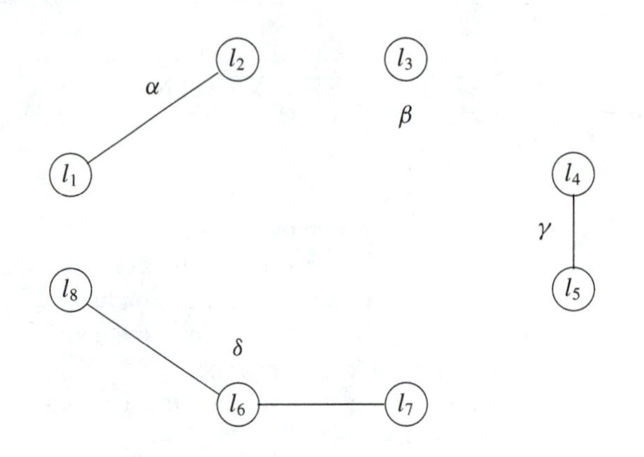

FIGURE 5.7

Graph G_C corresponding to the matrix of Table 5.1. Connected components are indicated by Greek letters.

TABLE 5.1

A binary matrix.

	c_1	c_2	c_3	c_4	c_5	c_6	c_7	c_8	c_9
l_1	1	1	0	1	1	0	1	0	1
l_2	0	1	1	1	1	1	1	1	1
l_3	0	1	0	1	1	0	1	0	1
l_4	0	0	1	0	0	0	0	1	0
l_5	0	0	1	0	0	1	0	0	0
l_6	0	0	0	1	0	0	1	0	0
l_7	0	1	0	0	0	0	1	0	0
l_8	0	0	0	1	1	0	0	0	1

Taking Care of a Component

To explain this part of the algorithm, we shall make use of the example matrix shown in Table 5.2 (assume it is just a section of some matrix with more than three rows).

Let us look first only at row l_1. If it were the only row in this component, all we would have to do is permute the columns such that all 1s become consecutive. Note that if there are k 1s in this row, we will have $k!/2$ possible permutations. We are not counting reversals as distinct solutions, because they do not matter for keeping 1s consecutive. In the case of row l_1 the 1s are in columns 2, 7, and 8. Then one set of possible permutations is 2, 7, 8, or 7, 2, 8 or 2, 8, 7. We can encode all these solutions by associating a set of possible columns for each element equal to 1 in a given row (we will call these sets the *column sets*). For just one row this set is set S defined above, and we describe the resulting configuration as follows:

TABLE 5.2

A section of a binary matrix.

	c_1	c_2	c_3	c_4	c_5	c_6	c_7	c_8
l_1	0	1	0	0	0	0	1	1
l_2	0	1	0	0	1	0	1	0
l_3	1	0	0	1	0	0	1	1

$$\begin{array}{lccccccc} & & & \{2,7,8\} & \{2,7,8\} & \{2,7,8\} & & \\ l_1 \rightarrow & \dots & 0 & 1 & 1 & 1 & 0 & \dots \end{array}$$

Now let us look at the following row, l_2. Because this row is in the same component as l_1 it must be the case that l_2 has 1s in some columns where l_1 also has 1s, and some other 1s where l_1 does not. Now we have to place l_2 with respect to l_1 in a consistent manner. We have two choices for this placement: Either put the 1s of l_2 to the left of those of l_1 or to the right. This means placing column 5 either to the left or to the right of columns 2 and 7. Suppose we place 5 to the left. The possible solutions, encoded by the sets above each column, now become

$$\begin{array}{lcccccccc} & & & \{5\} & \{2,7\} & \{2,7\} & \{8\} & & \\ l_1 \rightarrow & \dots & 0 & 0 & 1 & 1 & 1 & 0 & \dots \\ l_2 \rightarrow & \dots & 0 & 1 & 1 & 1 & 0 & 0 & \dots \end{array}$$

If these were the only two rows in this component we would have arrived at two possible permutations for its columns: 5, 2, 7, 8 and 5, 7, 2, 8. If we had chosen to place 5 on the right, we would obtain the reverse of these two permutations. This means that the direction we choose to place the second row does not matter, and this is a key observation.

Now let us consider the third row, l_3. Because it is in the same component as the other two, we know that in graph G_C we should find edge (l_3, l_1), or edge (l_3, l_2), or both. In our example both edges are present. Let us now place l_3 with respect to l_2. It seems that again we have the choice of placing l_3 to the left or to the right of l_2. But now we have to take into account the relation between l_3 and l_1 as well. The way to do this is by considering the number of elements in the intersections between S_1, S_2, and S_3. Let $x \cdot y = |S_x \cap S_y|$, for any two sets x and y (the *internal product* of rows x and y). If $l_1 \cdot l_3 < \min(l_1 \cdot l_2, l_2 \cdot l_3)$, row l_3 must go in the *same* direction that l_2 was placed with respect to l_1. If $l_1 \cdot l_3 > \min(l_1 \cdot l_2, l_2 \cdot l_3)$, then we must place l_3 in the *opposite* direction used to place l_2 with respect to l_1 (we cannot have equality; why?). In either case, if the component is C1P, it must be the case that l_1 and l_3 are properly placed; conversely, if l_1 and l_3 are not properly placed, then the component is not C1P. The reason for this lies on the fact that the only choice we made up until now was in the placing of l_2 with respect to l_1, and we know that both possibilities (left or right) result in the same solutions up to reversal.

In our example above, $S_3 = \{1, 4, 7, 8\}$. Then $1 \cdot 3 = 2$, $1 \cdot 2 = 2$, $3 \cdot 2 = 1$, which means we have to put l_3 to the right of l_2, as in the following figure:

			{5}	{2}	{7}	{8}	{1, 4}	{1, 4}		
$l_1 \rightarrow$	...	0	0	1	1	1	0	0	0	...
$l_2 \rightarrow$	...	0	1	1	1	0	0	0	0	...
$l_3 \rightarrow$	...	0	0	0	1	1	1	1	0	...

We now know that the right permutation for columns 2, 5, 7, and 8 is 5, 2, 7, 8. Other columns may be determined as we consider more rows. But how do we handle additional rows? As we saw, in placing the third row we had no choice of placement. The same is true for all other rows in the same component. All we have to do for a new row k is find two previously placed rows i and j such that there exist edges (k, i) and (i, j) in G_C, and proceed as we did with the three-row case above. Note how the algorithm preserves possible permutations of a component in the sets above each column. Considering that after placing the first two rows there are no other choices of placement for remaining rows, we see that this algorithm gives us *all* possible column permutations of a component having the C1P, up to reversal. In the example above, if the component had only three rows, the two possible permutations would be 5, 2, 7, 8, 1, 4 and 5, 2, 7, 8, 4, 1. As we have already mentioned, keeping track of all possible solutions is a desirable feature in the case of DNA mapping.

An implementation of the preceding algorithm is simple: Construct G_C and traverse it using depth-first search. When visiting a vertex invoke procedure *Place,* presented in Figure 5.8. To check whether the algorithm has found a C1P permutation, we can verify the consistency of the column sets as we try each new placement, as indicated in the last line of procedure *Place*. If we attempt to place a row and we find out that one of its columns should go, say, to the right, and the existing placement tells us that it should go to the left, then the component, and hence the matrix, does not have the C1P.

Algorithm *Place*
  input: u, v, w, vertices of $G_C = (V, E)$ such that $(u, v) \in E$ and $(v, w) \in E$.
  // v *and* w *can have value nil*
  output: A placement for row u, if possible
  if $v = nil$ **and** $w = nil$ **then**
    Place all 1s of row u consecutively
  else if $w = nil$ **then**
    Left- or right-place the 1s of u with respect to the 1s of v
    Record direction used
  else
    if $u \cdot w < \min(u \cdot v, v \cdot w)$ **then**
      Place u with respect to v using same direction used in v, w placement
      Record direction used
    else
      Place u with respect to v using opposite direction used in v, w placement
      Record direction used
  Check consistency of column sets.

FIGURE 5.8

The procedure to place a row given two others.

Let us analyze the running time of this algorithm. Building graph G_C takes $O(nm)$ time. We then process the n rows as sketched above, spending $O(m)$ per row to check consistency of column sets. Total time is thus $O(nm)$.

Joining Components Together

We now tackle the second part of the algorithm, component-joining. For this we need another graph, G_M, which will tell us how components fit together. Each component of the original matrix M will be a vertex in G_M. A directed edge will exist between vertex α and vertex β if the sets S_i for all $i \in \beta$ are contained in at least one set S_j of component α (we are making no distinction between a row j and its corresponding set S_j). The graph G_M corresponding to the components of the matrix from Table 5.1 is shown in Figure 5.9.

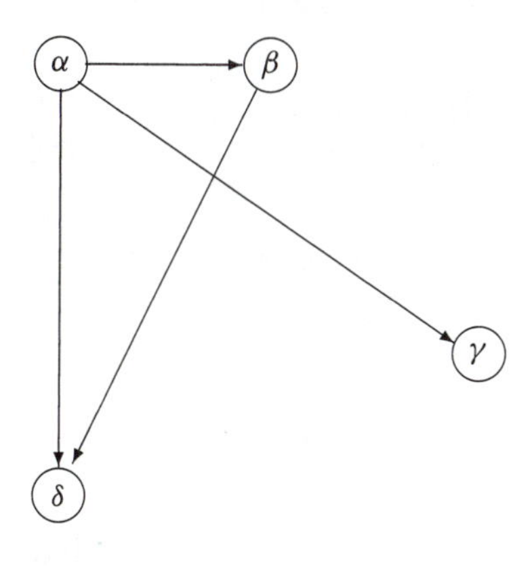

FIGURE 5.9

Graph G_M corresponding to the components of the matrix from Table 5.1.

From the discussion in the beginning of this section we know that the containment relations between a set of β and sets of α are the same for all sets of β. Let us make this more clear. Suppose set S_i belonging to component β is contained in set S_j belonging to another component α. Because the rows i and j are in different components, there is no row k in component α such that S_i is not contained in S_k and $S_i \cap S_k \neq \emptyset$. In other words, S_i is contained in some sets of α and is disjoint from the others. We claim that the exact same containments and disjunctions hold for all other sets from β. Suppose not and let us analyze one such exception. Suppose $i, l \in \beta$, $S_i \cap S_l \neq \emptyset$, $j \in \alpha$, and $S_i \subseteq S_j$, but $S_l \not\subseteq S_j$. In this case the only way l could be out of α is if $S_l \cap S_j = \emptyset$. But this cannot be, since S_i and S_l have at least one element in common and S_i is contained in S_j. We leave other cases to the reader.

The joining of components depends on the way sets in one component contain or are contained in sets from other components. Intuitively, we should process first components that have sets that are not contained anywhere else. Because containment is given by edge direction, we have to process G_M in topological ordering. (We can only do this if G_M is acyclic. Is it? Yes, and this is not difficult to prove.)

After having correctly permuted the columns of all components individually, we select all vertices in G_M without incoming edges, and freeze their columns. We then take the next vertex in topological order. Suppose we are following edge (α, β). This means that we must find some sort of "reference column" in component α that will tell us how to place the rows of β. Choose the row l from β that has the leftmost 1, and call the column where this 1 is c_β. We know that S_l is contained in some S_i of α but not in others. Find all rows from α that contain S_l, and find the leftmost column where all such rows have 1s (and call this column c_α). This is the reference column we are looking for, since we can now make c_α and c_β one and the same.

Let us illustrate this process with the matrix from Table 5.1 and the graph from Figure 5.9, and let us consider its components in topological ordering. One such ordering is $\alpha, \beta, \delta, \gamma$. Here are the rows of component α with columns permuted so that its 1s are consecutive:

$$\begin{array}{ccc|ccccc|cccc}
 & & \{1\} & & & \{2,4,5,7,9\} & & & & \{3,6,8\} & & \\
l_1 \rightarrow & \ldots & 1 & 1 & 1 & 1 & 1 & 1 & 0 & 0 & 0 & \ldots \\
l_2 \rightarrow & \ldots & 0 & 1 & 1 & 1 & 1 & 1 & 1 & 1 & 1 & \ldots
\end{array}$$

We will now join the next component, which is β, a singleton row:

$$\begin{array}{cc|ccccc|c}
 & & & & \{2,4,5,7,9\} & & & \\
l_3 \rightarrow & \ldots & 1 & 1 & 1 & 1 & 1 & \ldots
\end{array}$$

As should be clear, all we have to do is join l_3 with the rows from α, without any rearrangements:

$$\begin{array}{ccc|ccccc|cccc}
 & & \{1\} & & & \{2,4,5,7,9\} & & & & \{3,6,8\} & & \\
l_1 \rightarrow & \ldots & 1 & 1 & 1 & 1 & 1 & 1 & 0 & 0 & 0 & \ldots \\
l_2 \rightarrow & \ldots & 0 & 1 & 1 & 1 & 1 & 1 & 1 & 1 & 1 & \ldots \\
l_3 \rightarrow & \ldots & 0 & 1 & 1 & 1 & 1 & 1 & 0 & 0 & 0 & \ldots
\end{array}$$

The next component is δ. Here are its rows, with consecutive 1s:

$$\begin{array}{cccccccc}
 & \ldots & \{9,5\} & & \{4\} & \{7\} & \{2\} & \ldots \\
l_6 \rightarrow & \ldots & 0 & 0 & 1 & 1 & 0 & \ldots \\
l_7 \rightarrow & \ldots & 0 & 0 & 0 & 1 & 1 & \ldots \\
l_8 \rightarrow & \ldots & 1 & 1 & 1 & 0 & 0 & \ldots
\end{array}$$

We can see that in component δ it is l_8 that has the leftmost 1 (in either column 5 or 9). In component α both rows l_1 and l_2 contain l_8, and it is the second column (representing a bunch of possible columns) that has all 1s. Therefore that second column should coincide with columns 9 or 5 from component δ. This is the result:

		{1}	{9,5}		{4}	{7}	{2}	{3,6,8}			...
$l_1 \rightarrow$	...	1	1	1	1	1	1	0	0	0	...
$l_2 \rightarrow$	...	0	1	1	1	1	1	1	1	1	...
$l_3 \rightarrow$	...	0	1	1	1	1	1	0	0	0	...
$l_6 \rightarrow$	...	0	0	0	1	1	0	0	0	0	...
$l_7 \rightarrow$	...	0	0	0	0	1	1	0	0	0	...
$l_8 \rightarrow$	...	0	1	1	1	0	0	0	0	0	...

We finally glue in component γ. Its rows are

		{6}	{3}	{8}	
$l_4 \rightarrow$	...	0	1	1	...
$l_5 \rightarrow$	...	1	1	0	...

Joining with the rows above we obtain the final matrix with all 1s consecutive:

	{1}	{9,5}		{4}	{7}	{2}	{6}	{3}	{8}
$l_1 \rightarrow$	1	1	1	1	1	1	0	0	0
$l_2 \rightarrow$	0	1	1	1	1	1	1	1	1
$l_3 \rightarrow$	0	1	1	1	1	1	0	0	0
$l_6 \rightarrow$	0	0	0	1	1	0	0	0	0
$l_7 \rightarrow$	0	0	0	0	1	1	0	0	0
$l_8 \rightarrow$	0	1	1	1	0	0	0	0	0
$l_4 \rightarrow$	0	0	0	0	0	0	0	1	1
$l_5 \rightarrow$	0	0	0	0	0	0	1	1	0

Notice that one of the column sets is not a singleton, indicating that we have found two solutions. In this particular case the reason is that columns 5 and 9 are identical. But in general there may be multiple solutions that do not involve just the permutation of identical columns. For example, 1, 2, 7, 4, 9, 5, 8, 3, 6 is another solution for this same matrix. Multiple solutions may exist because G_M may allow different topological orderings, and this is compounded by the facts that each component may on its own have several solutions and each of them can be used in two ways (the permutation and its reversal).

We now analyze the running time of the algorithm. Topological sorting of G_M takes time $O(n + m)$. It is possible to preprocess the entries of M so that the queries needed when traversing G_M take constant time. For example, we can store for each row the column where its leftmost 1 is. Such preprocessing takes at most $O(nm)$. Therefore this is the time needed for joining components. Total time for processing each component was seen to be $O(nm)$ as well, so this is the running time of the whole algorithm.

Another algorithm exists that can find a C1P in time $O(n + m + r)$, where r is the total number of 1s in the matrix. In addition, the algorithm encodes in a compact way all possible solutions. This algorithm, however, is considerably more complicated than the algorithm we presented in this section. References are given in the bibliographic notes.

5.4 AN APPROXIMATION FOR HYBRIDIZATION MAPPING WITH ERRORS

In the previous section we studied an algorithm that solves the consecutive 1s problem. We saw that such a problem is a good model of hybridization mapping when there are no errors and when probes are unique. If errors are present, some other approach is needed, and that is the subject of this section.

Let us first examine the effect errors can have on a clones $\times$ probes binary matrix M. Suppose M is presented to us with the true column permutation. Given one row, if there are no errors, all its 1s will be consecutive. If a row corresponds to a chimeric clone, where two fragments were joined, we will see *two* blocks of 1s separated by some number of 0s (assuming no other errors are present in this row). We will call a consecutive block of 0s bordered by 1s a *gap*. Notice that this is different from the use of the term *gap* in other chapters of this book. We can thus say that a gap was created in this row because of the chimeric fragment. If, on another row, there is a false negative, the corresponding 0 may separate two blocks of 1s, creating another gap, as shown below:

$$\begin{array}{cccccccccc} 0 & 1 & 1 & 0 & 1 & 1 & 1 & 1 & 0 & 0 \\ & & & \uparrow & & & & & & \end{array}$$

A false negative

The gap will not be created if the probe was leftmost or rightmost for this clone. Finally, a false positive may split a block of 0s in two, thus possibly creating yet another gap. In this way we see that there is a close correspondence between errors and gaps in the matrix. Given the basic assumption that we want to avoid explaining gaps by experimental error as much as possible, a reasonable approach is to try to find a permutation where the total number of gaps in the matrix is minimum. Such an approach has the desirable property that, if there is a C1P permutation, it will have the minimum number of gaps. In other words, gap minimization can be seen as a generalization of the consecutive 1s problem. We have mentioned in Section 5.2.3 that some extensions to the C1P are NP-hard. Such is the case also with the gap minimization problem just sketched. However, for this particular NP-hard problem, we can use many special techniques to get approximate solutions that we can expect to be reasonably good, as we show next.

5.4.1 A GRAPH MODEL

It turns out that gap minimization is equivalent to solving a well-known graph problem. This is the *traveling salesman problem* (TSP), described in Section 2.2.

The input to this version of the TSP is a complete undirected weighted graph G. The vertices of G correspond to columns of the clones $\times$ probes binary matrix M; that is, they correspond to probes. For reasons that will soon become clear, we also have to add an extra column to M filled with zeros, and G must have the corresponding vertex. The weight on each edge of G is the number of rows where the two corresponding columns differ (this is also known as the *Hamming distance* between the rows). For example, in Table 5.3 we have an example of a binary matrix, and in Figure 5.10 we see the corresponding graph. We will now argue that a minimum-weight cycle in G corresponds to a column permutation in M with the least number of gaps.

TABLE 5.3

A clones × probes matrix with added column p_6^*.

	p_1	p_2	p_3	p_4	p_5	p_6^*
c_1	1	1	1	0	0	0
c_2	0	1	1	1	0	0
c_3	1	0	0	1	1	0
c_4	1	1	1	1	0	0

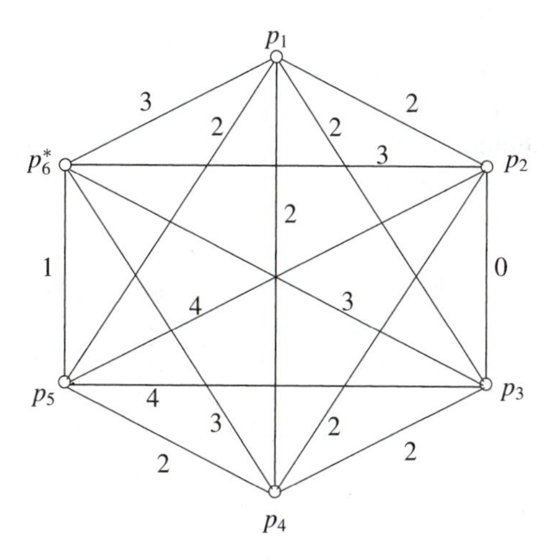

FIGURE 5.10

TSP graph for matrix of Table 5.3.

To see this, note that given a permutation of the columns, a gap in a row means that at a certain point we have a transition from 1 to 0 and further on we have a transition from 0 to 1. Thus for each gap we have two transitions and each gap contributes exactly 2 to the weight of the cycle corresponding to the given column permutation. However edge weights may also be increased by *extremal transitions,* that is, transitions between elements in extremal (1 or m) columns, and these do not correspond to gaps. To ensure that every row has a pair of extremal transitions we include an extra column of zeros in column $m + 1$. Without such a column, cycles in the graph correspond to permutations where we allow consecutive 1s to wrap around in each row, and we do not want this to happen. The relationship between cycles and permutations now becomes

$$\text{cycle weight} = \text{number of gap transitions} + 2n.$$

This means that for a given n minimizing cycle weight is the same as minimizing the number of gaps.

What we have just done is to show how we can reduce the gap minimization problem to the TSP. It is well known that the TSP is an NP-hard problem, so in principle we have not accomplished much. However, a large array of techniques is available to solve or approximate traveling salesman problems, and these techniques can now be used in this context. The existence of such techniques is not in itself enough to give us confidence that solving traveling salesman problems will give us the true probe permutation. What we need are *guarantees* that such solutions are, in some sense, close to the true solutions. We shall show one such guarantee in Section 5.4.2.

Before that, let us go back to the gap minimization problem and present an example of the ideas outlined at the end of Section 5.2.4. We have defined above a function that, given an input matrix, returns the total number of gaps in the matrix. We have further argued that obtaining a permutation that has the minimum value for this function (or approximately minimum) will help us find the true column permutation. Since we do not have a guarantee that the true permutation will be among the solutions we find, we should look for other functions that might also be helpful. The idea is that by carefully defining several such optimization functions and developing algorithms for them we will increase the likelihood of hitting upon the true solution. In particular, it is reasonable to expect that the true solution will be in the intersection of all solution sets. This will only be the case, however, if each function represents one property that true solutions do have or are likely to have.

Here is an example of another optimization function. One possible drawback of gap minimization is that in a permutation with a minimum value for this function one or a few rows may have many gaps, while others may have none. Having many gaps in one row is undesirable, since it would mean that one clone was subject to many more errors than other clones, which contradicts laboratory experience. Therefore we could try to minimize the number of gaps *per row*. We leave as Exercise 15 how to show that we can still use the preceding graph model. The resulting graph problem is known as *the bottleneck traveling salesman problem,* which is also NP-hard.

5.4.2 A GUARANTEE

In this section we give a formal proof that the TSP approach outlined in the last section will give us, with high probability, the true permutation. The proof we present depends on two basic assumptions: that the number of probes is sufficiently large, and that the mapping process obeys a certain mathematical model. This model seems to be a good representation of what actually happens in large mapping projects. We describe the model next.

First we assume that the DNA molecule we are dealing with is so long that we may think of it as an interval on the real line, extending from 0 to N. The clones are subintervals of this long interval, and we assume that all of them have the same length; for convenience each clone is one unit long. To simplify the exposition, we will speak of *clone permutations* rather than probe permutations. The unit length assumption makes them equivalent. This means that we will be looking for consecutive 1s in columns, not in rows, and that each vertex in the associated TSP will correspond to a clone. We use each clone's left endpoint as a clone locator. We, of course, do not know the precise position of each clone along the molecule; and because we are dealing with hybridization, all we will be able to determine is relative clone order.

A critical feature of this model is clone distribution along the target DNA. We will assume that each clone's position is an independent random variable, that clone locators are distributed uniformly over $[0, N-1]$, and that the clones cover the interval $[0, N]$ (that is, for every subinterval I of $[0, N]$, there always exists at least one clone C such that $C \cap I \neq \emptyset$).

Another important aspect of the model is probe distribution. We will *not* assume that each probe is unique; instead, we will assume that each probe occurs rarely along the target DNA. More formally, we will say that the occurrences of a given probe obey a Poisson process with rate λ. Moreover, the Poisson process of any one probe is independent of all the others. This part of the model lets us immediately obtain an expression for the probability of a specific probe hybridizing to a clone. The expression is

$$\Pr\{\text{a given probe occurs } k \text{ times in a given clone}\} = e^{-\lambda}\frac{\lambda^k}{k!}. \tag{5.1}$$

This expression can be obtained from any textbook formula for Poisson processes, using the fact that our clones are unit length intervals.

This completes the model description. We shall now argue that, given a clones $\times$ probes binary matrix, the row permutation given by solving the associated TSP is a good approximation to the true permutation, in the following precise sense: The probability that both permutations are the same tends to 1 as the number of probes increases. Note that the number of probes is *fixed* for a given instance of the problem. We are just claiming that in larger and larger instances of the problem (and we are measuring size here by number of probes) the TSP permutation will be the same as the true permutation with higher and higher probability.

To prove this claim, we must argue in terms of graph weights, or more appropriately, clone distances. As we mentioned, the weight of each edge of the graph associated with input matrix M is called the Hamming distance between its two endpoints (clones). We denote by h_{ij} the Hamming distance between clone i and clone j. We can also think of the *true* distance between clones. Denoting a clone's coordinates by l (left) and r (right) we can define this distance to be

$$t_{ij} = |l_j - l_i| + |r_j - r_i| = 2|l_j - l_i|, \tag{5.2}$$

given that clones are all of the same size.

Suppose now that we *knew* all true distances. Then it is clear that the largest such distance would give us the clones that are farthest apart, which is to say, the clones that occur at opposite ends of the interval $[0, N]$. The next largest such distance gives us another similar pair that occurs between the previous two, and so on. This means that given the true distance we are able to obtain the true clone permutation, which is not surprising. However, we have distances h_{ij} and not distances t_{ij}. But because we are trying to obtain only the true *relative* order of clones, it would suffice if we could say that, given any four clones i, j, r and s, $h_{ij} < h_{rs}$ implies that $t_{ij} < t_{rs}$ and vice versa. The reason is that some notion of *order* between clone distances was all we needed to place clones relative to each order using the true distances. If we prove that the probability that $h_{ij} < h_{rs} \iff t_{ij} < t_{rs}$ tends to 1 as the number of probes increases, we will have the result we need.

Let us look at a pair of clones i and j, and let us find the probability that a certain probe p contributes to their Hamming distance. This will happen if probe p hybridizes to i but not to j or vice versa. Referring to Equation (5.1) we see that the probability that probe p does *not* occur on clone j is $e^{-\lambda}$, since $k = 0$. On the other hand, the probability of probe p occurring at least once in clone i is the same as the complement of the probability of probe p *not* occurring in that section of i that does not overlap j. If this overlap is z_{ij}, we obtain the following result:

$$\Pr\{p \text{ hybridizes to } i \text{ and not to } j \text{ or vice versa}\} = p_{ij} = 2e^{-\lambda}(1 - e^{-\lambda(1-z_{ij})}). \quad (5.3)$$

This essentially means that there is a well-defined probability of this event happening. We can now take into account all probes, and each of them will have a certain probability of hybridizing to clones i and j. Assume we have m probes, and consider h_{ij}/m. This represents the mean contribution of each probe to the Hamming distance between clones i and j. (Note that a probe's contribution to the Hamming distance is either 0 or 1.) If the number of probes is large, we can invoke the law of large numbers and say that h_{ij}/m approaches p_{ij}. Or, for any fixed small positive real number ϵ, that

$$\Pr\left\{\left|\frac{h_{ij}}{m} - p_{ij}\right| > \epsilon\right\} \to 0$$

as $m \to \infty$. Note now that since clones are unit length, the true distance between two overlapping clones i and j with an intersection measuring z_{ij} is given by $t_{ij} = 2 - 2z_{ij}$ (see Equation 5.2). This allows us to substitute t_{ij} for z_{ij} in Equation (5.3), effectively showing us that $t_{ij} > t_{rs}$ if and only if $p_{ij} > p_{rs}$, for pairs of clones i, j and r, s. But if this is the case, then we can say that

$$\Pr\left\{\left|t_{ij} - h_{ij}\right| > \epsilon\right\} \to 0.$$

This implies that $h_{ij} < h_{rs} \iff t_{ij} < t_{rs}$, which is what we wanted to prove.

5.4.3 COMPUTATIONAL PRACTICE

In this section we present some actual results of computational tests with algorithms for the hybridization mapping problem. Before presenting these results, however, we must make considerations regarding the way the results of such tests can be interpreted. Some of these considerations are valid for many problems in computational biology.

Initially we shall look at the input data using another graph. By doing so it will become clearer what we can expect of any algorithm that tries to obtain the true probe permutation. This graph will also be used in the next section.

We define the *hybridization graph* H as a bipartite graph (U, V, E) that is built using information from the hybridization matrix: Clones are the vertices of the U partition, and probes are the vertices of the V partition. There is an edge between two vertices if the corresponding probe hybridized to the corresponding clone. In Figure 5.11 we can see the bipartite graph H that was built based on matrix M shown in Table 5.3, but excluding the all-zeros column.

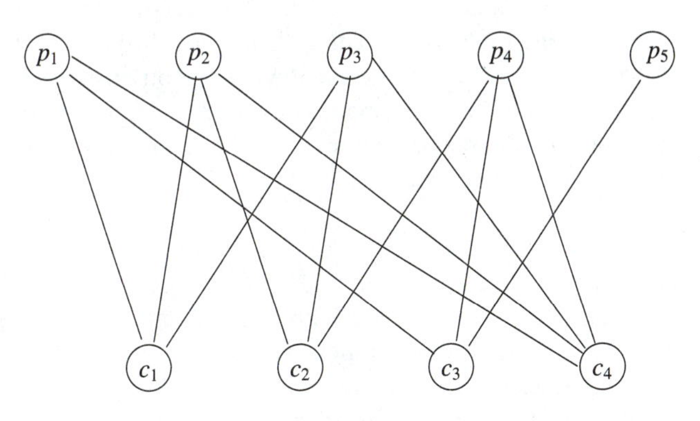

FIGURE 5.11

Hybridization graph H corresponding to hybridization matrix from Table 5.3, without the added column.

The first thing to notice is that H may not be connected, even if all entries in the hybridization matrix are correct. If this is the case, then no matter how good our algorithm is, we will not be able to tell the relative order between probes that belong to different connected components; the information to do so is simply not present in the hybridization matrix. A connected component may be as simple as a singleton vertex, meaning that there is a probe that did not hybridize to any clone or a clone that was not hybridized to any probe. Another observation is that there may be *redundant probes,* or probes that hybridize to exactly the same set of clones. This could happen if the probes, although different, hybridize to parts of the target DNA that are close together. It could also happen if certain clones for that particular DNA stretch are missing, leaving some probes without any positive hybridization results.

Connected components of H show up when we solve the corresponding consecutive 1s problem. Redundant probes can also be easily seen in the hybridization matrix: They are columns that have exactly the same 1s and 0s. But if there are errors we may get wrong information regarding the number and structure of connected components of H and whether a probe pair is redundant. We say that errors may *mask* these properties. So notice the difficult situation that errors create: The input matrix without errors may lack information necessary to find the true permutation (for example, the errorless H has several components), and we may have a lot of trouble just to recognize that lack of information (because our errorful H has only one component). Assuming matrix in Table 5.3 is errorless, probes p_2 and p_3 are redundant, but if there were a false positive between clone c_3 and probe p_3 we would not be able to recognize that.

Given this situation, it is clear that evaluation of a mapping algorithm is a difficult task (in addition to mapping itself!). We will now take a look at how we can evaluate such algorithms assuming that we somehow know the correct answer to any mapping problem. This can be done, for example, if we use a computer program to generate artificial instances of mapping problems, simulating experimental errors. If such instances are faithful to real instances, it should be clear from the above discussion that the input matrix may lack information to enable an algorithm to determine the true probe order. Therefore we should try to evaluate how "close" the solution found by a particular mapping algorithm is to the true probe order. The question now becomes: How should we define closeness in this context? At the moment there is no consensus on how to do this. However, to give some idea of the performance of current mapping algorithms in practice, we will present as an example one definition that has appeared in the literature. It is a reasonable definition, but even if it becomes widely accepted, it may still undergo some refinements.

We will measure a mapping algorithm by the fraction of *strong adjacencies* reported by it. Strong adjacencies are defined in terms of the number b of blocks of consecutive 1s present in a hybridization matrix with a given probe permutation $\pi = p_1, p_2, \ldots, p_m$. We analyze the effect of *translocations,* which are operations that reverse the order of a set of consecutive probes. We say that two adjacent probes p_i and p_{i+1} represent a strong adjacency if placing these probes apart by any translocation increases b in each row. When such an increase takes place we have some evidence (albeit not conclusive) that probes p_i and p_{i+1} should stay adjacent in all solutions.

Based on this concept, we can define the *strong adjacency cost* of a given permutation. This is given by the formula

$$100\left(\frac{1}{m-1}\sum_{0}^{m-2}\delta_i\right), \tag{5.4}$$

where $\delta_i = 1$ if p_i and p_{i+1} is a strong adjacency in the true permutation but these probes are not adjacent in the proposed permutation, and $\delta_i = 0$ otherwise. Note that the cost is given as a percentage. Good permutations should have low strong adjacency cost.

With this definition we are finally able to give the reader an idea of algorithm performance. Table 5.4 presents the strong adjacency cost of two algorithms. One of them is "random": A random probe permutation is selected. The other is based on the TSP approach; this is how it works. Given the TSP graph the algorithm builds a cycle by choosing pairs of vertices and making them adjacent on a path. The pair (u, v) chosen at each iteration must fulfill the following conditions: If both u and v already belong to paths, and these paths are different, the paths can be joined only if u and v are endpoints in their respective paths (that is, they each have just one neighbor); and they must be the closest among all qualifying pairs. After all vertices are on the same path, that path is closed forming a cycle. This solution is then submitted to another algorithm, which tries to improve the solution by applying another heuristic.

In Table 5.4 we can see that the TSP-based approach (called "greedy") performs well compared to "random." The paper from which these results were obtained presents the performances of three other, much more sophisticated algorithms, and the results are similar to those shown above for "greedy TSP." This can be seen as a point in favor of the TSP approach, but in a sense it is yet another measure of how difficult the mapping problem is. The table also shows that in the presence of false negatives the solution of "greedy TSP" was fairly poor. It is fair to assume that results would be even worse if all

TABLE 5.4

Strong adjacency costs for two algorithms on matrices with different kinds of errors. Error rates are indicated in the heading of each column (only one type of error per column). Coverage in all cases is 10, where coverage is the ratio between the total length of all clones and target DNA length.

	C1P 0	*Chimerism* 0.5	*False Positives* 0.04	*False Negatives* 0.32
Greedy TSP	1.9	0.9	16.0	28.3
Random	86.4	89.7	94.4	94.9

kinds of errors were combined in the same instance. This motivates our next section, in which we present a heuristic that appears to be robust in the presence of false negatives.

5.5 HEURISTICS FOR HYBRIDIZATION MAPPING

As the previous sections have shown, mapping is a difficult problem, and no general and good algorithms for it have been found. As a consequence, what we see in practice is that researchers resort to various heuristics to help them arrive at a solution. In this section we present two such heuristics that have yielded good results in hybridization mapping projects.

5.5.1 SCREENING CHIMERIC CLONES

As noted above, chimeric clones occur with high frequency in clone libraries, and their presence brings serious problems to any mapping algorithm. In this section we present a simple heuristic that tries to split chimeric clones into fragments. Such a heuristic is very useful as a *screening procedure,* which can be used as a preprocessing step before employing more sophisticated techniques.

The idea is very simple: if a clone is chimeric (and let us assume it is composed of only two fragments), the probes that hybridize to one of its fragments should not be related to the probes hybridizing to the other fragment. The key here is of course the concept of "relatedness". This is made concrete by looking at the following graph.

Take clone i and the set of probes that hybridize to it, P_i. We create a graph $H_i = (P_i, E)$ for every clone i. We create an edge between two probes from P_i if they hybridize to a clone other than i. If the resulting graph is connected, we say that i is not chimeric. If it has more than one component, we say that i is chimeric and we replace i by new "artificial" clones, each new clone given by a connected component of H_i. This

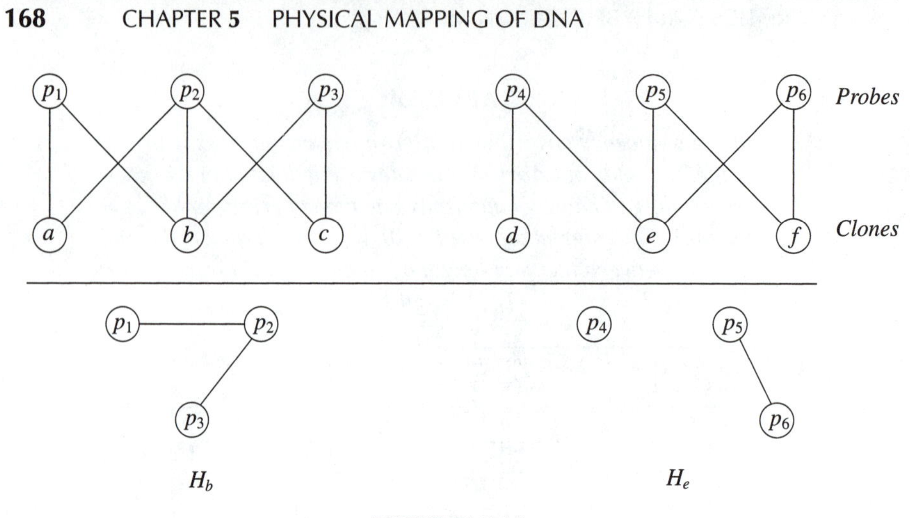

FIGURE 5.12

Above the line we show the hybridization graph H for some clones × probes matrix. Below the line at left is graph H_b for clone b. This clone is probably not chimeric, because H_b is connected. At right is graph H_e for clone e. This clone could be chimeric, because H_e is not connected.

method can be refined by requiring that an edge exists between probes p and q in H_i only if p and q hybridize to at least k other clones, where k is a parameter depending on the particular problem at hand. See Figure 5.12 for an example.

Experience has shown that this heuristic behaves well. On the other hand it may consider a clone chimeric when in fact it is not. Therefore, another useful heuristic would be one that combines two clones that are actually one. See Exercise 22.

5.5.2 OBTAINING A GOOD PROBE ORDERING

The heuristic presented here is more ambitious than the one we saw in the previous section. It aims at actually solving the problem, that is, obtaining a permutation of the probes. The idea is to estimate for every probe p the number of probes to its left and the number of probes to its right by looking at the hybridization graph H (defined in Section 5.4.3). We can then sort the probes using this estimate and thus obtain one "good" permutation.

Given a probe we will be able to count the number of probes to its left and to its right if we can somehow split the other probes into *two* separate components: one left component and one right component. Probes near the ends of the target DNA may not split the others in two; therefore, the probes for which we can obtain two components will be called *splitters*. The method for detecting splitters is described next.

Given a probe p build a set of vertices S_p (whose elements are clones and probes) by including p in S_p and every other probe that shares a clone with p. Include in S_p all clones that are incident to any probe in S_p. Now remove all vertices in S_p from H. We will say that p is a splitter if the resulting graph has exactly two components. Using this method, the heuristic for obtaining a good probe ordering is described in Figure 5.13.

```
Algorithm Probe Permutation Heuristic
    input: clones × probes hybridization matrix
    output: a "good" permutation for the probes
    Identify splitters as described in the text
    for every splitter i do Determine the components A_i and B_i
    for each probe p do Initialize l_p and r_p with zero
    Select a pair (A_k, B_k) arbitrarily
    for each probe p do
        for each pair of components (A_i, B_i) do
            if p ∈ A_i then
                if |A_i ∩ A_k| > |A_i ∩ B_k| then
                    Increment l_p
                else
                    Increment r_p
            else if p ∈ B_i then
                if |B_i ∩ A_k| > |B_i ∩ B_k| then
                    Increment l_p
                else
                    Increment r_p
    Sort probes in decreasing order of l_p − r_p.
```

FIGURE 5.13

Heuristic to obtain a "good" probe permutation.

For each probe p, the algorithm keeps two counters, l_p and r_p, recording the number of left and right regions that contain p. These counters are initialized with zero and are incremented using the splitters. It is not obvious how to do this, because although we know the two components A_i and B_i of a splitter i, we cannot tell which one goes to the left or to the right. To solve this difficulty we rely on a fixed arbitrary splitter k that we use as a reference. We assume that A_k lies to the left of B_k. Then, given another splitter i, a component X of i (either A_i or B_i) is the leftmost one when $|X \cap A_k| > |X \cap B_k|$. Once we have the final counts, we sort probes so that the ones with higher left bias $l_p - r_p$ come first. A simple improvement to this heuristic is in the choice of splitter k. The more "central" this splitter, the better should the results be.

As with the heuristic of the previous section, experience has shown that this one performs well. In particular, it is relatively robust with respect to false negatives. In addition, one can envisage this heuristic also used as delivering its result to other algorithms or heuristics that could try to improve the solution.

SUMMARY

In this chapter we studied two methods that yield data for DNA mapping: digestion by restriction enzymes and hybridization experiments. One way to use restriction enzyme data is to measure and compare the corresponding fragment lengths. This leads to the double digest and partial digest problems. We showed that the double digest problem is NP-complete. Data about fragments can also be used to characterize a fragment's fingerprint. In this case the order of fragments can be reconstructed by determining overlaps between fragments based on fingerprints. This can be modeled in various ways by interval graphs, but most models result in NP-complete problems.

In hybridization mapping, we use as primary data a clones $\times$ probes hybridization matrix. If we place the stringent requirements on this matrix that probes are unique and that there are no errors, the resulting problem can be solved in polynomial time. All we have to do is to determine whether the matrix has the consecutive ones property. We presented one algorithm for this problem. When errors are present, we must resort to approximations, and we showed how mapping can be reduced to one version of the traveling salesman problem. This reduction, coupled with another model for the hybridization process, yields a guarantee that a solution to the TSP will be very close to the solution of the mapping problem. The algorithmic difficulties of mapping motivate the search for algorithm evaluation criteria. We presented one such criterion and showed that a TSP-based approach behaves well under this criterion compared to other more sophisticated approaches. We closed the chapter by presenting two heuristics that aid in screening errors and in recovering a good probe ordering.

EXERCISES

1. Find an alternative solution to the problem in Figure 5.2.

2. Given the following results of double digest experiments, try to find a solution:

Enzyme A: 4, 5, 7, 8, 12.

Enzyme B: 3, 4, 4, 6, 9, 10.

Enzyme $A + B$: 1, 2, 3, 3, 4, 4, 4, 4, 5, 6.

3. Given the following results of a partial digest experiment, try to find a solution: 2, 3, 7, 8, 9, 10, 11, 12, 17, 18, 19, 21, 26, 29.

4. Design an exhaustive search algorithm for solving the DDP problem.

5. Explain why two restriction sites for two different restriction enzymes cannot coincide.

⋆ 6. A necessary condition for a graph G to be interval is that any cycle in G containing four or more vertices has a *chord*, that is, an edge joining two nonconsecutive vertices belonging to the cycle. Prove this fact.

7. Show that the condition Exercise 6 is not sufficient.

8. Show that an interval graph where no interval is properly contained in any other is equivalent to an interval graph in which all intervals have the same length.

9. Execute the C1P algorithm from Section 5.3 on the input matrix given in Table 5.5 and obtain a permutation for the columns that leaves the 1s consecutive in each row.

10. Characterize binary matrices that have the C1P for both rows and columns.

TABLE 5.5

The input binary matrix for Exercise 9.

	1	2	3	4	5	6	7	8	9	10
1	0	1	0	0	0	0	0	0	0	0
2	1	1	0	1	0	1	1	1	1	1
3	1	0	0	0	0	0	0	1	0	0
4	0	1	0	0	0	1	0	0	1	0
5	0	0	0	1	0	1	1	0	1	0
6	0	1	0	1	0	1	1	0	1	0
7	1	0	0	1	1	0	1	1	0	1
8	1	0	1	0	1	0	0	0	0	0
9	0	1	0	1	0	1	1	0	1	0
10	0	1	0	0	0	0	0	0	1	0

⋆⋆ **11.** Characterize interval graphs using the C1P. That is, given a graph G, propose a binary matrix M based on G such that G is an interval graph if and only if M has the C1P.

⋆ **12.** Based on the answer to Exercise 11 design an efficient algorithm to recognize interval graphs.

13. In the C1P algorithm of Section 5.3, prove that the containment relations between a set of component β and sets of another component α are the same for all sets of β. It is a case-by-case proof; one case was already given in the text.

14. Prove that the graph G_M defined in Section 5.3 is acyclic.

15. Develop the model related to gap minimization per row alluded to at the end of Section 5.4.1.

16. Prove that the decision version of the problem of obtaining a column permutation in a binary matrix such that the total number of blocks of consecutive 1s is minimum is NP-complete. *Hint:* Reduce from the Hamiltonian path problem.

17. Run the TSP greedy algorithm sketched at the end of Section 5.4.3 on the graph of Figure 5.10.

18. Theoretical and practical results show that it is a little easier to solve TSP problems if distances obey the constraints given in Section 3.6.1. Does the Hamming distance obey those constraints?

◇ **19.** Hybridization experiments sometimes are not conclusive: The probe may or may not have hybridized to that clone. Assume that the result of each experiment is a real number between 0 and 1 that gives the degree of confidence that the hybridization actually occurred. Propose an algorithm to find the true probe permutation based on such data.

⋆ **20.** Propose an algorithm for hybridization mapping based on the minimum spanning tree heuristic for the TSP problem. A description of this heuristic can be found, for example, in [152].

21. In hybridization mapping sometimes we have the information that, among all probes, some of them hybridize to ends of clones, and we know which ones. What kind of matrix results from this information? How would you reformulate the problem to be solved? Is it any easier than the problem without endprobe information?

22. Propose a heuristic that tries to combine clones that have been wrongly separated by the screening heuristic described in Section 5.5.1. Your heuristic should be used as a post-processing step; that is, it should receive as input one possible ordering for the probes obtained in a previous step.

23. In the heuristic to obtain a good probe ordering, presented in Section 5.5.2, show how the choice of a bad central splitter will put another splitter in the wrong position in the ordering.

24. Fragment assembly and hybrization mapping are similar problems. Are any of the techniques from Chapter 4 applicable to hybridization mapping?

25. One of the main institutions working on the physical map of the human genome is the Whitehead Institute for Biomedical research from MIT. Check out their Web site at http://www-genome.wi.mit.edu/ and find out about the current status of the Human Genome Project.

BIBLIOGRAPHIC NOTES

A nice overview of mapping problems was given by Karp [111]. A good survey of restriction mapping was presented by Lander [119]. In the same paper Lander described algorithms for genetic linkage mapping, a topic that we have not covered and that uses data coming from important genetic markers known as restriction fragment length polymorphisms (RFLPs).

Double digest mapping has been studied by Goldstein and Waterman [73] and by Schmitt and Waterman [169], among others. The proof that DDP is NP-complete is taken from [73]. In [169] it is shown that the number of solutions to DDP may increase exponentially with the target DNA length. In this paper some hope is placed on a general treatment based on equivalence classes of fragments, but this hope is unfortunately not fulfilled by the extensive study made by Pevzner [160], who concluded that a multiple digest approach should be pursued. The partial digest problem was studied by Skiena and Sundaram [174], who propose a branch-and-bound algorithm to solve it. They report promising experimental results.

Waterman and Griggs [200] present an algorithm for the following problem from restriction site mapping. The target DNA is digested separately by two enzymes. Then overlap information between the fragments of each digest is obtained. The problem is to discover the order of the restriction sites based on this information.

Computational complexity analyses of physical mapping appear in [72, 59, 75]. Section 5.2.2 was based on these papers, and they contain the NP-completeness proofs alluded to in that section. Kaplan, Shamir, and Tarjan [108] present an algorithm for the following problem, related to physical mapping: Transform the input graph into a proper interval graph by adding as few edges as possible. A proper interval graph is one in which no interval properly contains another (or, equivalently, one in which all intervals have equal length). They present an algorithm that runs in linear time if the maximum number of edges to be added is a fixed k.

The algorithm for the consecutive 1s problem presented in Section 5.3 is from Fulkerson and Gross [67]. A much more sophisticated algorithm was presented by Booth and Lueker [27]. It runs in time $O(n+m+r)$ and encodes in a compact manner all possible solutions (r is the total number of 1s in the matrix). However, the algorithm is cumbersome and attempts have been made to provide better alternatives. See, for instance, the work by Hsu [97] and by Meidanis and Munuera [136].

Mapping by hybridization with unique probes was studied by Greenberg and Istrail [78, 79] and by Alizadeh et al. [9]. The paper [79] is an excellent description of many aspects of the problem, and several algorithms are presented. The results from Section 5.4.3 were taken from there, and Section 5.2.4 owes much to that paper as well. A number of other techniques are described in [9] and the heuristics from Section 5.5 came from this paper. Grigoriev, Mott, and Lehrach [80] also describe a heuristic for chimeric clone detection. Alizadeh et al. [8] studied hybridization mapping where probes are not unique; Section 5.4 was mainly based on that paper. The Poisson process model for clone distribution is from Lander and Waterman [120], the same model presented in Section 4.1.2 under the topic of coverage. This model is also used in mapping projects to determine the number of clones necessary for a desired coverage.

A software package for physical mapping with a graphic interface is described by Soderlund and Burks [177], and in the same paper they mention software for automatic generation of test instances.

CHAPTER

6

PHYLOGENETIC TREES

In this chapter we describe the problem of reconstructing phylogenetic trees. This is a general problem in biology. It is used in molecular biology to help understand the evolutionary relationships among proteins, for example. We present models and algorithms for two basic kinds of input data: characters and distances.

Modern science has shown that all species of organisms that live on earth undergo a slow transformation process through the ages. We call this process *evolution*. One of the central problems in biology is to explain the evolutionary history of today's species and, in particular, how species relate to one another in terms of common ancestors. This is usually done by constructing trees, whose leaves represent present-day species and whose interior nodes represent hypothesized ancestors. These kinds of trees are called *phylogenetic trees*. One example is shown in Figure 6.1. According to this tree human beings and chimpanzees are genetically closer to each other than to the other primates in the tree. This means that they have a common ancestor that is not an ancestor of any of the other primates. The tree also shows a similar situation between gibbons and siamangs.

The problem with phylogenetic tree construction is that generally we do not have enough data about distant ancestors of present-day species; and even if we did, we could not be 100% sure that a particular fossil belongs to a species that *really* is the ancestor of two or more current species. Therefore we have to *infer* the evolutionary history of current organisms, and *re-create* their phylogenetic tree, generally using as primary data comparisons between today's species. For this reason the tree shown in Figure 6.1 is not necessarily the true one; it is simply a hypothesis.

Phylogenetic trees, or more simply, **phylogenies**, have been proposed since the last century for all sorts of groups of organisms. The methods used up to the 1950s were generally based on researchers' experience and intuition. Gradually mathematical formalisms were introduced and these were incorporated into numerical methods. Today there is an immense body of work on the mathematics and algorithmics of phylogeny reconstruction. There are also software packages offering a host of programs to aid scientists in re-creating phylogenies.

It is not our aim in this chapter to survey computational methods for phylogeny reconstruction; a whole book would be required for this. Our goal is rather to study some

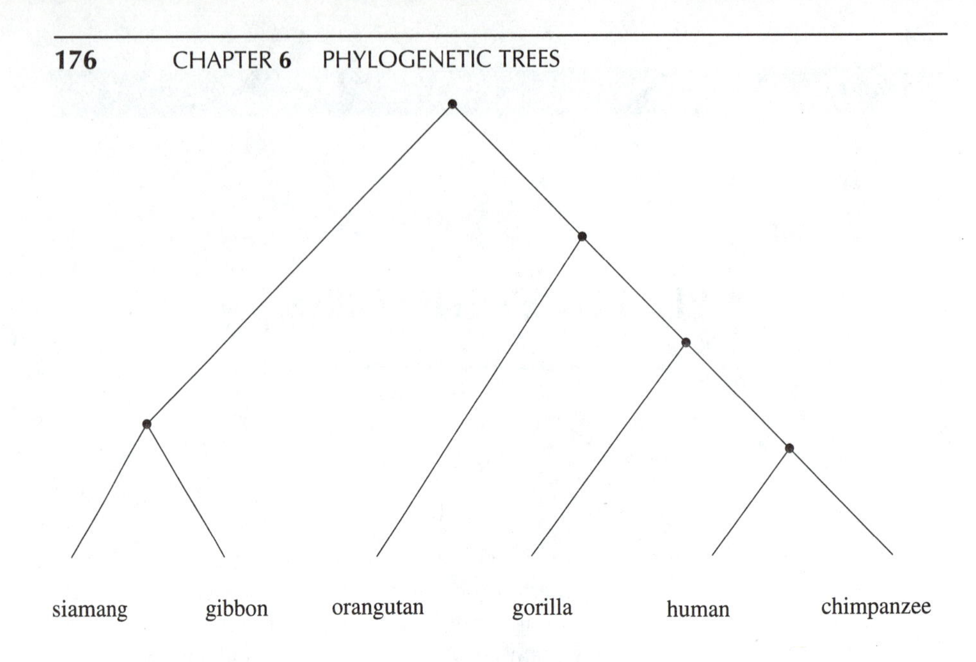

FIGURE 6.1

A phylogenetic tree for some primates.

problems of this area that are amenable to algorithmic solution. The reader should be aware that this is a very narrow and biased treatment, considering the extent and diversity of the literature on this topic. Pointers to more general texts on the subject are given in the bibliographic notes for this chapter.

In evolutionary biology, we can build phylogenetic trees for species, populations, genera, or other *taxonomical units*. (Taxonomy is the "science of classification." The term *numerical taxonomy* is used to describe the collection of numerical methods used for any kind of grouping. This chapter's topic can be seen as a subfield of numerical taxonomy.) Because nucleic acids and proteins also evolve, we can also build phylogenies for them, which explains why molecular biologists are also interested in methods for phylogeny reconstruction. To keep the exposition as general as possible we will use the term **object** for the taxonomical units for which we want to reconstruct a phylogeny.

Let us recall the definition of trees from Section 2.2. A tree is an undirected acyclic connected graph. We distinguish between its exterior nodes, or *leaves,* and the *interior nodes*. Leaves have degree one, whereas interior nodes have degree greater than one. To the leaves we associate the objects under study. Thus we can say that a leaf is *labeled* by an object or a set of objects.

Scientists are interested in two main aspects of phylogenetic trees. One is its *topology,* that is, how its interior nodes connect to one another and to the leaves. (Some authors use the term *branching pattern* for this.) Another important aspect is the *distance* between pairs of nodes, which can be determined when the tree edges are weighted. This distance is an estimate of the evolutionary distance between the nodes. Depending on many assumptions that we will not discuss, the distance between an interior node (ancestral object) to a leaf (present-day object) can be interpreted as an estimate of the time it took for one to evolve into the other. But the reader should be aware evolutionary distance in general is not equivalent to elapsed time.

Another important feature of a phylogenetic tree is whether it has a root. A tree with a root implies ancestry relationships between interior nodes. However, in many of the problems that we will study there is not enough information in the data to allow us to determine the root; in those cases the reconstructed tree will be unrooted.

We mentioned that phylogenies are reconstructed based on comparisons between present-day objects. We can classify input data for phylogeny reconstruction into two main categories:

1. *Discrete characters,* such as beak shape, number of fingers, presence or absence of a molecular restriction site, etc. Each character can have a finite number of *states*. The data relative to these characters are placed in an objects × characters matrix, which we call **character state matrix**.
2. *Comparative numerical data,* which we call *distances* between objects. The resulting matrix is called the **distance matrix**.

Each of these categories gives rise to different methods for phylogeny reconstruction. One significant difference between the categories is that character state matrices are in general not square matrices, since the number of objects need not be equal to the number of characters. On the other hand, distance matrices are triangular matrices, because we have a distance for each pair of objects and these distances are symmetric. In the following sections we present representative algorithms for several of these methods. There is also a third category of *continuous character data,* but we will not cover this case.

6.1 CHARACTER STATES AND THE PERFECT PHYLOGENY PROBLEM

We start this section with some general remarks about assumptions on characters and the meaning of states in interior tree nodes.

The basic assumption regarding characters is a somewhat obvious one: that the characters being considered are "meaningful" in the context of phylogenetic tree reconstruction. The task of defining what is meaningful and checking that the chosen characters are meaningful according to this definition rests largely with the scientist, or "user" of the tree. Nevertheless, we briefly mention two assumptions that underlie all character-based phylogeny reconstruction methods.

The first is that characters can be inherited independently from one another; this is crucial for all algorithms we will discuss. Another assumption is that all observed states for a given character should have evolved from one "original state" of the nearest common ancestor of the objects being studied. Characters that obey this assumption are called *homologous*. This is also important when the objects are biological sequences. In such cases, a character will be a position in the sequence, and its states, in the case of

DNA, will be A, T, C, and G. If we are comparing position, say, 132 of sequence s to position 722 of sequence t, we should make sure that these two positions owe their present state to the same position in an ancestral sequence.

Regarding interior nodes, we recall that they represent hypothetical ancestor objects. The algorithms we describe assign specific states to these nodes, but the set of states of an interior node does not necessarily have biological significance. By this we mean that the characterization of interior nodes should not be taken as a back-prediction of the features of ancestral objects; the main goal is always the grouping of present-day objects, and the reconstruction of ancestor objects is just a means to this end. Nevertheless we have to be cautious about the assignment of states to interior nodes, in order to obtain a tree as close as possible to the true tree.

Having made these introductory remarks, we pass on to a formal definition of the *character state matrix*. We define the character state matrix as a matrix M with n rows (objects) and m columns (characters). Thus M_{ij} denotes the state the object i has for character j. There can be at most r states per character and states are denoted by non-negative integers. A given row of this matrix is the *state vector* for an object. Since we will also assign states to interior nodes in the tree, we will also say that interior nodes have associated state vectors. An example of a character state matrix is seen in Table 6.1.

TABLE 6.1

A character state matrix.

	Character				
Object	c_1	c_2	c_3	c_4	c_5
A	1	1	0	0	0
B	0	0	1	0	1
C	1	1	0	0	1
D	0	1	1	1	0
E	1	1	0	0	1

When trying to create a phylogeny from a character state matrix, we encounter some difficulties. The first arises when two or more objects have the same state for the same character. Almost all methods for phylogeny reconstruction are based on the assumption that objects that share the same state are genetically closer than objects that do not. However there exists the possibility that two objects share a state but are not genetically close. A case in point is the presence of wings in bats and birds. Such phenomena are called *convergence* or *parallel evolution*. We observe in nature that such cases are rare; therefore, we will deal with this difficulty in the definition of our tree reconstruction problems in the following manner: Convergence events should not happen, or their number should be minimized.

The second difficulty has to do with the relationships among different states of the same character. Let us illustrate this problem using the matrix shown in Table 6.1. Suppose that there was an ancestral object X from which objects A and B evolved. Which state should we assign to X relative to character c_1? Note that $c_1 = 1$ for A and $c_1 = 0$ for B. If we make $c_1 = 0$ for X we are saying that 0 is the *ancestral state* and that 1 is the *derived state*. Assume we have made such a choice, and now suppose that objects C and D have as ancestral object $Y \neq X$. Further assume we have decided that Y's state for c_1 is 1. In this case notice that object D presents a *reversal* to the ancestral state 0 for character c_1. In this particular case, since the characters are binary, we could interpret this situation in terms of gains and losses of the character. That is, A has acquired the character with respect to its ancestor X, whereas D has lost it with respect to its ancestor Y. As in the case of the difficulty discussed in the previous paragraph, we observe in nature that such reversals are uncommon. Therefore we will handle reversals the same way we did with convergence events: We will require reversals not to happen, or that their number be minimized. Notice that we may not know beforehand which state is ancestral and which is derived. If we do not, we will have to choose, and then remain consistent with this choice in order to avoid or minimize reversals.

In the previous example we dealt with a binary character. Because in general there can be r states for a given character, the relationships between them may be more complex. Depending on how much we know about these relationships, characters can be classified as **ordered** or **unordered**. For an unordered character we assume nothing about the way states can change; that is, any state can change into any other. In the case of ordered characters extra information is known. For example, a given character with four states may have the following *linear* order: $3 \leftrightarrow 1 \leftrightarrow 4 \leftrightarrow 2$. This means, among other things, that in the reconstructed tree there should be no change of state 3 directly into state 4; state 1 should always be an intermediate. In other words, if there is a node with state 3 for the given character, there should be no edge linking it to another node with state 4. Characters may also be *partially ordered,* in which case there is a *derivation tree* that shows how states are derived from one another by means of a tree. Note that even with ordered characters we are not saying anything about the *direction* in which state changes must take place; characters for which the direction of change is known are said to be **directed**. In the literature unordered characters are also known as *qualitative characters* and ordered characters as *cladistic characters*. Directed characters are also known as *polar characters*. Intuitively, ordered characters provide us with much more information on how we should build the phylogeny, and this intuition is useful in understanding some computational complexity results below.

If we want to altogether avoid convergence events and reversals, the considerations above require that the desired tree T have the following property: for each state s of each character c, the set of all nodes u (leaves and interior nodes) for which the state is s with respect to c must form a subtree of T (that is, a connected subgraph of T). This means that the edge e leading to this subtree is uniquely associated with a transition from some state w to state s. A phylogeny with this property is a *perfect phylogeny*. We are now ready to state the central problem of phylogeny reconstruction based on character state matrices. This problem is known as the *perfect phylogeny problem:*

PROBLEM: PERFECT PHYLOGENY

INSTANCE: A set O with n objects, a set C of m characters, each character having at most r states (n, m, r positive integers).

QUESTION: Is there a perfect phylogeny for O?

For the matrix shown in Table 6.1, assuming 0 is ancestral and 1 is derived, such a phylogeny does not exist, as will be shown in the next section. For the matrix shown in Table 6.2 a perfect phylogeny does exist and is shown in Figure 6.2. A character labeling an edge in the figure indicates that a transition from state 0 to state 1 takes place along that edge, such that the subtree below the edge contains all objects that have state for 1 that character and no others.

TABLE 6.2

Another character state matrix.

	Character					
Object	c_1	c_2	c_3	c_4	c_5	c_6
A	0	0	0	1	1	0
B	1	1	0	0	0	0
C	0	0	0	1	1	1
D	1	0	1	0	0	0
E	0	0	0	1	0	0

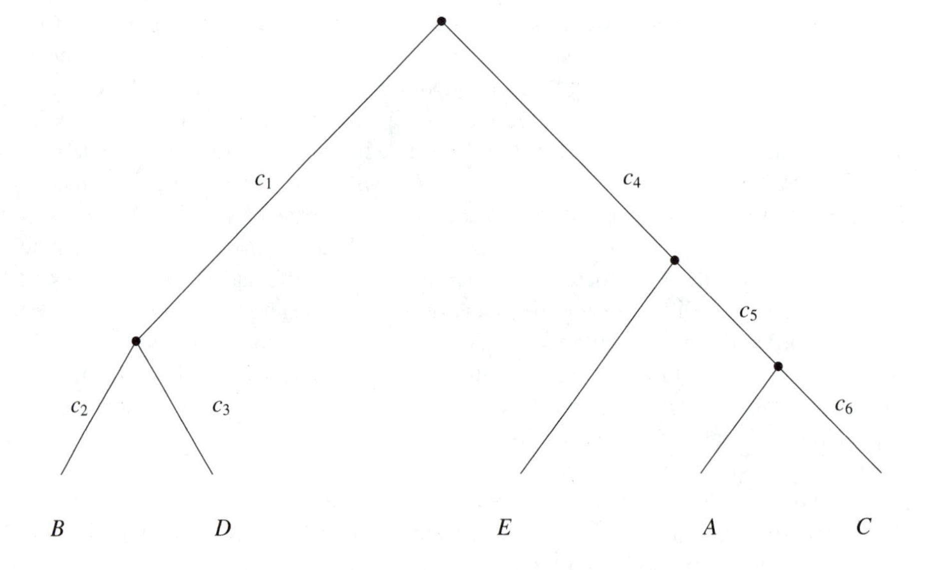

FIGURE 6.2

A phylogeny corresponding to the matrix in Table 6.2. Edges where state transitions take place are labeled by the corresponding character.

Whenever a set of objects defined by a character state matrix admits a perfect phylogeny, we say that the defining characters are **compatible**. This notion of compatibility will be later refined for the special cases of the problem that we will study.

Having defined the problem, the first question we would like to ask is this: Is there an efficient algorithm for solving it? We shall answer this question in a moment, but first let us have an idea of how many different trees can we build for n objects. Suppose we want only unrooted binary trees. Recall that we are interested in trees where objects are the leaves; hence in general it does matter which object is what leaf. For three objects there is only one tree, shown in Figure 6.3. For four objects there are three trees, as shown in Figure 6.4. It can be shown (see Exercise 2) that in general there are $\prod_{i=3}^{n}(2i - 5)$ trees for n objects. This number grows faster than $n!$, so this rules out any algorithm that builds all trees trying to find one that is a perfect phylogeny. We would like an algorithm for this problem that runs in time proportional to a polynomial in n, m, and r.

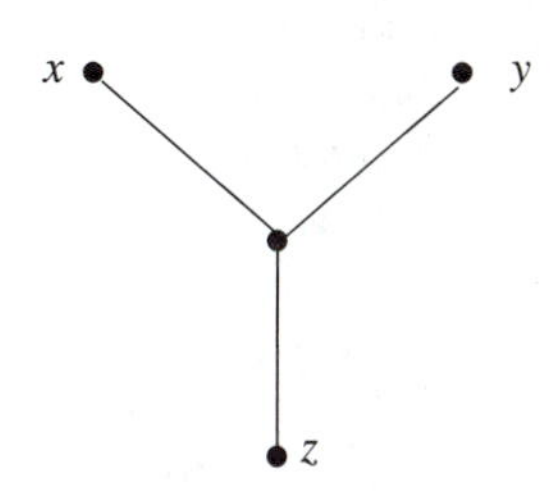

FIGURE 6.3

The unique tree for three objects.

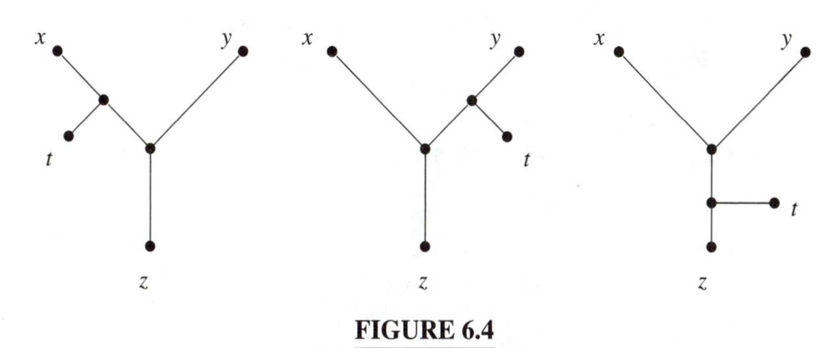

FIGURE 6.4

The three trees for four objects.

It turns out that whether the perfect phylogeny problem is computationally hard depends on the nature of the characters. If they are unordered, we have bad news: The problem is NP-complete. But if they are ordered, then it can be efficiently solved. Having these two facts as background, we now proceed to study in more detail some special cases of the problem that represent a sample of algorithmic results for character state matrices.

BINARY CHARACTER STATES

6.2

First we examine the special case of the perfect phylogeny problem in which characters are binary, such as those from Tables 6.1 and 6.2. Notice that for this case the distinction between ordered and unordered characters depends on knowing whether the characters are directed. This means that if we know that state 0 is ancestral and 1 is derived (or vice versa) then the character is ordered; otherwise it is unordered. As we will see, with binary characters (ordered or unordered) the perfect phylogeny problem can be solved efficiently; we present in this section an algorithm that runs in time $O(nm)$.

We present the algorithm as working in two phases. (In an actual implementation these two phases can be combined, but for presentation purposes this separation is better.) In the first phase the algorithm *decides* whether the input matrix M admits a perfect phylogeny. If it does, then the second phase of the algorithm will construct one possible phylogeny.

In the algorithm to be presented we assume that state 0 is ancestral and state 1 is derived. We will see later on how to drop this restriction. Notice that because we are dealing with directed binary characters, a rooted tree T that is a perfect phylogeny for input matrix M will have the following property, already mentioned in the previous section: To every character in input matrix M there corresponds an edge in T, and this edge marks the transition from state 0 to state 1 for that character. Such edges will be labeled by their respective characters. The root always has character state vector $(0, 0, \ldots, 0)$. This means that when we traverse the path linking an object i in a leaf to the root the edge labels found along the way correspond to characters for which object i has state 1. (See Figure 6.2.)

In the special case under study, there is a very simple necessary and sufficient condition that enables us to tell whether a given matrix M admits a perfect phylogeny. To state this condition, we first need a few definitions.

Each column j of M is a character. Thus we use without distinction the terms *column* and *character*. Each row i of M is an object, and we also use without distinction the terms *row* and *object*.

DEFINITION 6.1 For each column j of M, let O_j be the set of objects whose state is 1 for j. Let $\overline{O}_j$ be the set of objects whose state is 0 for j.

The simple condition is the following:

LEMMA 6.1 A binary matrix M admits a perfect phylogeny if and only if for each pair of characters i and j the sets O_i and O_j are disjoint or one of them contains the other.

Proof. Let us assume that M admits a perfect phylogeny. Because M is binary, we know that to each character i we can uniquely associate an edge (u, v) in the tree. Moreover, the subtree rooted at v (assuming it is the deeper of the two nodes) contains all nodes having state 1 for character i; and any node having state 0 for character i does not belong to this subtree. Suppose now that there are three objects A, B, and C such that

$A, B \in O_i$, $C \notin O_i$ and $B, C \in O_j$, $A \notin O_j$. This means that, according to character i, A and B belong to the same subtree but C does not. However, according to character j, B and C do belong to the same subtree, which is a contradiction.

Let us now suppose that all pairs of columns of M satisfy the condition stated in the lemma. We will show inductively how to build a rooted perfect phylogeny. Assume we have only one character, say 1. This character partitions the objects into two sets, $A = O_1$ and $B = \overline{O}_1$. Create a root and then node a for set A and node b for set B. Link node a to the root with an edge labeled by 1. Link node b to the root by an unlabeled edge. As a final step, we split each child of the root into as many leaves as there are objects in them. The tree thus created is clearly a perfect phylogeny; this is our base case. Now assume we have built a tree T for k characters, but without having executed the final step (that is, there are no leaves; nodes still contain sets of objects). We now want to process character $k + 1$. This character also induces a partition in the object set, and using it we should be able to get tree T' from T. We will be able to do this as long as the partition induced by character $k + 1$ separates objects belonging to the *same* node. If this were not the case, we would be forced to label *two* edges with label $k + 1$ and the resulting phylogeny would not be perfect. But such a situation cannot happen. Suppose it did; that is, suppose character $k + 1$ separates objects belonging to the nodes a and b. Because these are different nodes, there must be a character i that led them to be in different nodes in the first place. It is the case that $O_i \cap O_{k+1} \neq \emptyset$, because either the objects in a or those in b belong to O_i. But it is also the case that O_i does not contain nor is contained in O_{k+1}, because either the objects in a or those in b are *not* in O_i. But this would contradict our hypothesis, and the lemma is proved. ■

From this lemma it is easy to see why the character state matrix shown in Table 6.1 does not admit a perfect phylogeny: Columns c_1 and c_5 do not satisfy the lemma's condition. We say that those characters are not **compatible**. The notion of compatibility between characters can here be given a precise meaning and is just another way to state Lemma 6.1: A collection of binary directed characters is compatible if and only if they are pairwise compatible, where pairwise compatibility is given by the set relationships in the lemma's statement.

We mentioned that we would present the algorithm for the binary case in two phases: In the first we decide whether the matrix admits a perfect phylogeny, and in the second we construct one, if possible. Lemma 6.1 already gives us an algorithm for the decision phase: Simply consider each column in turn and check whether it is compatible with all others. Each such check takes $O(n)$ and we have to make $O(m^2)$ checks, resulting in an algorithm with running time $O(nm^2)$. From the lemma's proof it is also possible to develop an algorithm for the construction phase, with the same time bounds. Our aim, however, is to present an algorithm running in time $O(nm)$.

How can we attain this goal? For the decison phase, the algorithm just sketched scans all remaining columns each time a new column is checked. To get the efficiency we want, we should scan all columns just *once,* perhaps imposing some order on the columns. Which order should this be? One idea is to process first characters for which a maximum number of objects has state 1. Intuitively, once we have processed this character we can "forget" about it, as all other characters are either subsets of it or are disjoint from it. And that is exactly what we do. The algorithm is shown in Figure 6.5.

Let us now show that the algorithm is correct. First note that if $O_j \subseteq O_k$, then k

```
Algorithm Perfect Binary Phylogeny Decision
  input: binary matrix M
  output: TRUE if M admits a perfect phylogeny, FALSE otherwise
  // Assumes all columns are distinct
  Sort columns based on the number of 1s in each column using radix-sort
      (columns with more 1s come first)
  // Initialize auxiliary matrix L
  for each L_ij do
      L_ij ← 0
  // Compute L
  for i ← 1 to n do
      k ← −1
      for j ← 1 to m do
          if M_ij = 1 then
              L_ij ← k
              // k is the rightmost column to the left of j such that
              // M_ik = 1. If no such column exists, k = −1
              k ← j
  // check columns of L
  for each column j of L do
      if L_ij ≠ L_lj for some i, l and both L_ij and L_lj are nonzero then
          return FALSE
  return TRUE
```

FIGURE 6.5

An algorithm that decides whether a collection of directed binary character states admits a perfect phylogeny.

is to the left of j in M, that is, $k < j$. Suppose that M has a perfect phylogeny but the algorithm tells us otherwise. That is, M has column j such that $L_{ij} = k$ and $L_{lj} = k' < k$. This means (looking now at column k) that element M_{lk} is zero but element M_{ik} is 1. Thus $O_k \cap O_j \neq \emptyset$. On the other hand neither set is a subset of the other: O_j is not a subset of O_k because M_{lk} is zero and M_{lj} is one; and O_k is not a subset of O_j because k is to the left of j. Therefore columns k and j do not satisfy Lemma 6.1 and this is a contradiction.

We now argue that if the algorithm tells us that there is a perfect phylogeny, then this is true. We will show that for every pair of columns of M Lemma 6.1 is satisfied. Consider an arbitrary column j and the column k to the left of j such that $O_j \subseteq O_k$ and k is maximum (that is, k is the common nonzero value of column j of L). By this definition, any column p between k and j has the property that $O_p \cap O_j = \emptyset$. If $k > 0$ we can apply the same argument to show that the intersection between O_j and all columns q between k and k' (where k' is the common nonzero value of column k in L) is also empty. We thus showed that for an arbitrary column j the corresponding O_j is contained in some sets to its left and has an empty intersection with the others, which was exactly what we set out to prove.

The running time of the algorithm is $O(nm)$ because the radix-sorting of M takes time $O(nm)$ ($O(n)$ per column), and the other steps can easily be done in the same time bound.

The Construction Phase

The idea in the construction is to process objects in turn, building the tree as we go. We start with a tree with a single node: the root. For each object, we look at the characters for which its state is 1. Whenever we see that there is no edge so far labeled with such a character, we create a new node and link it to the current node by an edge labeled by this character. The new node becomes the current node. If there already exists an edge labeled with this character, we simply move on to the next character, letting the current node be the endpoint of this edge. The algorithm is described formally in Figure 6.6. Notice that the tree obtained is rooted, but we can unroot it simply by removing the root and connecting its children. Let us now show that the tree constructed in this algorithm is a perfect phylogeny for M, and that the running time is $O(nm)$.

```
Algorithm Perfect Binary Phylogeny Construction
    input: binary matrix M with columns sorted in nonincreasing order by number of 1s
    output: perfect phylogeny for M
    Create root
    for each object i do
        curNode ← root
        for j ← 1 to m do
            if M_ij = 1 then
                if there already exists edge (curNode, u) labeled j then
                    curNode ← u
                else
                    Create node u
                    Create edge (curNode, u) labeled j
                    curNode ← u
        Place i in curNode
    for each node u except root do
        Create as many leaves linked to u as there are objects in u
```

FIGURE 6.6

An algorithm that constructs a perfect phylogeny from directed binary characters.

We first recall that each character in M must correspond to one and only one edge of the tree, and that is precisely what the algorithm does. In addition, in a path from the root to a leaf (object) i we should traverse the edges corresponding to the characters for which i has state 1, and no others, and again that is what happens. The running time is $O(nm)$ given that we look at each element from M exactly once, and we spend constant time per element. Applying the algorithm to the matrix in Table 6.2 we obtain the phylogeny shown in Figure 6.2.

We have mentioned that the case where the characters are unordered can also be solved efficiently. In fact, the same algorithm can be used, provided a simple transformation is performed. When characters are unordered, think of the states as denoted by the letters a and b. The transformation is the following: For each character, determine the state that occurs more often. The majority state becomes 0 and the other 1. If both occur with equal frequency, then we can choose either one to be 0 and the other to be 1. Now we can use the algorithm above. The reason why this works derives from the following fact:

Fact: There exists a perfect phylogeny for a collection of objects on two binary unordered characters i and j (that is, these two characters are compatible) if and only if both O_i, O_j and $\overline{O}_i$, $\overline{O}_j$ obey the conditions stated in Lemma 6.1.

We leave it as an exercise to show that the preceding transformation preserves the compatibility between unordered characters as required by the fact above.

We now close this section with two observations. The first is that a simple argument shows that there cannot be a faster algorithm than the one described. It should be clear that any algorithm that solves this problem has to look at least once at every element in the matrix, and this gives us a lower bound of $\Omega(nm)$. This means that the algorithm presented above is optimal. However, if the input is given as a list of characters for each object, there are algorithms that solve the problem and run in time proportional to the total lengths of such lists, which is $O(n+m)$. The second observation is that polynomial-time algorithms are known for unordered characters whenever the number of states is fixed. For details on both observations, see the bibliographic notes at the end of this chapter.

6.3 TWO CHARACTERS

Here we study another special case of the perfect phylogeny problem. We allow the more general situations in which characters can be unordered and have an arbitrary number of states. This time, however, we place a restriction on the maximum *number* of characters; in particular we study the case when only *two* characters are allowed.

As was the case for binary states, we will see that having only two characters makes the problem computationally easy. We present in what follows a very simple algorithm for this problem, but one that makes use of nontrivial results from graph theory. We begin by stating some definitions.

DEFINITION 6.2 A *triangulated graph* is an undirected graph in which any cycle with four or more vertices has a *chord,* that is, an edge joining two nonconsecutive vertices of the cycle.

Recall that a subtree is a connected subgraph of a tree, and that the set of tree nodes that have the same state for a given character make up a subtree in a perfect phylogeny. The connection between triangulated graphs and the perfect phylogeny problem is given by the following theorem:

THEOREM 6.1 To every collection of subtrees $\{T_1, T_2, \ldots, T_l\}$ of a tree T there corresponds a triangulated graph and vice versa.

We will not present the proof of this theorem, but we will describe the graph that should correspond to the given collection of subtrees. This graph turns out to be a convenient tool to help us determine whether a set of characters is compatible (i.e., if there is a perfect phylogeny for the corresponding objects). As will become clear shortly, we will be able to say that a set of characters is compatible if and only if the graph built according to the theorem can be triangulated. To understand this graph we need some more definitions.

DEFINITION 6.3 An *intersection graph* for a collection C of sets is the graph G that we get by mapping each set in C to a vertex of G, and linking two vertices in G by an edge if the corresponding sets have a nonempty intersection.

The graph specified in the proof of Theorem 6.1 is an intersection graph. To describe it, we first present another graph, which we call *state intersection graph,* or SIG, and which represents an intermediate step. Given a character state matrix, its SIG is obtained in the following way. For each state of each character we create a vertex in the SIG. Note that to each state, and thus vertex, there corresponds a set, the set of objects that have that state for that character. We then create an edge between vertices u and v of the SIG if and only if the corresponding sets have a nonempty intersection. This is why we call such a graph a state intersection graph. An example can be seen in Figure 6.7, which was obtained from the matrix in Table 6.3.

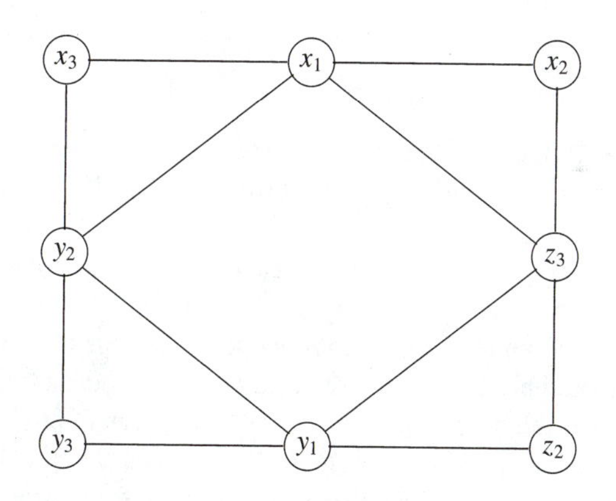

FIGURE 6.7

The SIG corresponding to the matrix in Table 6.3. The subscript of each vertex label is the color of the vertex.

Note that the graph mentioned in Theorem 6.1 is built based on a set of *subtrees*. The graph we described in the previous paragraph is built based on a set of objects, which are *leaves* in the phylogeny. To arrive at the graph mentioned in the theorem, we must *enlarge* the SIG (by adding new edges) to account for the interior nodes. It is in this

TABLE 6.3

A matrix of objects × characters.

	Character		
Object	c_1	c_2	c_3
A	x_1	x_2	z_3
B	x_1	y_2	x_3
C	y_1	y_2	y_3
D	y_1	z_2	z_3

enlarged graph that we will check for the triangulated property. In fact, we will try to add edges to the SIG with the hope of making it triangulated. However we cannot add edges at will; we must take into account the relationships between objects according to each character. To explain this we need one more definition.

DEFINITION 6.4 Given a graph $G = (V, E)$ with a coloring c on V, we say that G can be *c-triangulated* if there exists a triangulated graph $H = (V, E')$, such that $E \subseteq E'$ and c is a valid coloring for H. In other words, any edge present in E' but not in E must link two vertices with different colors.

Now observe that the set of characters of a character state matrix defines a coloring for the SIG, as an object cannot have two different states for the same character. We thus obtain a new formulation for Theorem 6.1 that makes clear the connection between the SIG and the phylogeny.

THEOREM 6.2 A character state matrix M, with a character set defining a coloring c, admits a perfect phylogeny if and only if its corresponding SIG can be c-triangulated.

In the example of Figure 6.7 the color of each vertex is given by the character that is part of the label of the vertex. To triangulate this graph, all we need do is create an edge between vertices y_2 and z_3; this is possible because these two vertices have different colors [but we would not be able to create edge (x_1, y_1) because they have the same color]. This means that the matrix in Table 6.3 admits a perfect phylogeny.

Note the generality of Theorem 6.2: nothing is said about fixed number of characters. Recalling the fact that the perfect phylogeny problem for unordered characters is in general NP-complete, you should not be surprised to learn that c-triangulation of graphs is also an NP-complete problem. However, if M has only two characters, the SIG will have only two colors, which gives us the following result.

THEOREM 6.3 A character state matrix M with only two characters admits a perfect phylogeny if and only if its corresponding SIG is acyclic.

Proof. First note that if the SIG is acyclic it is already c-triangulated, and thus, by Theorem 6.2, M admits a phylogeny. To show the only-if part, it suffices to use a classical result from graph theory, which states that a graph can be colored with two colors if and only if all its cycles are of even length. From Theorem 6.2 we know that the SIG can be 2-triangulated, which means that it can be colored with two colors. It follows that the SIG cannot have odd-length cycles. A triangle is a cycle of odd length. Hence the only way that the SIG can satisfy the 2-triangulation property is if it is acyclic. ■

With the above theorem we now have a simple and efficient algorithm to check whether a two-character state matrix admits a phylogeny, since a test for acyclicity takes $O(|V|+|E|)$ on a graph with $|V|$ vertices and $|E|$ edges. In our case the number of vertices is $O(n)$ (since there are only two characters, there can be at most $2n$ states) and the number of edges is also $O(n)$. Hence the algorithm runs in linear time. The reconstruction of the perfect phylogeny T, if one exists, proceeds as follows. Create an auxiliary graph $G = (V, E)$ such that there exists a vertex in V for each edge in the SIG. In this way, to a vertex $v \in V$ corresponding to edge (x, y) in the SIG there also corresponds a set L_v of objects: the intersection between the x set and the y set (recall that a vertex in the SIG corresponds to a set of objects). Create an edge between two vertices $u, v \in V$ if $L_u \cap L_v \neq \emptyset$. The desired tree is any spanning tree of G (which means that the tree may not be unique), with the objects in L_v added as leaves to each node v. The resulting tree is unrooted and can be obtained in time $O(n)$. This construction applies only to the two-character case. If there are more characters, the nodes in the tree should correspond to maximal complete subgraphs (that is, maximal *cliques*) in the triangulated SIG. This fact comes from the proof of Theorem 6.1, which we omitted, and which also explains how to connect the nodes thus obtained to build the tree.

As an example of the two-character tree construction see Figure 6.8, which is the tree obtained from the matrix in Table 6.4. We can verify that this tree is a perfect phylogeny by checking that all objects with the same state for a given character form a subtree. For example, objects A, C, F, and H all have state x_1 for character c_1; and indeed they form a subtree. By contrast, E and G do not form a subtree. See also Exercise 9.

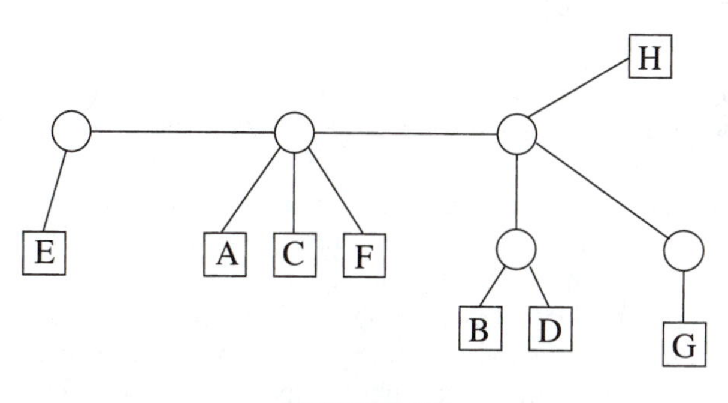

FIGURE 6.8

The reconstructed tree from data in Table 6.4.

The results presented provide the foundation for a much more general result: An algorithm that runs in polynomial time for any fixed value of m (in other words, m appears as an exponent in the running time function). A description of this result is beyond our scope, but a reference is given in the bibliographic notes.

TABLE 6.4

An example of a two-character state matrix.

	Character	
Object	c_1	c_2
A	x_1	x_2
B	y_1	y_2
C	x_1	x_2
D	y_1	y_2
E	w_1	x_2
F	x_1	x_2
G	z_1	y_2
H	x_1	y_2

6.4 PARSIMONY AND COMPATIBILITY IN PHYLOGENIES

We have seen in the previous sections that even though the perfect phylogeny problem is NP-complete for unordered characters, the fixed-parameter special cases are solvable in polynomial time. Fixed parameter algorithms might seem good enough in practice, but the problem is that real character state matrices are unlikely to admit perfect phylogenies. The reason is twofold: Experimental data in biology always carries errors, and the assumptions made by the algorithms mentioned (no reversals and no convergence) sometimes are violated. For these reasons other approaches for reconstructing phylogenies are necessary.

Ignoring errors but working around the assumptions, two approaches suggest themselves. The first is to allow reversal and convergence events, but to try to *minimize* their occurrence. This is known as the *parsimony criterion*. The second is to insist in avoiding reversals and convergence, but to exclude characters that cause such "problems." It is obviously a good idea to try to exclude as few characters as possible. This means that in this approach we try to find a maximum set of characters such that we can find a perfect phylogeny for them (or, in other words, a maximum set of characters that are compatible). This is known as the *compatibility criterion*.

Both criteria result in optimization problems. Such problems are in general harder than decision problems, such as the perfect phylogeny problem considered in Section 6.1. Because we know that deciding perfect phylogeny for unordered characters is an NP-complete problem, the problems just sketched should be at least just as hard; and so they are. But we might have better luck in the case of *ordered* characters, since, as noted in Section 6.1, for them the perfect phylogeny problem *is* polynomially solvable. Unfortunately we have no such luck, and both criteria outlined above result in NP-hard problems for ordered characters as well. Below we discuss these problems in more detail.

When using the parsimony criterion, we try to minimize reversal and/or convergence events, as already mentioned. In general this means that we want to minimize the number of *state transitions* (that is, edges) in the reconstructed phylogeny. If we allow reversals but forbid convergence, we get what is known as the *Dollo parsimony criterion*. If, on the other hand, we forbid reversals but minimize convergence events we get the *Camin-Sokal parsimony criterion*. Both criteria result in NP-hard problems. The proofs are based on the Steiner tree problem, a classic optimization problem in graphs.

The concept of parsimony also appears in the literature in connection with a different problem. Given a set of objects *and* the tree through which they are related, assign states to the interior nodes such that a minimum number of state transitions occur from node to node. This is a problem for which efficient algorithms exist depending on the exact characterization of the problem. We will not cover such algorithms; we give references at the end of this chapter.

We now turn our attention to the compatibility criterion and present the proof that the associated decision problem is NP-complete. The proof uses the *clique problem,* which is well-known to be NP-complete. In fact, we show that clique and compatibility are equivalent. This means that enumerative algorithms for clique can be used for the compatibility problem, and there are phylogeny software packages that take advantage of this. We are of course very limited in the size of problems that we can solve by this approach, but it seems to be useful nevertheless.

The version of the compatibility problem to be shown NP-complete is the following:

PROBLEM: COMPATIBILITY
INSTANCE: a character state matrix M with n objects and m directed binary characters, and a positive integer $B \leq m$.
QUESTION: Is there a subset L of the characters that satisfies Lemma 6.1 with $|L| \geq B$?

Notice that we are looking at a very restricted form of the problem, as characters are binary and directed. With such restrictions, the corresponding notion of compatibility is precisely that given by Lemma 6.1. As mentioned, the NP-completeness proof uses the clique problem, which can be formally characterized as follows:

PROBLEM: CLIQUE
INSTANCE: Graph $G = (V, E)$, and positive integer $K \leq |V|$.
QUESTION: Does G contain a subset $V' \subseteq V$ with $|V'| \geq K$ such that every pair of vertices in V' is linked by an edge in E?

As we have said we want to show that COMPATIBILITY is NP-complete and that it is equivalent to CLIQUE. As described in Section 2.3, this task requires that we show that both problems are in class NP, and that each can be transformed in polynomial time to the other. We are already assuming that CLIQUE is an NP-complete problem, so it is immediate that it belongs to NP. The other parts are given in the following theorems.

THEOREM 6.4 COMPATIBILITY is NP-complete.

Proof. Given a solution to COMPATIBILITY, we can easily check that it in fact solves the problem by using the polynomial-time algorithm described in Section 6.2. Thus we see that COMPATIBILITY is in NP. We now show how to transform CLIQUE to COMPATIBILITY in polynomial time.

Given an instance $G = (V, E)$ of CLIQUE, we construct an instance M of COMPATIBILITY as follows. We let $m = |V|$; that is, for every vertex v_i in V we create character

i in M. The number of objects (rows) of M is given by $n = 3m(m-1)/2$; that is, we create three objects for every possible pair of characters. We now have to fill in M. For every pair (v_i, v_j) such that it is *not* an edge in E we create three objects r, s, and t in M with the following values for the corresponding elements of M: $M_{ri} = 0$, $M_{si} = 1$, and $M_{ti} = 1$; and $M_{rj} = 1$, $M_{sj} = 1$, and $M_{tj} = 0$. It is easy to see that with such values, characters i and j will not be compatible (see Lemma 6.1). The remaining elements of M should be zero. The transformation just given can clearly be accomplished in polynomial time.

We now have to show that G contains a clique V', with $|V'| \geq K$ if and only if M contains a compatible character subset L, with $|L| \geq K$. First assume that such a clique exists. Then to every edge of this clique there corresponds a pair of characters in M, such that whenever one of them has state 1 for an object, the other has state 0 (or both have 0). All these pairs therefore satisfy Lemma 6.1 and hence are compatible. Now assume that L exists, $|L| \geq K$. Then to every pair of characters of L there corresponds a pair of vertices in V linked by an edge. All these pairs taken together form a clique of size greater than or equal to K. ■

THEOREM 6.5 COMPATIBILITY can be transformed to CLIQUE in polynomial time.

Proof. Given an instance M of COMPATIBILITY, obtain an instance $G = (V, E)$ of CLIQUE as follows. For each character of M create a vertex in V. Join two vertices v_i and v_j if and only if characters i and j are compatible. The checking of compatibility between pairs of characters can be easily done in polynomial time. Lemma 6.1 ensures that G contains a clique of size B or more if and only if M contains a subset of compatible characters whose size is also greater than or equal to B. ■

6.5 ALGORITHMS FOR DISTANCE MATRICES

In this section we consider the problem of reconstructing trees based on comparative numerical data between n objects. The basic input here is an $n \times n$ matrix M, whose element M_{ij} is a nonnegative real number, which we call the *distance* between objects i and j. Such distances come from comparisons between the objects. For example, if the objects are DNA sequences, we could use the notion of distance between sequences described in Section 3.6.1.

The first problem we will consider can be informally stated as follows. Given distance matrix M, we would like to build an edge-weighted tree where each leaf corresponds to only one object of M and such that distances measured on the tree between leaves i and j correspond exactly to the value of M_{ij}. When such a tree can be constructed, we say that the distances in M are *additive*.

We will see that in practice distance matrices are rarely additive and that trying to minimize the deviation from additivity leads to NP-hard problems. This prompts our second problem, in which the tree to be constructed uses as input *two* matrices instead of one. These matrices give us lower and upper bounds on the pairwise distances between objects, and the problem is to find a tree that fits "between" these bounds.

6.5.1 RECONSTRUCTING ADDITIVE TREES

When discussing distances we need the concept of a *metric space*. We have already seen this concept in Section 3.6.1, where it was called *distance*. We repeat the definition here for convenience. A metric space is a set of objects O such that to every pair $i, j \in O$ we associate a nonnegative real number d_{ij} with the following properties:

$$d_{ij} > 0 \quad \text{for } i \neq j, \tag{6.1}$$

$$d_{ij} = 0 \quad \text{for } i = j, \tag{6.2}$$

$$d_{ij} = d_{ji} \quad \text{for all } i \text{ and } j, \tag{6.3}$$

$$d_{ij} \leq d_{ik} + d_{kj} \quad \text{for all } i, j, \text{and } k \text{ (the \textit{triangle inequality})}. \tag{6.4}$$

We will require our input matrix M to be a metric space. This means in particular that it will be a symmetric matrix (from Property 6.3). We now must define exactly what we mean by constructing a tree based on M.

Because M is an $n \times n$ matrix, our tree must have n leaves. The leaves are nodes with degree one; all other nodes must have degree three; such a tree is therefore unrooted. All edges in the tree have nonnegative weight. (This requirement coupled with the degree requirement means that rather than have an interior node with degree four, we will split it into two nodes of degree three attached by an edge of length zero.) The *weight of a path* between nodes u and v on the tree is the sum of the weights of the edges that form the path. The final requirement is that the weight of the path between any two leaves i and j must be equal to M_{ij}. If such a tree T can be found, we say that M and T are additive.

Given this framework, the basic questions we have are: Given matrix M which is a metric space, is it also additive? And if it is, how do we build the corresponding additive tree? The answer to the first of these questions is given by the following lemma.

LEMMA 6.2 A metric space O is additive if and only if given any four objects of O we can label them i, j, k, and l such that

$$d_{ij} + d_{kl} = d_{ik} + d_{jl} \geq d_{il} + d_{jk}.$$

We will not present the proof of this elegant result (which is known as the *four point condition*); please see the bibliographic notes. Rather, we will concentrate on the answer to the second question: an algorithm to build an additive tree. This is a polynomial-time algorithm, and the best way to understand it (or to design it) is through induction.

Let us assume that the condition of Lemma 6.2 has already been verified for matrix M. Suppose now that the number n of objects is 2 (call them x and y). Then it is easy to build a tree: It is simply an edge with weight M_{xy} joining nodes x and y. Now suppose we add a third object, z. How should we include it in the tree? It is clear that we should obtain a tree that looks like the one in Figure 6.9. The problem is to discover where exactly between nodes x and y the central node (call it c) should be located; or in other words,

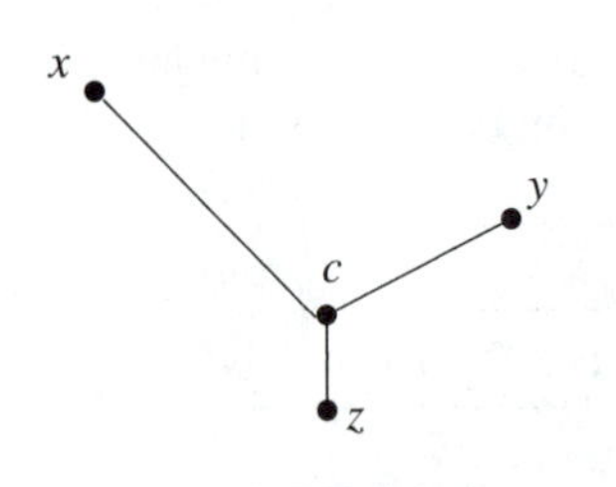

FIGURE 6.9

The additive tree for three objects.

what the lengths of edges (z, c), (x, c), and (y, c) should be. Denoting the length of edge (i, j) by d_{ij}, we can formulate the following equations:

$$M_{xz} = d_{xc} + d_{zc} \tag{6.5}$$

$$M_{yz} = d_{yc} + d_{zc} \tag{6.6}$$

On the other hand, we know that

$$d_{yc} = M_{xy} - d_{xc}. \tag{6.7}$$

If we now subtract Equation (6.6) from (6.5), and replace d_{yc} using (6.7), we get

$$d_{xc} = \frac{M_{xy} + M_{xz} - M_{yz}}{2}. \tag{6.8}$$

Notice that thanks to the triangle inequality (Equation 6.4), $d_{xc} \geq 0$. We can proceed similarly for the other two unknowns and get

$$d_{yc} = \frac{M_{xy} + M_{yz} - M_{xz}}{2} \tag{6.9}$$

$$d_{zc} = \frac{M_{xz} + M_{yz} - M_{xy}}{2}. \tag{6.10}$$

We now have an additive tree corresponding to the three objects. Notice that this tree is unique, just as in the two-object case. Now let us add a fourth object, w. (The reason we are looking at so many base cases of the induction is that we will use them later to show that the additive tree is unique for *any* number of objects.) Object w will create another interior node that we will call c_2. The problem now is to decide whether c_2 should split edge (x, c), (y, c), or (z, c). We choose any two of the nodes already on the tree, say x and y, and apply Equations (6.8), (6.9), and (6.10) with z replaced by w. If we find that node c_2 should either split (x, c) or (y, c), we are done. But if we find that c_2 coincides with node c, this means that edge (z, c) is the one that should be split, and we have to apply the same equations again, this time using z and replacing x (or y) by w. What if there are two additive trees for four objects? In this case, we have the choice of placing c_2 in two different edges, for example (x, c) and (z, c). But this in turn means that there are two different trees for the three objects x, z, and w, which contradicts our earlier result that for any three objects there is always only one tree. So we conclude that for four objects there is again only one tree.

We now assume we have built a tree for k objects. The process of adding object $k+1$ to the tree is similar to what we did for the fourth object. We select any two objects x

and y that are already on the tree and compute the position of the new interior node. If this new node does not coincide with any existing node, we are done. That is, we know exactly the edge that must be split in order to attach object $k+1$ to the tree. Otherwise, the split position falls on an existing node, say u. Because u is an interior node, we know that there is a subtree hanging from it (it could be a subtree with just one leaf). We choose now any object belonging to this subtree, say r, and apply Equations (6.8), (6.9), and (6.10) on objects x (or y), r, and object $k+1$. We repeat this process until the right split position is found.

This algorithm clearly builds an additive tree if one exists. For every object that we add to the tree, we may have to check all other already placed objects, spending constant time per check. This means that in the worst case the algorithm runs in time $O(n^2)$.

Example 6.1 Let us build an additive tree for the objects shown in Table 6.5. We begin by creating edge (A, B) with 63 length units. We now add object C. By applying Equations (6.8), (6.9), and (6.10) we find that a new node x_1 should be created at a distance of 39 units from A and 24 units from B. We then add new edge (x_1, C), 55 units long. Now let us consider object E. If we apply the equations using A and C as objects already on the tree we find that E should go in the same subtree as B. Therefore we have to apply the equations to E, B, and (say) A. Doing this we find the position of new node x_2, 18 units from x_1 and 6 units from B. Going on in this fashion, we obtain the final tree shown in Figure 6.10.

TABLE 6.5

Input matrix M for Example 6.1.

	B	C	D	E	F	G
A	63	94	111	67	23	107
B		79	96	16	58	92
C			47	83	89	43
D				100	106	20
E					62	96
F						102

We now close the uniqueness argument, showing that for $n > 4$ both the topology and the edge lengths of the additive tree are unique. We first look at the topology. Any three leaves determine one unique interior node, and this node in turn determines a unique partition $\{P_1, P_2, P_3\}$ of all leaves. If there are two distinct topologies, then there must be a set of three leaves, say x, y, and z, that determine partition $\{P_1, P_2, P_3\}$ in the first tree and partition $\{Q_1, Q_2, Q_3\}$ in the second tree, such that these two partitions are different. Then it must be the case that one of the leaves, say x, belongs to P_1 and to Q_2, but $P_1 \neq Q_2$. This in turn means that there must be some other object, say w, that belongs to P_1 but does not belong to Q_2. But this is equivalent to saying that there are two different trees for the four objects x, y, z, and w, which we know cannot be true.

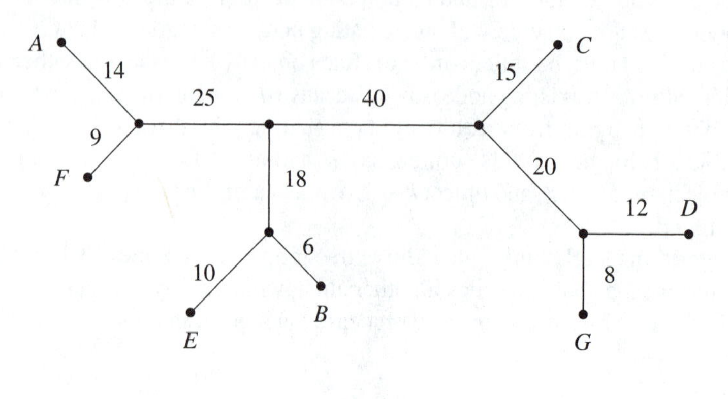

FIGURE 6.10

Additive tree for Example 6.1 with edge lengths indicated.

Now we prove that edge lengths are the same. Any edges leading to a leaf x must have unique length, because the tree determined by x and any two other objects in the tree is unique. Interior edges are also unique, since they partition the leaves into sets P_1, P_2, P_3, and P_4. Taking a leaf from every set we obtain four objects, and again we know that with them we can only build one tree; hence the edge length of the interior edge must be unique.

We have presented a simple and nice algorithm for reconstructing additive trees. However, this result has limited use in practice, because real-life distance matrices are rarely additive. This situation is not necessarily due to errors in the distance measurements. It is possible to construct examples of biological sequences related by a tree where the ancestor sequences undergo multiple changes in the same site (for example, a particular DNA base might change from A to C and then to G). In other words, the record of earlier changes is destroyed by later changes. In addition, convergent evolution can also happen. Allowing these possibilities may result in pairwise distances that are not additive. Such a scenario is particularly probable in the case of protein sequences.

These facts lead us to generalize the additive tree problem. One possibility is the following: Obtain a tree that is as close as possible to an additive tree. Unfortunately it has been shown that under several notions of "closeness" the resulting problems are NP-hard. But there is another way of approaching the problem that *is* tractable; that is the topic of the next section.

⋆ 6.5.2 RECONSTRUCTING ULTRAMETRIC TREES

Pairwise distance measures sometimes carry an uncertainty. Suppose this uncertainty can be quantified so that each measurement is expressed in terms of an interval. This interval defines a lower bound and an upper bound for the true distance. From such data we can obtain two distance matrices, M^l and M^h, containing, respectively, pairwise distance lower bounds and pairwise distance upper bounds. The problem we have now is to reconstruct an evolutionary tree such that the distances measured on the tree fit "between" these two input matrices. In other words, let d_{ij} be the distance measured on the tree between objects i and j. Then the following inequality should hold:

$$M^l_{ij} \leq d_{ij} \leq M^h_{ij}. \tag{6.11}$$

We call Equation (6.11) the *sandwich constraints*. We will show in this section that if we impose one additional requirement on the desired tree this problem can be solved efficiently. This requirement is that the tree be *ultrametric*. A tree is ultrametric when it is additive and can be rooted in such a way that the lengths of all leaf-root paths are equal. The biological meaning of this requirement is that the objects being studied have evolved at equal rate from a common ancestor.

Because the ultrametric property is a special case of the additive property, real-life matrices are also rarely ultrametric. On the other hand, by requiring the tree simply to be "between" matrices M^l and M^h the problem becomes more realistic; and as we will see the algorithm that solves it is quite interesting. It is also the most involved algorithm we present in this chapter. For this reason the presentation will follow this structure: First we present a few definitions and concepts; then we present the algorithm itself, but in a high-level view; next we detail the algorithm, explaining how every part is to be implemented, and giving a running time analysis; finally we prove that the algorithm works.

The first concept we need is the interpretation of a distance matrix as an undirected edge-weighted graph. Given an $n \times n$ distance matrix M, we can interpret it as a complete graph on n vertices, where the weight of edge (i, j) is given by M_{ij}. This interpretation is used extensively below. We assume that M_{ij} is defined for every pair i, j, so that the corresponding graph is always connected. We let $M_{ii} = 0$ for every i. With this interpretation we will talk about graphs G^l and G^h, which correspond, respectively, to matrices M^l and M^h defined above. We refer to edge weights by the function $\mathcal{W}$, as in $\mathcal{W}(e)$ or $\mathcal{W}(a, b)$, where e and (a, b) are edges.

Another concept we will need is that of a minimum spanning tree (MST). This concept is described in Section 2.3. The reader should be careful not to confuse the tree we want to reconstruct (a phylogenetic ultrametric tree) with the minimum spanning tree, which is just a tool to help solve the problem. We distinguish between the two by always referring to the ultrametric tree by U and to the MST by T.

The MST T is defined over graph G^h. We compute T because its edges will direct the way tree U is built. In particular, given any two nodes a and b in T, the largest-weight edge in the unique path from a to b in T is called the *link* of a and b and denoted by $(a, b)_{\max}$. The link is used in the following crucial function definition on edges of T:

DEFINITION 6.5 The *cut-weight* of an edge e of the minimum spanning tree of G^h is given by

$$CW(e) = \max\{M^l_{a,b} | e = (a, b)_{\max}\}.$$

Let us understand the meaning of this definition. Given matrix M^h, we can compute the minimum spanning tree T of the corresponding graph G^h. Each edge $e = (a, b) \in T$ is the link of at least one pair of vertices of T (namely, a and b); it may be the link of more pairs. Now take all pairs of vertices for which edge e is link; and choose the largest entry of *the other matrix* (i.e., M^l) among these pairs. This is the cut-weight of e.

An intuition about the role of cut-weights helps in understanding the algorithm. Consider leaf nodes a and b in U, such that their distance d_{ab} is the largest interleaf distance in U. It is clear that d_{ab} should be greater than the largest entry in M^l, which we denote by $\max^l$. But on the other hand, d_{ab} need not be larger than $\max^l$. Therefore we should have $d_{ab} = \max^l$, and a should belong to one subtree of the root and b to the other. We now use T to determine a partition of the remaining nodes between the two subtrees. The properties of T and the definition of cut-weights ensure that $M^l_{xy} \leq d_{xy} \leq M^h_{xy}$ for all objects x and y in different subtrees, and that the resulting tree is ultrametric. All this will be shown in the proofs of Lemma 6.4 and Theorem 6.6. We are now ready to give a first sketch of the algorithm, as follows:

1. Compute a MST T of G^h;

2. For each edge $e \in T$ compute $CW(e)$;

3. Build ultrametric tree U.

The first step is simple, since we can use well-known algorithms and spend $O(n^2)$ time. The other two steps are more involved, and we examine each in detail.

Cut-Weight Computation

The basic information we need in order to compute cut-weights is $(a, b)_{\max}$ for every pair of vertices a and b. We find these edges by building yet another tree, which we denote by R. This is a rooted binary tree, with the objects in the leaves. In every interior node of R we store an edge of T. Our aim is the following: for every pair of objects a and b, the interior node of R that is the *lowest common ancestor* of a and b contains $(a, b)_{\max}$. For example, at the root of R we store the largest-weight edge of T. The reason is that for every pair of objects a and b, such that they are on opposite sides of T with respect to the largest-weight edge, precisely this edge is $(a, b)_{\max}$. We could use this observation to design an algorithm that works in a top-down fashion: First we create the root of R and store in it the largest-weight edge e; then we do the same thing recursively first considering nodes on one side of e (whose root will be the left child of the overall root), and then considering the nodes on the other side of e (whose root will be the right child of the overall root). However, it is more efficient to build R from the bottom up, as is done in the algorithm shown in Figure 6.11. This algorithms uses a *disjoint-set forest* data structure (see Section 2.3), represented by routines *MakeSet, FindSet,* and *Union.* The running time analysis is as follows. Sorting the edges takes $O(n \log n)$ time. Building R takes $O(n\alpha(n, n))$ time (because of the disjoint-set forest data structure). Hence total time is $O(n \log n)$.

With R it becomes easy to compute cut-weights. The basic operation that we have to apply on it is the lowest common ancestor operation for any two leaves. This is a classic computer science problem, and algorithms exist that can preprocess a tree with n leaves in time $O(n)$ so that future lowest common ancestor queries take constant time. The resulting pseudo-code is shown in Figure 6.12. The time spent on computing cut-weights can now be summed up: it is $O(n^2)$, since we process every pair of objects, spending constant time per pair.

```
Algorithm Construction of R
    input: minimum spanning tree T of G^h
    output: tree R
    for each object i do
        MakeSet(i)
        Create one-node tree for i
    Sort edges of T in nondecreasing order of weights
    for each edge e = (a, b) ∈ T in this order do
        A ← FindSet(a)
        B ← FindSet(b)
        if A ≠ B then
            // join trees
            r_a ← tree that contains a
            r_b ← tree that contains b
            create tree R
            R.edge ← e
            R.leftChild ← r_a
            R.rightChild ← r_b
            Union(A, B)
    return R
```

FIGURE 6.11

Construction of tree R using disjoint-set operations.

```
Algorithm Cut-weight Computation
    input: matrix M^l, trees T and R
    output: cut-weight array CW
    Pre-process R for lowest common ancestor (lca) queries
    for each edge e ∈ T do
        CW[e] ← 0
    for each pair of objects a and b do
        e ← lca(R, a, b)
        if M^l_ab > CW[e] then
            CW[e] ← M^l_ab
    return CW
```

FIGURE 6.12

Computation of cut-weights.

Building U

We now present the final part of the algorithm, where we build the ultrametric tree U itself. As we will see, the construction process is essentially the same as the one we used to build R, but instead of using edge weights, we use edge cut-weights. We use cut-weights to sort the edges of T, and then, as we build U out of smaller trees, we use the cut-weights to determine the edge weights of U. The algorithm is shown in Figure 6.13.

```
Algorithm Construction of U
    input: trees R and T, array CW
    output: tree U
    for each object i do
        MakeSet(i)
        Create one-node tree for i
        height[i] ← 0  // auxiliary array
    Sort edges of T in nondecreasing order of cut-weights
    for each edge e = (a, b) ∈ T in this order do
        A ← FindSet(a)
        B ← FindSet(b)
        if A ≠ B then
            // join trees
            u_a ← tree that contains a
            u_b ← tree that contains b
            create tree U
            U.leftChild ← u_a
            U.rightChild ← u_b
            height[U] ← CW[e]/2
            W(u_a, U) ← height[U] − height[u_a]
            W(u_b, U) ← height[U] − height[u_b]
            Union(A, B)
    return U
```

FIGURE 6.13

Construction of ultrametric tree U.

At this point the reader may think that the algorithm is rather mysterious. In a moment we will present a proof that it works, and this proof contains some of the intuition behind the algorithm. Before that, however, let us complete the running time analysis. The time spent in building U is the same as the one for building R: $O(n \log n)$. Therefore, the total time of the algorithm is: $O(n^2)$ for the MST computation, plus $O(n^2)$ for the cut-weights computation, and plus $O(n \log n)$ for constructing U. The $O(n^2)$ term dominates the other two and that is the final running time.

Example 6.2 We will work out an example based on input matrices M^l and M^h as shown in Table 6.6. Graph G^h and one of its minimum spanning trees T appear in Figure 6.14. Tree R appears in Figure 6.15. The cut-weights we can derive from R are:

$$CW[B, D] = 1,$$

$$CW[A, C] = 2,$$

$$CW[A, E] = 3,$$

$$CW[A, D] = 4.$$

The resulting ultrametric tree is shown in Figure 6.16.

TABLE 6.6

Input matrices M^l and M^h for Example 6.2

M^l					M^h			
	B	*C*	*D*	*E*	*B*	*C*	*D*	*E*
A	3	2	4	3	7	3	6	5
B		4	1	1		10	4	8
C			3	3			8	5
D				1				7

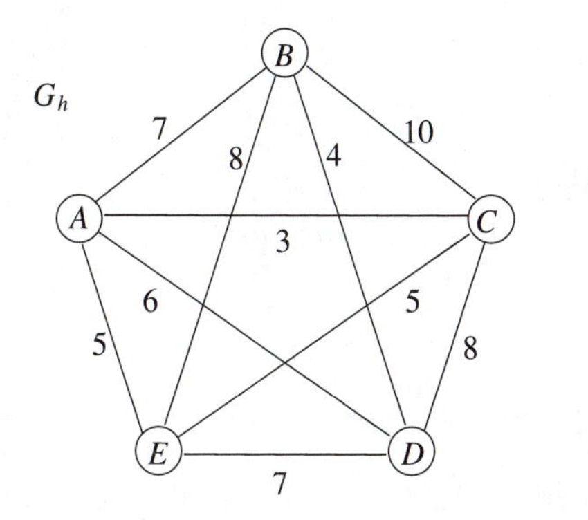

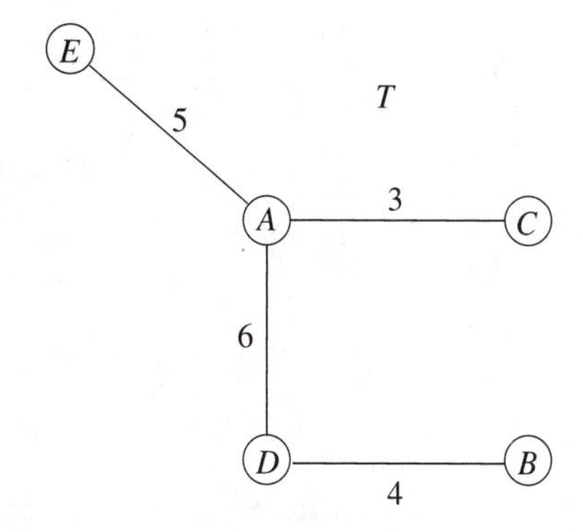

FIGURE 6.14

Graph G^h and MST T for Example 6.2

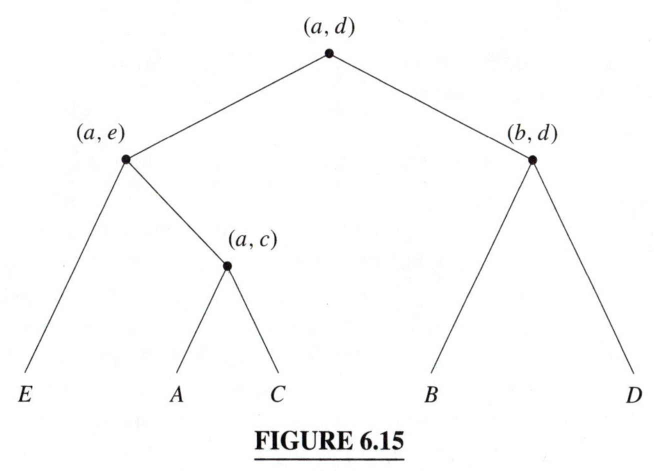

FIGURE 6.15

Tree R for Example 6.2.

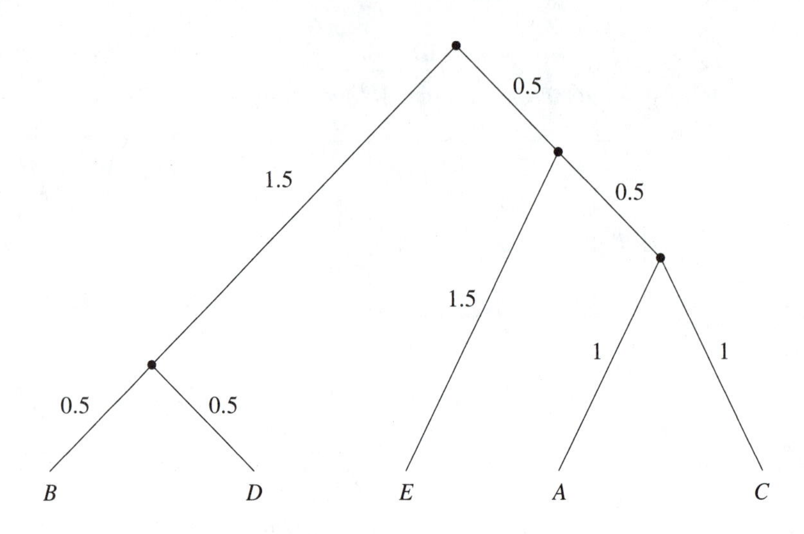

FIGURE 6.16

Ultrametric tree U for Example 6.2. Edge labels are distances between nodes.

A Proof That the Algorithm Works

To prove that the preceding algorithm is correct, we need to know more about ultrametric distances. We have mentioned that in an ultrametric tree the root-leaf distance is the same for all leaves. Although such a notion is easy to grasp, we need an alternative characterization of ultrametric distances, given by the following lemma.

LEMMA 6.3 An additive matrix M is ultrametric if and only if in the corresponding complete weighted graph G, the largest-weight edge in any cycle is not unique.

We leave the proof of this lemma as Exercise 19. We pass on to the first evidence that the algorithm works as claimed.

LEMMA 6.4 The tree U constructed by the algorithm is ultrametric.

Proof. Create graph G from U: It is simply the complete graph having as vertices the n objects and having as edge weights the distances assigned between objects by the algorithm. Now consider a cycle C in this graph, and the largest-weight edge (a, b) on this cycle. The weight of this edge is a cut-weight of some edge $e \in T$ (because that is the way the algorithm works). In the algorithm, when e is considered, one of two things can happen: Either a and b already belong to the same subtree (and hence their respective subtrees were joined before) or they belong to different subtrees. In the first case, because the nodes already belong to the same subtree, their distance in U is already set; hence it must be the case that there is another edge in C of equal weight to e, which was considered before e. This satisfies Lemma 6.3. In the second case, consideration of e led to the joining of a and b for the first time. And because e is a largest-weight edge of C, no other pair of vertices belonging to C was joined *after* e was examined. But in this case some other pair of vertices x and y from C was also joined for the first time. Otherwise a and b would be leaves of U and would remain with their distances to the other vertices of C unset, which we know is not the case. Therefore the weight of (x, y) in C is $CW(e)$, and this again satisfies Lemma 6.3. ■

We still have to prove that the distances obey the sandwich constraints. We will do this through a theorem that in fact lies at the heart of the algorithm.

THEOREM 6.6 An ultrametric tree U that lies between matrices M^l and M^h exists if and only if for every pair of objects a and b it is true that $M^l_{ab} \leq \mathcal{W}((a, b)_{\max})$.

Proof. Suppose U is ultrametric and obeys the sandwich constraints, but there is a pair of objects a, b that does not satisfy the inequality above. Then $M^l_{ab} > \mathcal{W}((a, b)_{\max})$. But by definition $(a, b)_{\max}$ is a largest-weight edge in a path from a to b in MST T. Hence, for (x, y) any edge on the path from a to b in T, we get the following string of inequalities:

$$M^h_{ab} \geq d^U_{ab} \geq M^l_{ab} > \mathcal{W}((a, b)_{\max}) \geq M^h_{xy} \geq d^U_{xy} \geq M^l_{xy}.$$

Now consider the cycle created in T by including edge (a, b). [Edge (a, b) cannot be in T because in that case $(a, b)_{\max} = (a, b)$ and $M^l_{ab} > \mathcal{W}((a, b)_{\max}) = M^h_{ab}$, which clearly cannot happen.] This is a cycle where (a, b) is a unique largest-weight edge (due to the central strict inequality above), which is a contradiction.

The converse of the theorem was already partly proved in Lemma 6.4. It only remains to show that the distances computed by the algorithm obey the sandwich constraints. We will examine first the case of M^l. Consider an arbitrary pair of vertices a, b, and let $e \in T$ be the edge that led a and b to be first joined in U. We have two cases. Suppose $e = (a, b)_{\max}$. Then the distance between a and b in U is $CW(e) \geq M^l_{ab}$, by the definition of cut-weights. If $e \neq (a, b)_{\max}$, then $CW((a, b)_{\max}) \geq CW(e)$, and $(a, b)_{\max}$ is an edge not yet considered by the algorithm. But it is easy to show that when the algorithm joins a pair u, v in U all edges on the unique path from u to v in T (including e) have already been considered, which is a contradiction.

We now prove that the upper bounds are also respected. Consider again an arbitrary pair of vertices a, b, and let $e \in T$ be the edge that led a and b to be first joined in U. Now take the pair p, q such that $(p, q)_{\max} = e$ and such that $CW(e) = M^l_{pq}$. That is, p, q is the pair that was used to determine $CW(e)$. The pair p, q satisfies the condition of the theorem, that is, $M^l_{pq} \leq \mathcal{W}((p, q)_{\max})$. But $\mathcal{W}((p, q)_{\max}) = M^h_e \leq M^h_{ab}$. The last inequality follows because T is a minimum spanning tree and e is the on the unique path from a to b in T. ■

This finishes our discussion of algorithms for distance matrices. We note that the algorithms presented in this section are still mainly of theoretical interest, in part because they make too many assumptions about the input data. Practitioners generally rely on various heuristics for reconstructing trees based on distances, and one example is given in Exercise 14.

6.6 AGREEMENT BETWEEN PHYLOGENIES

In the past sections we have considered several different methods for phylogeny reconstruction; in the specialized literature it is possible to find many more. In practice it occurs quite often that two different methods applied on the same data (or at least on the same set of objects) yield different trees (in the topological sense). Sometimes even the same method on the same data (but arranged differently) yields different trees . In those cases, which tree should a researcher choose? This is a difficult and intensely debated question. We introduce the reader to this area by presenting in this section a simple related problem and its algorithmic solution.

The problem is the following: Suppose we are given two different rooted trees T_1 and T_2 for the same set of n objects, such that every interior node has exactly two children (except the root, which may have only one). We would like to know whether T_1 and T_2 *agree,* where agreement is given by the following definition.

DEFINITION 6.6 We say that a tree T_r *refines* another tree T_s whenever T_r can be transformed into T_s by contracting selected edges from T_r. Two trees T_1 and T_2 *agree* when there exists a tree T_3 that refines both.

The term *tree compatibility* is also used in the literature. We chose the term *agreement* to avoid confusion with the compatibility concept studied in previous sections. Intuitively this notion of agreement means that whatever information is contained in T_1 and not in T_2 may be added to T_1 without causing a contradiction, and vice versa. Examples of trees that agree are shown in Figures 6.17 and 6.18; the tree that refines both is shown in Figure 6.19. Two trees that do not agree are shown in Figure 6.20. These examples are so simple that a mere eye inspection can tell us whether the trees shown agree; of course it is much more of a problem when the trees have dozens or hundreds of leaves.

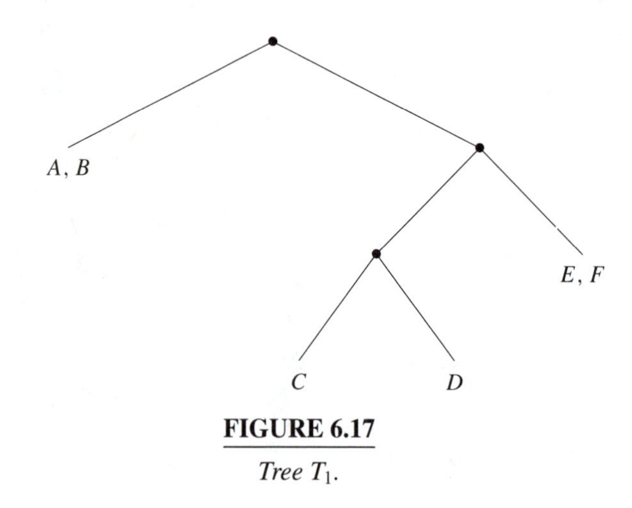

FIGURE 6.17

Tree T_1.

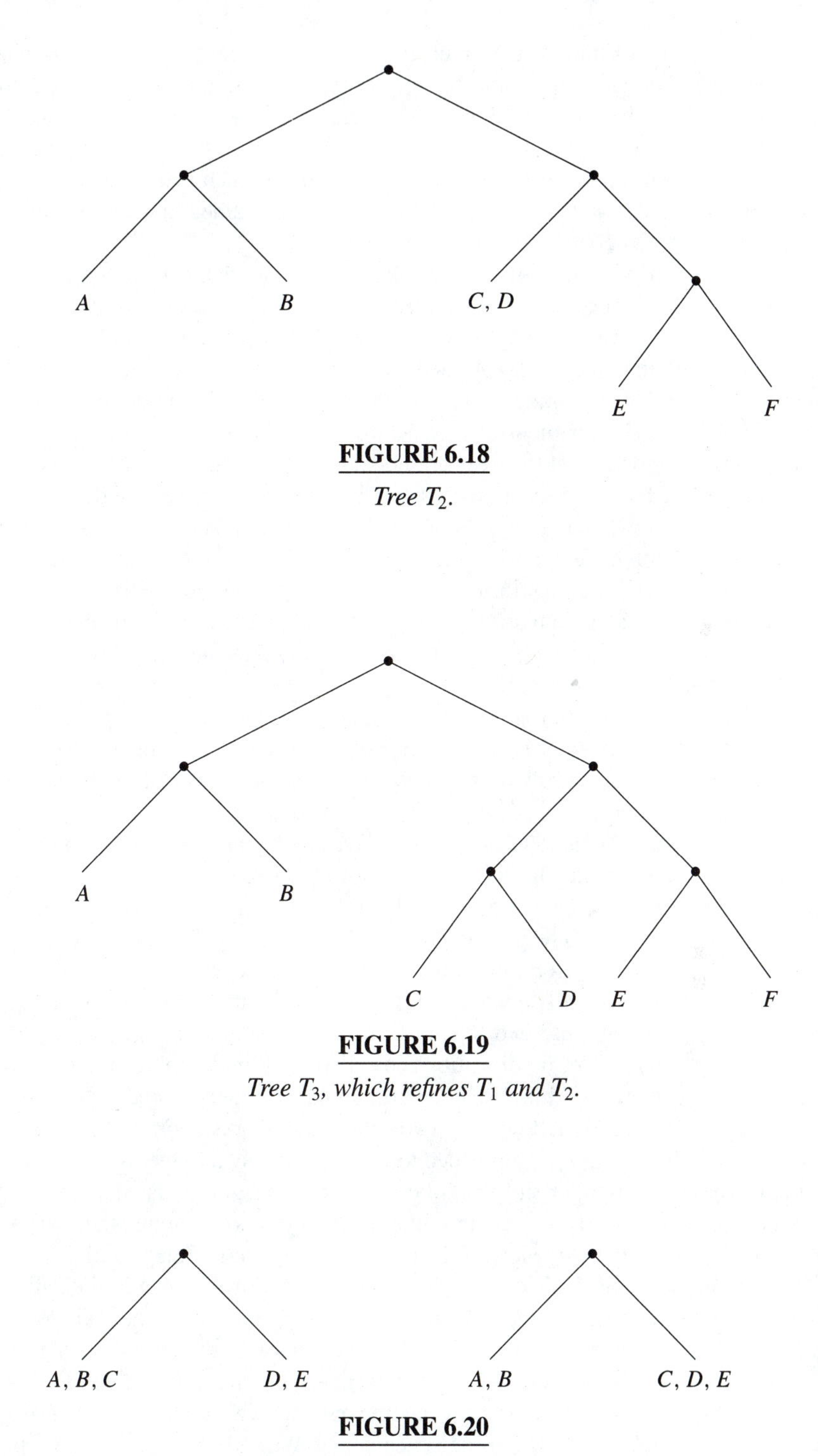

FIGURE 6.18

Tree T_2.

FIGURE 6.19

Tree T_3, which refines T_1 and T_2.

FIGURE 6.20

Two trees that do not agree.

In the foregoing examples, the input trees have leaves containing several objects. This is the general case, but as a first step let us first deal with the case where each leaf contains or is labeled by only one object. In this case what we are really trying to verify is whether trees T_1 and T_2 are *isomorphic*. Two trees T_1 and T_2 are isomorphic when there is an one-to-one correspondence between their nodes such that for every pair u, v of corresponding nodes, $u \in T_1$ and $v \in T_2$, the objects contained in leaves below u are the same as the objects contained in leaves below v.

Although graph isomorphism is a difficult problem, binary tree isomorphism can be solved by a simple algorithm. The basic idea is to compare both trees from the bottom up, removing nodes as we go. Once we have determined that two pairs of sibling leaves (one pair in T_1 and the other in T_2) are isomorphic, we can remove them, and mark the parent of each pair as containing a new, "artificial" object. After removing all leaves, their parents become leaves themselves, and we repeat the process until we have processed all nodes or found that the trees are not isomorphic. The algorithm relies on the following data structures and functions: a set S, from which we can add and remove specific elements (objects), and test for emptiness; a function $object(u)$, which returns the object associated with node u; a function $sib(u)$, which returns the sibling node of node u; a function $parent(u)$, which returns the parent of node u; a marking of nodes that can tell whether a node is a leaf; an array L_1 indexed by objects that gives us the leaf node in tree T_1 where object i appears; and an analogous array L_2. The algorithm is shown in Figure 6.21.

The algorithm can be seen as a careful comparison of the input trees T_1 and T_2 node by node, and hence its correctness is straightforward. The running time is $O(n)$, for all we are doing is traversing trees with $O(n)$ nodes a constant number of times and spending constant time per node.

We now consider the general case where leaves in the input trees may contain several objects. The idea in the algorithm is similar to the isomorphism test above: Carefully compare the two trees node by node. However, instead of removing nodes from the trees, we inspect T_1 with respect to T_2 and possibly modify T_1 using the extra information found on T_2. As an example, consider the leafnode containing objects A and B in tree T_1 from Figure 6.17. A quick look at tree T_2 in Figure 6.18 tells us that A and B should actually be split into two separate nodes, and we modify T_1 accordingly. We try to perform these kinds of modifications considering all nodes of T_1, and then we do the same with T_2. In the algorithm we keep the number of objects stored at each subtree, which enables us to quickly tell whether two subtrees are not isomorphic. If the number is the same in both T_1 and T_2, we proceed to compare the actual objects stored in each. After both trees have been possibly modified we apply an isomorphism test to the (possibly) modified T_1 and the (possibly) modified T_2. This test is essentially the same as the one in Figure 6.21. A detailed view of the algorithm is given in Figure 6.22.

The running time analysis of this algorithm is as follows. In the initialization we spend time $O(n)$ traversing the trees a constant number of times. In the main loop, notice that once nodes are declared inactive, they remain so for the rest of the algorithm, and each active node is examined at most once. Checking that nodes contain the same objects also requires traversing parts of each tree only once. So up to here we have spent time $O(n)$. Finally there is the isomorphism checking. We must modify the isomorphism algorithm presented earlier, because in the input trees leaves may contain more than one object. First, the function $object(u)$ must return all objects associated with leaf u (see

Algorithm *Binary Tree Isomorphism*
  input: Trees T_1 and T_2
  output: TRUE if T_1 and T_2 are isomorphic, FALSE otherwise
  // *Initializations*
  $k \leftarrow n$ // *k counts the number of objects plus "new" objects*
  By appropriate depth-first searches in tree T_1 **do**
    if node u is a leaf **then**
      $L_1[object(u)] \leftarrow u$
      if $sib(u)$ is a leaf **then**
        $add(S, u)$
  By appropriate depth-first searches in tree T_2 **do**
    if node u is a leaf **then**
      $L_2[object(u)] \leftarrow u$
  // *Main loop*
  while $S \neq \emptyset$ **do**
    $u \leftarrow remove(S)$
    $j \leftarrow object(u)$
    $v \leftarrow L_2[j]$
    $y \leftarrow sib(u)$
    $w \leftarrow sib(v)$
    if w is not a leaf **or** $object(w) \neq object(y)$ **then**
      return FALSE
    $p_1 \leftarrow parent(u)$
    $p_2 \leftarrow parent(v)$
    Remove nodes u, v, w, y from respective trees
    Mark p_1 and p_2 as leaves
    $k \leftarrow k + 1$
    $L_1[k] \leftarrow p_1$
    $L_2[k] \leftarrow p_2$
    // p_1 *and* p_2 *now contain "new object"* k
    if $sib(p_1)$ is leaf **then**
      $add(S, p_1)$
  // *end* **while**
  return TRUE

FIGURE 6.21

Isomorphism testing of rooted binary trees.

initialization part in Figure 6.21). Then in the main loop of the algorithm the tests to check whether two nodes are labeled by the same object must be modified to check for set equality. We leave the implementation of these modifications such that total time spent in this checking is $O(n)$ as Exercise 18. The overall running time of the algorithm is thus $O(n)$.

This finishes our brief introduction to the subject of phylogeny agreement. More sophisticated problem formulations can be given, and some references to work in this area are presented in the bibliographic notes.

```
Algorithm Rooted Binary Tree Agreement
    input: Trees T1 and T2
    output: TRUE if T1 and T2 agree, FALSE otherwise
    By an appropriate depth-first search in tree T1 do
        if node u is leaf then
            for every object i stored in leaf u do
                L1[i] ← u
                K1[i] ← number of objects stored in leaf u
        else J1[u] ← number of objects in this subtree
    By an appropriate depth-first search in tree T2 do
        if node u is leaf then
            for every object i stored in leaf u do
                L2[i] ← u
                K2[i] ← number of objects stored in leaf u
        else J2[u] ← number of objects in this subtree
    for each object i do
        active[i] ← TRUE
    // we now examine tree T2 with respect to T1 and possibly modify T1
    for each object i do
        if active[i] and K1[i] > K2[i] then
            // T2 contains more information about object i
            w ← L2[i]
            repeat
                w ← parent(w)   // we go up in the tree
            until J2[w] ≥ K1[i]
            // We have reached the root w of a subtree of T2 that has
            // at least as many objects in it as the leaf of i in T1
            if J2[w] ≠ K1[i] then return FALSE
            Traverse subtree of T2 rooted at w checking that
                every object of L1[i] is in a leaf of this subtree
            if this test fails then return FALSE
            Replace node L1[i] in T1 with subtree of T2 rooted at w
            for each object j stored at L1[i] do
                Update L1[j]
                active[j] ← FALSE
    // End of outermost for
    // here we must repeat what was done above, but this time examining T1
    // with respect to T2 and possibly modifying T2 in the process
    // (similar code omitted)
    Run algorithm Binary Tree Isomorphism on T1 and T2
    return TRUE if trees are isomorphic and FALSE otherwise.
```

FIGURE 6.22

An algorithm to test the agreement of two rooted binary trees.

SUMMARY

Phylogenetic trees are an important tool to help in the understanding of relationships between objects that evolve through time, in particular molecular sequences. Building a tree requires that objects be compared, for which there are two basic criteria. One is the usage of discrete characters, in which we select a set of characters and we list for each object the states that that object has for each character. All such lists taken together form a character state matrix, from which a tree may be constructed. The idea is that objects that share more states should be closer in the tree. Another criterion is distance between objects, where the concept of distance depends on the objects being grouped. Again here the idea is that objects that are close in terms of distance should be close in the tree as well.

If the comparison data is in a character state matrix, and if we impose certain restrictions on the desired tree, reconstruction can be cast as the perfect phylogeny problem. Depending on the characters this problem can be NP-complete. We presented algorithms for some special cases that admit efficient algorithms. One was for binary states and the other was for the case of only two characters. In this last case we saw that graph theory concepts greatly facilitate algorithm development, being also the basis for more general cases.

Real data almost never support a perfect phylogeny. Hence we must relax the restrictions on the desired tree, and that leads us to optimization problems like the most parsimonious tree, or the largest set of compatible characters. We saw that these problems are NP-hard, which means that algorithms for these problems either perform exhaustive searches or try somehow to approximate the optimum value.

When objects are compared through distances, we may require that the observed distance be equal to the distance measured on the reconstructed tree. This is the additive tree problem, and we presented a simple algorithm that solves it. But again here, real data rarely support additive trees. A way around this was the idea of building a tree where distances obey lower and upper bounds. When we impose the additional restriction that the tree be ultrametric — all leaf-root paths must have equal length — it is possible to devise an efficient algorithm, which we described.

Because there are so many methods for building phylogenies, the ability to compare different trees for the same set of objects is desirable. We presented a simple algorithm that determines a tree given two others, incorporating information from both input trees in the output tree. The algorithm produces the agreement tree as long as the two input trees agree in a very precise sense.

EXERCISES

1. Phylogenetic trees have many applications. Two of them are for grouping languages and medieval manuscripts. Try to apply the concepts of ordered and unordered characters and distances to these applications.

2. Show that the number of labeled unrooted binary trees is $\prod_{i=3}^{n}(2i-5)$. (*Hint:* Start by finding how many interior nodes and interior edges such a tree has. Then use induction, by noting what happens when you add one more object to the tree by making it branch out from an existing edge.)

3. Propose a linear representation for a rooted phylogenetic tree. Such a representation can be used, for example, in the output of a phylogeny computer program.

4. The algorithm in Figure 6.5 assumes that all columns are distinct. How would you modify the algorithm to handle columns that are not distinct?

5. The algorithm for binary states (Section 6.5) was presented in two phases. How would you merge them into just one phase?

6. Notice the similarity between Lemma 6.1 and the conditions used in the algorithm for the consecutive ones problem in Section 5.3. Show that binary character matrices that admit a perfect phylogeny have the C1P for columns. Can you design an algorithm that will find the row permutation that makes 1s consecutive, thus taking advantage of the special nature of this matrix?

7. Show that the transformation outlined at the end of Section 6.2 to change unordered binary characters into ordered binary character works, using the fact mentioned there.

8. Suppose we have the following character state matrix:

	c_1	c_2
A	0	1
B	1	1
C	1	0

According to the results for binary characters (Section 6.2) this input does not have a perfect phylogeny, in that c_1 and c_2 are not compatible. But, according to the results for two characters (Section 6.3), the input does have a perfect phylogeny, because the SIG is acyclic. Which interpretation is correct?

9. If a character state-based phylogeny is perfect, we should be able to assign to each of its edges a character state transition. Assign such transitions to the edges of the tree in Figure 6.8.

⋆⋆ 10. Design an algorithm for the perfect phylogeny problem when the number of states in each character is at most three.

⋆⋆ 11. Design an algorithm for the perfect phylogeny problem when the number of characters is at most three.

⋆ 12. Design an algorithm for the perfect phylogeny problem when the number of characters is at most three *and* the number of states in each character is at most three.

13. Suppose we have an algorithm that finds maximum cliques in graphs, and we want to use it to solve the general character state compatibility problem. By general we mean that character states are not necessarily binary. The proof of Theorem 6.5 shows how to do this for the case of binary characters. What should be changed for the case of general characters?

14. A well-known heuristic for building a phylogenetic tree using distance data is as follows. Select the pair of objects closest to each other, say A and B. Make them children of new object O, and set each edge length to $d(A, B)/2$. Now remove A and B from the distance matrix, and add O, setting the distance between O and every other object i to $(d(A, i)+d(B, i))/2$. Find the closest pair again, and repeat the process until all objects are in the tree. (When linking two objects by a new interior node we have to take into account the edge leghths that may already exist below the objects being linked.) Apply this heuristic to the matrix in Table 6.5. Does it find an additive tree?

15. In the presentation of the algorithm for constructing additive trees, we assumed that the input matrix M is additive. Suppose we do not know whether M is additive. What should we do?

⋆⋆ 16. Suppose that in the additive tree problem we know that interior nodes have bounded degree. Show that in this case it is possible to obtain an algorithm running in time $O(n \log n)$.

17. Prove that ultrametric distances can also be characterized in the following way: For any three objects i, j and k, we have $M_{ij} \leq \max\{M_{ik}, M_{jk}\}$.

18. Modify the tree isomorphism algorithm in Figure 6.21 to account for the possibility that leaves may contain more than one object.

19. Prove Lemma 6.3.

BIBLIOGRAPHIC NOTES

As we mentioned at the beginning of this chapter, our coverage of phylogenetic tree algorithms is narrow relative to the vastness of the field. In particular, we have not covered algorithms based on statistical analyses of the problem, commonly known as maximum likelihood methods. We now provide some references for the interested reader.

A good starting point for more details on phylogeny reconstruction for molecular data is the excellent review paper by Swofford and Olsen [182]. Another good, but older reference is Felsenstein [60]. The molecular evolution book by Nei [147] has a useful review in Chapter 11. Penny, Hendy, and Steel [159] present a short survey.

The perfect phylogeny problem for unordered characters mentioned in Section 6.1 was shown to be NP-complete by Bodlaender, Fellows, and Warnow [25], and independently by Steel [179]. The algorithm of Section 6.2 is due to Gusfield [82], and simplified by Waterman [199]. McMorris [132] showed how to convert unordered binary characters to directed binary characters.

As mentioned in the text, whenever the number of characters is fixed, the perfect phylogeny problem can be solved in polynomial time. This discovery is due to Agarwala and Fernández-Baca [4], and the running time of their algorithm is $O(2^{3r}(nm^3 + m^4))$. Algorithms have also been proposed for particular values of r. For $r = 3$ there are two algorithms: by Dress and Steel [52], which runs in $O(nm^2)$ time, and by Kannan and Warnow [107], which runs in $O(n^2 m)$ time. The algorithm by Kannan and Warnow works for the case where $r = 4$ as well. This is an important algorithm because it can be applied to DNA sequences aligned without gaps. In this case we can interpret each position in the alignment as a character that can have one of four states (A, C, G, or T).

Agarwala, Fernández-Baca, and Slutzki [5] and Meidanis and Munuera [135] presented algorithms for binary phylogeny when the input is given as lists of characters for each object rather than as a matrix.

The presentation used in Section 6.3 was based on the paper by Warnow [194]. Theorem 6.1 is due to Buneman [29]; Table 6.3 and Figure 6.7 were adapted from there. McMorris, Warnow, and Wimer [133] described an algorithm for fixed number of characters running in time $O(r^{m+1}m^{m+1} + nm^2)$. Three papers describing algorithms for the case $m = 3$ are Bodlaender and Kloks [26], Kannan and Warnow [106], and Idury and Schäffer [101].

The proof that the parsimony criterion yields an NP-complete problem appears in Day, Johnson, and Sankoff [43]; they used a previous important result relating phylogenies to the Steiner tree problem due to Foulds and Graham [64]. The survey by Swofford and Olsen mentioned above supplies more pointers to the parsimony literature. Swofford and Maddison [181] provided a parsimony algorithm for the assignment of ancestral character states under the another criterion called the Wagner parsimony criterion. The NP-completeness proof for the compatibility criterion presented in Section 6.4 is due to Day and Sankoff [44].

The four-point condition that appears in Section 6.5.1 was proved by Buneman (as cited in [202]) and independently by Dobson [49]. The additive tree reconstruction algorithm presented in Section 6.5.1 is by Waterman, Smith, Singh, and Beyer [202]. Culberson and Rudnicki [40] and Hein [90] present faster algorithms for the case when it is known that interior nodes have bounded degree. The problem of obtaining the tree that best approximates the additive tree under various metrics was shown to be NP-complete by Day [42]. The algorithm for sandwich ultrametric trees is due to Farach, Kannan, and Warnow [56]. In this paper the authors further show that the requirement that the tree be ultrametric is essential, because finding an additive tree that satisfies the sandwich constraints is an NP-complete problem. They also present other complexity results concerning additive and ultrametric trees. Agarwala and others [3] have used the result on ultrametric trees to present an approximation algorithm for the additive tree problem with a performance guarantee. They have further shown that obtaining approximations better than a certain threshold is an NP-hard problem.

The algorithms presented in Section 6.6 are due to Gusfield [82]. Warnow [195] presented a more general algorithm for the case where the trees are not necessarily binary. She also showed how this algorithm could be used in a branch-and-bound algorithm for the unordered character perfect phylogeny problem. A recent result on agreement of trees by Amir and Keselman [14] showed that obtaining a subtree that agrees with k input trees while maximizing the number of objects contained in it is already NP-hard for $k = 3$. They presented a polynomial-time algorithm for the case when one of the trees has bounded degree, and this result was later improved by Farach, Przytycka, and Thorup [57].

When the objects under study are molecular sequences, it is common to start the phylogeny reconstruction using as input a multiple alignment of the sequences. On the other hand, it is very useful to have a phylogeny for the sequences when aligning them. This situation led a number of researchers to propose algorithms that perform both tasks simultaneously. One version of this problem is known as "tree alignment," and a brief description is given in Section 3.4.3. Some pointers for this topic are given in the bibliographic notes of Chapter 3. Additional references that give heuristic solutions to the combined problems of multiple alignment and phylogeny reconstruction are Hein [89, 91, 92] and Vingron and von Haeseler [190].

Two popular phylogeny reconstruction software packages are PHYLIP (phylogeny inference package), by Felsenstein [61], and PAUP (phylogenetic analysis using parsimony), by Swofford (a reference to this package is given in [182]). Both provide many ways of building phylogenies.

CHAPTER

7

GENOME REARRANGEMENTS

In previous chapters we have seen how differences in genes at the sequence level can be used to infer the evolutionary relationship among related species. There are cases, however, when comparisons of corresponding genes in two or more species yield less information than comparisons of larger portions of the genome, or even of the entire genome. The comparison of these larger portions is done at a higher level. Instead of looking at the sequences, we compare the positions of several related genes and try to determine gene rearrangement operations that transform one genome into the other. A rich mathematical theory has been developed to model these situations, and this chapter presents an introductory view of this theory.

7.1 BIOLOGICAL BACKGROUND

When comparing entire genomes across species, we want to compute a form of *distance* between them, but we want to use a set of basic operations different from those seen in Chapter 3. We are no longer interested in point mutations such as substitutions, deletions, and insertions, but rather in larger mutations. Pieces of chromosomes can be exchanged in various ways, affecting sections consisting of a significant percentage of the chromosome. A piece can be moved or copied to another location, or it can leave one chromosome and land in another. Such movements are called **genome rearrangements**.

In some cases, for instance, when studying chloroplast or mitochondrial genomes, a single type of rearrangement event — the *reversal* — appears to be almost the only way to effect evolutionary change. It makes sense in these cases to measure the evolutionary distance between two genome organizations by the number of reversals needed to transform one of them into the other.

As an example, consider the chloroplast genomes of alfafa and pea, two related plant species, depicted in Figure 7.1. There each labeled arrow denotes a *block*. A block here is

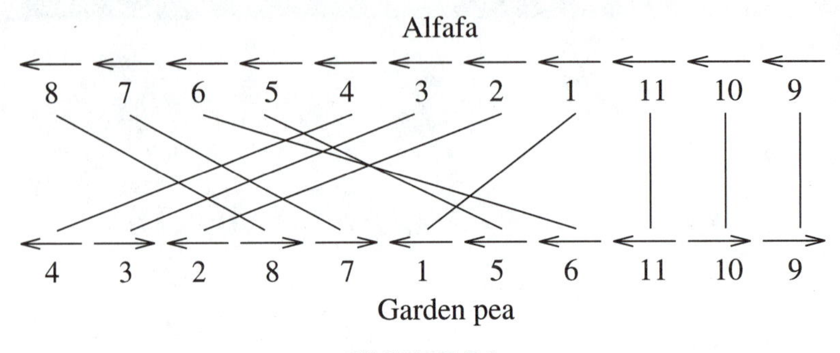

FIGURE 7.1

Relation between the chloroplast genomes of alfafa and garden pea. (Adapted from [150].)

a section of the genome possibly containing more than one gene, which is transcribed as a unit. The arrow denotes the fact that blocks have orientation. The reason that blocks are oriented has to do with the particular strand from where transcription takes place, as was explained in Section 1.4.2. Two blocks in different genomes have the same numerical label if they are *homologous,* that is, if they contain the same genes. Notice finally that the distances between genes are not always the same in both genomes.

The reversal operation for oriented blocks can be defined as follows. It operates in a contiguous segment of blocks. It inverts the order of the affected blocks and also flips their arrows. Given this definition, we are interested in the following question: What is the minimum number of reversals that leads from one genome to the other? One possible solution is given in Figure 7.2. The first line represents the alfafa genome, the last line represents the garden pea genome, and each line except the first is obtained from the previous one by a reversal involving the underlined blocks. But, how do we know that no shorter series of reversals exists to accomplish the required transformation? We would like to have a polynomial-time algorithm to give us such an answer, and that is precisely the topic of the this chapter.

The reason that we want a shortest possible series of reversals is the same *parsimony assumption* given in other sections of this book that deal with minimization problems. We assume that Nature always finds paths that require a minimum of change, and therefore, if we want to know how alfafa could have changed into pea or vice versa, we should try to find a minimum series of reversals that accomplishes this task.

A similar rearrangement problem without the orientation can be conceived. Sometimes the information that we have about genes is not accurate enough to ascertain their relative orientation, but it does give out their relative order. In this case, we would look for a shortest series of reversals that transforms one block order into the other, without regard to orientation. For instance, suppose that we did not know the orientations of the blocks in Figure 7.1. In that case a series of three reversals would be enough to solve the problem, as shown in Figure 7.3.

The next sections explore both versions of the problem, which is known as **sorting by reversals**. It is interesting to note that while there is a polynomial-time algorithm to find a minimum series of reversals in the oriented case, the unoriented version has been shown to be NP-hard.

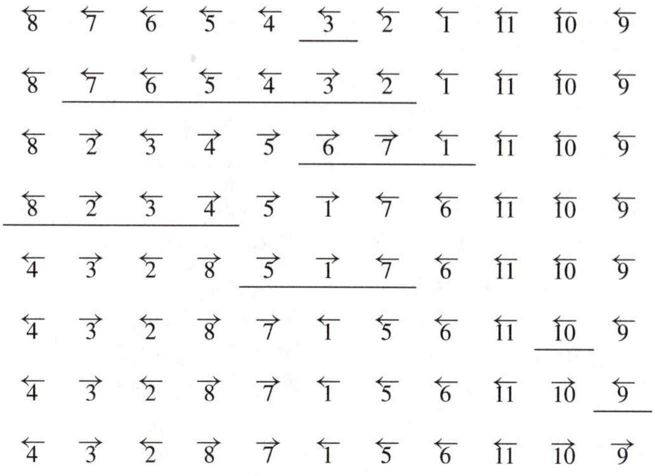

$$\overleftarrow{8}\quad \overleftarrow{7}\quad \overleftarrow{6}\quad \overleftarrow{5}\quad \overleftarrow{4}\quad \underline{\overleftarrow{3}}\quad \overleftarrow{2}\quad \overleftarrow{1}\quad \overleftarrow{11}\quad \overleftarrow{10}\quad \overleftarrow{9}$$

$$\overleftarrow{8}\quad \underline{\overleftarrow{7}\quad \overleftarrow{6}\quad \overleftarrow{5}\quad \overleftarrow{4}\quad \overrightarrow{3}\quad \overleftarrow{2}}\quad \overleftarrow{1}\quad \overleftarrow{11}\quad \overleftarrow{10}\quad \overleftarrow{9}$$

$$\overleftarrow{8}\quad \overrightarrow{2}\quad \overleftarrow{3}\quad \overrightarrow{4}\quad \overrightarrow{5}\quad \underline{\overrightarrow{6}\quad \overrightarrow{7}\quad \overleftarrow{1}}\quad \overleftarrow{11}\quad \overleftarrow{10}\quad \overleftarrow{9}$$

$$\underline{\overleftarrow{8}\quad \overrightarrow{2}\quad \overleftarrow{3}\quad \overrightarrow{4}}\quad \overrightarrow{5}\quad \overrightarrow{1}\quad \overleftarrow{7}\quad \overleftarrow{6}\quad \overleftarrow{11}\quad \overleftarrow{10}\quad \overleftarrow{9}$$

$$\overleftarrow{4}\quad \overrightarrow{3}\quad \overleftarrow{2}\quad \overrightarrow{8}\quad \underline{\overrightarrow{5}\quad \overrightarrow{1}\quad \overleftarrow{7}}\quad \overleftarrow{6}\quad \overleftarrow{11}\quad \overleftarrow{10}\quad \overleftarrow{9}$$

$$\overleftarrow{4}\quad \overrightarrow{3}\quad \overleftarrow{2}\quad \overrightarrow{8}\quad \overrightarrow{7}\quad \overleftarrow{1}\quad \overleftarrow{5}\quad \overleftarrow{6}\quad \overleftarrow{11}\quad \underline{\overleftarrow{10}}\quad \overleftarrow{9}$$

$$\overleftarrow{4}\quad \overrightarrow{3}\quad \overleftarrow{2}\quad \overrightarrow{8}\quad \overrightarrow{7}\quad \overleftarrow{1}\quad \overleftarrow{5}\quad \overleftarrow{6}\quad \overleftarrow{11}\quad \overrightarrow{10}\quad \underline{\overleftarrow{9}}$$

$$\overleftarrow{4}\quad \overrightarrow{3}\quad \overleftarrow{2}\quad \overrightarrow{8}\quad \overrightarrow{7}\quad \overleftarrow{1}\quad \overleftarrow{5}\quad \overleftarrow{6}\quad \overleftarrow{11}\quad \overrightarrow{10}\quad \overrightarrow{9}$$

FIGURE 7.2

A solution to the problem of Figure 7.1.

$$8\quad 7\quad \underline{6\quad 5\quad 4\quad 3\quad 2\quad 1}\quad 11\quad 10\quad 9$$

$$\underline{8\quad 7\quad 1\quad 2\quad 3\quad 4}\quad 5\quad 6\quad 11\quad 10\quad 9$$

$$4\quad 3\quad 2\quad \underline{1\quad 7\quad 8}\quad 5\quad 6\quad 11\quad 10\quad 9$$

$$4\quad 3\quad 2\quad 8\quad 7\quad 1\quad 5\quad 6\quad 11\quad 10\quad 9$$

FIGURE 7.3

A solution to the same problem, but with unoriented blocks.

7.2 ORIENTED BLOCKS

We start the study of the oriented version of the problem with an example. Suppose we have two genomes divided in homologous blocks as indicated in Figure 7.4. Each of these blocks represents a contiguous segment that is well conserved in both genomes. In this example, a series of three reversals does the job and is the shortest possible as we show below. There are many other series that work, but all of them have three or more reversals.

Figure 7.5 displays one possible solution, with labels on each block showing the action of reversals. The arrows above the labels indicate the orientation of homologous blocks. Because the labels are arbitrary, we use the numerals from 1 to n ($n = 5$ in this

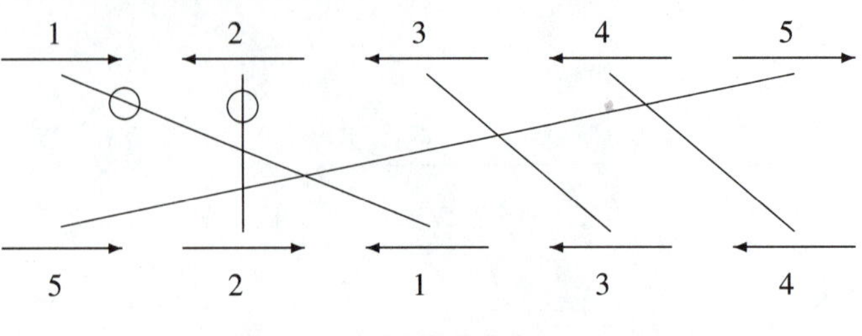

FIGURE 7.4

Two genomes with homologous blocks. Arrows represent the blocks in each genome. Homologous blocks are linked by lines. A small circle around a line indicates that the block has different orientations in the two genomes.

$\overrightarrow{1}\ \overleftarrow{2}\ \underline{\overleftarrow{3}\ \overleftarrow{4}\ \overrightarrow{5}}$

$\overrightarrow{1}\ \overleftarrow{2}\ \overleftarrow{5}\ \underline{\overrightarrow{4}\ \overrightarrow{3}}$

$\underline{\overrightarrow{1}\ \overleftarrow{2}\ \overleftarrow{5}}\ \overleftarrow{3}\ \overleftarrow{4}$

$\overrightarrow{5}\ \overrightarrow{2}\ \overleftarrow{1}\ \overleftarrow{3}\ \overleftarrow{4}$

FIGURE 7.5

A solution to the problem of the previous figure. Each line is obtained from the previous one by a reversal involving the underlined elements.

example) and assign them in the natural order to the upper genome. With this, the labels in the lower genome are uniquely determined. Each block gets the same number as its homologous block, and the arrows reflect the actual orientations. In the figure, reversals are indicated by a line running under the affected blocks. The initial permutation is changed step by step to the final arrangement.

How do we know that no shorter series of reversals exists that accomplishes the required transformation? In the case of this particular setting there is an easy argument that confirms that less than three reversals are not enough. The argument relies on the concept of a *breakpoint*. A breakpoint is a point between consecutive labels in the initial permutation that must necessarily be separated by at least one reversal in order to reach the target permutation. In other words, these two consecutive labels are not consecutive in the target, or, even if they are, the orientations are not the same in a relative sense. In addition, a breakpoint exists before the first label if it is not the first label — with the same orientation — in the target. Figure 7.6 shows the breakpoints for our initial permutation relative to the target in Figure 7.4. Each breakpoint is indicated by a bullet (•). For instance, the location between labels $\overleftarrow{2}$ and $\overleftarrow{3}$ is a breakpoint because in the target blocks $\overleftarrow{2}$ and $\overleftarrow{3}$ are not consecutive. If no reversal separates these labels, they will

$$\bullet\ \overrightarrow{1}\ \overleftarrow{2}\ \bullet\ \overleftarrow{3}\ \overleftarrow{4}\ \bullet\ \overrightarrow{5}\ \bullet$$

FIGURE 7.6

Each • is a breakpoint.

remain together until the end and will be then in the wrong order. In contrast, $\overrightarrow{1}$, $\overleftarrow{2}$ is *not* a breakpoint because this is exactly the way these two blocks appear in the target. Our initial permutation has four breakpoints altogether.

A reversal can remove at most two breakpoints, because it cuts the permutation in exactly two locations. The target permutation has zero breakpoints and it is the only permutation with this property. Hence we need at least two reversals to fix the four breakpoints in our example. But a more careful inspection reveals that no single reversal can remove two breakpoints from our initial permutation. Indeed, to remove the first breakpoint, the one before $\overrightarrow{1}$, we would have to move $\overrightarrow{5}$ to the first position, and even this is not enough. The only reversal that can bring $\overrightarrow{1}$ there also flips its arrow, and the resulting permutation still has a breakpoint before the first position. The label is in its correct position but the orientation is wrong.

A similar argument convinces us that there is no way of removing the last breakpoint with one reversal. Our only hope of removing two breakpoints relies on the two central ones. However, the only reversal affecting these two locations does not remove any breakpoint! So, we know now that the first reversal will decrease the number of breakpoints by at most 1. Then we will have at least three breakpoints left, and another two reversals are needed, for a total of at least three reversals. This shows that the solution presented earlier is optimal.

7.2.1 DEFINITIONS

Genome rearrangements is a topic that requires us to be very careful in our definitions and to employ a coherent notation. Let us first define what an oriented permutation is. Given a finite set $\mathcal{L}$ of *labels*, of oriented labels from $\mathcal{L}$, where

$$\mathcal{L}^o = \bigcup_{a \in \mathcal{L}} \{\overleftarrow{a}, \overrightarrow{a}\}.$$

The set $\mathcal{L}^o$ has twice as many elements as $\mathcal{L}$. For $x \in \mathcal{L}^o$, denote by $|x|$ the element of $\mathcal{L}$ obtained by removing x's arrow. Thus, $|\overleftarrow{a}| = |\overrightarrow{a}| = a$ for any $a \in \mathcal{L}$. Let n be the number of elements of $\mathcal{L}$. An *oriented permutation* over $\mathcal{L}$ is a mapping $\alpha : [1..n] \mapsto \mathcal{L}^o$ such that for any label $a \in \mathcal{L}$ there is exactly one $i \in [1..n]$ with $|\alpha(i)| = a$. Informally, a permutation "picks" for each index i an oriented label $\overleftarrow{a}$ or $\overrightarrow{a}$ from $\mathcal{L}^o$, but if $\overleftarrow{a}$ is picked for some index, $\overrightarrow{a}$ cannot be picked by any other index and vice versa. There being $n = |\mathcal{L}|$ indices guarantees that for each $a \in \mathcal{L}$ either $\overleftarrow{a}$ or $\overrightarrow{a}$ will be picked exactly once.

We represent a permutation α as a list of the elements $\alpha(1), \alpha(2), \ldots, \alpha(n)$. Usually we just list the elements separated by blanks, as in the previous section, but sometimes we use commas and parentheses to avoid ambiguities. In most of cases our set $\mathcal{L}$ will be just the first n positive integers: $1, 2, \ldots, n$. Thus, for instance, for $n = 6$, the following is an example of a permutation over $\mathcal{L} = \{1, 2, \ldots, 6\}$:

$$\alpha = (\overleftarrow{2}, \overleftarrow{3}, \overrightarrow{1}, \overrightarrow{6}, \overleftarrow{5}, \overleftarrow{4}). \tag{7.1}$$

So, in the above example, $\alpha(1) = \overleftarrow{2}$, $\alpha(2) = \overleftarrow{3}$, and so on. The *identity* permutation is the permutation I such that $I(i) = \overrightarrow{i}$ for all i between 1 and n.

A **reversal** is an operation that transforms one permutation into another by reversing the order of a contiguous portion of it, and at the same time flipping the signs of the elements involved. Let i and j be two indices with $1 \leq i, j \leq n$. A reversal affecting elements $\alpha(i)$ through $\alpha(j)$ is indicated by $[i, j]$. The reversal $[i, j]$ is actually a mapping from $\{1, 2, \ldots, n\}$ into itself, and it can be *composed* with an oriented permutation α to give another mapping from $\{1, 2, \ldots, n\}$ to $\mathcal{L}^o$, that is, another permutation. The *composition* of mappings α and ρ, denoted by $\alpha\rho$, is the mapping defined by

$$(\alpha\rho)(k) = \alpha(\rho(k)).$$

For instance, applying the reversal [2, 4] to the permutation α of Equation (7.1) results in

$$\alpha[2, 4] = (\overleftarrow{2}, \overleftarrow{6}, \overleftarrow{1}, \overrightarrow{3}, \overleftarrow{5}, \overleftarrow{4}).$$

In general, a reversal $\rho = [i, j]$ changes a permutation α into $\alpha\rho$ defined by

$$\alpha[i, j](k) = \begin{cases} \overline{\alpha(i + j - k)} & \text{if } i \leq k \leq j \\ \alpha(k) & \text{otherwise,} \end{cases}$$

where $\bar{l}$ is label l with its arrow flipped. There are $n(n + 1)/2$ different reversals in a setting involving n blocks. This includes the unitary reversals $[i, i]$ that just flip the arrow of $\alpha(i)$.

Our main interest in this section is in the problem of *sorting by reversals*. Given two oriented permutations α and β over the same label set $\mathcal{L}$, we seek the minimum number of reversals that will transform α into β. In other words, using the notation just introduced, we seek a series of reversals $\rho_1, \rho_2, \ldots, \rho_t$, with as few reversals as possible, such that

$$\alpha\rho_1\rho_2 \ldots \rho_t = \beta.$$

The number t is called the *reversal distance* of α with respect to β and is denoted by $d_\beta(\alpha)$. Oriented permutation α is called the *initial* permutation and β is the *target*. This process is called *sorting α with respect to β*.

Notice that while $t = d_\beta(\alpha)$ is fixed for a given pair α, β, the actual series of reversals that sort α is generally not unique. Usually we will be interested in finding a reversal ρ that "makes progress" toward β. Such a reversal is called a **sorting reversal** of α with respect to β; it is characterized by the property

$$d_\beta(\alpha\rho) < d_\beta(\alpha),$$

which is equivalent to

$$d_\beta(\alpha\rho) = d_\beta(\alpha) - 1$$

because the distances of α and $\alpha\rho$ cannot differ by more than one.

It is important to observe that the reversal distance is symmetric, that is, $d_\beta(\alpha) = d_\alpha(\beta)$. This stems from the fact that any reversal has no net effect if applied twice. In other words, for every reversal ρ, $\rho\rho$ has no effect. Therefore, if

$$\alpha\rho_1\rho_2\ldots\rho_t = \beta,$$

then, composing both sides of this equation successively with $\rho_t, \rho_{t-1}, \ldots, \rho_1$, we get

$$\alpha = \beta\rho_t\rho_{t-1}\ldots\rho_1.$$

We conclude that the same reversals transform β into α, but they have to be applied in reverse order. Thus, the sorting reversals of β with respect to α are not necessarily the same as the sorting reversals of α with respect to β.

We will abbreviate $d_\beta(\alpha)$ to just $d(\alpha)$ when the target β is clear from the context. Likewise, if this is the case, we will omit β when talking about sorting reversals, breakpoints, and so on.

7.2.2 BREAKPOINTS

Let us now formalize the concept of a breakpoint. Given a permutation α, the first thing we have to do in order to define breakpoints is to consider the *extended* version of α, which is obtained from α by adding a new, artificial label L (for *left*) before the first real label $\alpha(1)$ and also another new label R (for *right*) after the last label $\alpha(n)$. Thus, for the permutation α of Equation (7.1), the extended version is

$$(L, \overleftarrow{2}, \overleftarrow{3}, \overrightarrow{1}, \overrightarrow{6}, \overleftarrow{5}, \overleftarrow{4}, R). \tag{7.2}$$

A *breakpoint* of α with respect to β is a pair x, y of elements of $\mathcal{L}^o$ such that xy appears in the extended version of α but neither xy nor the reverse pair $\overline{y}\,\overline{x}$ appear in extended β. For instance, if α is the permutation in (7.2) and β is the identity, then breakpoints are $L\overleftarrow{2}$, $\overleftarrow{2}\overleftarrow{3}$, $\overleftarrow{3}\overrightarrow{1}$, $\overrightarrow{1}\overrightarrow{6}$, $\overrightarrow{6}\overleftarrow{5}$, and $\overleftarrow{4}R$. The pair $\overleftarrow{5}\overleftarrow{4}$ is *not* a breakpoint, because its reverse pair $\overrightarrow{4}\overrightarrow{5}$ appears in β.

The number of breakpoints of a permutation α is denoted by $b_\beta(\alpha)$ or just by $b(\alpha)$ if β is clear from the context. In our example, $b(\alpha) = 6$.

Notice that artificial labels L and R never move. So, L is involved in a breakpoint only if the first labels of α and β are different, or are equal but with different orientations. An analogous observation holds for R.

A Lower Bound

Let β be fixed. As we mentioned earlier, a reversal can remove at most two breakpoints from a permutation. In our notation,

$$b(\alpha) - b(\alpha\rho) \leq 2, \tag{7.3}$$

for any permutation α and any reversal ρ. This observation leads immediately to a lower bound on the reversal distance as follows. Let $\rho_1, \rho_2, \ldots, \rho_t$ be a series of reversals (not necessarily optimal) that converts α into β. Then

$$\alpha\rho_1\rho_2\ldots\rho_t = \beta,$$

so

$$b(\alpha\rho_1\rho_2\ldots\rho_t) = b(\beta) = 0. \tag{7.4}$$

Notice that β has zero breakpoint with respect to itself. But, by Equation (7.3)

$$\begin{array}{rcll} b(\alpha) & - & b(\alpha\rho_1) & \leq 2, \\ b(\alpha\rho_1) & - & b(\alpha\rho_1\rho_2) & \leq 2, \\ & & & \vdots \\ b(\alpha\rho_1\ldots\rho_{t-1}) & - & b(\alpha\rho_1\ldots\rho_t) & \leq 2. \end{array}$$

Adding these inequalities, and taking Equation (7.4) into account, we get

$$b(\alpha) \leq 2t.$$

This is true for any sorting of α. Now if we apply this result to a shortest series, then $t = d(\alpha)$ and we have

$$\frac{b(\alpha)}{2} \leq d(\alpha).$$

This is a lower bound for $d(\alpha)$, that is, a quantity that specifies a minimum value under which we can be sure $d(\alpha)$ will not lie. If we ever find a series of reversals with the size of a lower bound for a particular permutation α, we know we have found a solution in this case. However, this lower bound is not always achievable. For instance, we saw at the beginning of Section 7.2 a pair of permutations α and β with $b_\beta(\alpha) = 4$, hence the lower bound is 2, but $d_\beta(\alpha) = 3$.

A surprising fact in this context is that there are reversals that remove two breakpoints but are not sorting reversals, that is, they lead us in the wrong way as far as sorting α is concerned. See Exercise 5.

7.2.3 THE DIAGRAM OF REALITY AND DESIRE

The lower bound $d(\alpha) \geq b(\alpha)/2$ is not very tight. For a significant portion of the permutations it is far from being an equality. In this section we derive a better lower bound, based on a structure called the reality–desire diagram of a permutation with respect to another, which will be defined shortly.

To acquire some intuition about this structure, let us try to determine what makes a reversal remove breakpoints from α. A general situation is depicted below, where the reversal is indicated by the two vertical lines:

$$\cdots \quad x \mid y \quad \cdots \quad z \mid w \quad \cdots.$$

After the indicated reversal is performed, we end up with the sequence

$$\cdots \quad x \quad \overline{z} \quad \cdots \quad \overline{y} \quad w \quad \cdots,$$

where, recalling from above, $\overline{l}$ means the label l with flipped arrow. Assuming that the breakpoint xy was removed, we have that either $x\overline{z}$ is in β or $z\overline{x}$ is in β. However, xy was a breakpoint, so xy and $\overline{y}\,\overline{x}$ do not appear in β, although xy appears in α.

When we have a breakpoint, it is caused by two labels that are adjacent in α but that do not "want" to be adjacent. They want to be adjacent to other labels. Thus, we observe a deviation between the *reality*, that is, the neighbor a certain label has in α, and its *desire*, that is, the neighbor this same label has in β. Actually labels are oriented entities, so they have a specific desire for each one of their ends. An oriented label can be viewed as a battery with charged terminals indicated by positive and negative signs, as in Figure 7.7.

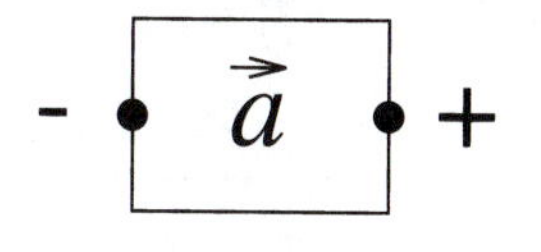

FIGURE 7.7

Oriented label as battery. The positive terminal is always at the tip of the arrow, while the negative terminal is at the tail.

Let us then analyze the particular reality and desire situation in the situation above. We draw each label as a box. In addition, we draw "reality" lines linking label ends that are in contact in the present permutation, and we draw "desire" lines linking label ends that will be in contact in the target permutation. The result for the case under study is shown in Figure 7.8. Notice that the removal of a breakpoint resulted in a desire line linking the same terminals; that is, desire became reality after this reversal. In this particular case we do not know which terminals are positive or negative, but this is not the most relevant issue.

The above considerations motivate us to construct the **reality and desire diagram** of a permutation α with respect to β, denoted $RD_\beta(\alpha)$, as follows. We want to represent each label by two nodes, so that we can keep track of the orientation of the labels. For each label $l \in \mathcal{L}$ we include two nodes $\overleftarrow{l}$ and $\overrightarrow{l}$ in $RD(\alpha)$. The node $\overleftarrow{l}$ represents the tail of the arrow and $\overrightarrow{l}$ is the head. We also include two extra nodes L and R to be

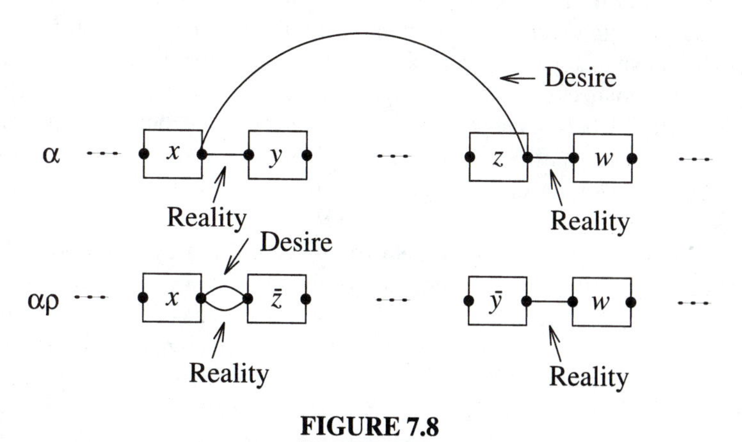

FIGURE 7.8

Reality and desire lines in breakpoint removal.

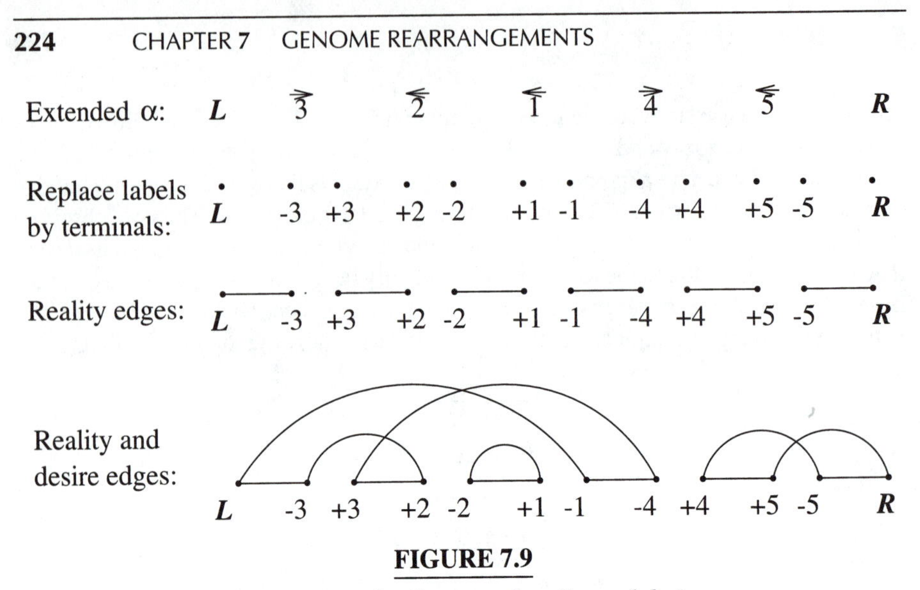

FIGURE 7.9

Construction of a diagram of reality and desire.

able to record the reality and desire of end labels.This is the same for all permutations.

The diagram $RD(\alpha)$ has two types of lines or *edges:* the reality edges and the desire edges. Reality edges connect terminals of consecutive labels in permutation α. Also, L is connected to the first terminal and the last terminal is linked to R. Desire edges connect terminals according to permutation β. Finally we arrange all terminal nodes around a circle, with L and R occupying the top two positions, L to the left of R, and the other nodes following α counterclockwise. Figure 7.9 displays a step-by-step construction of the reality and desire diagram of the permutation $\overrightarrow{3}\,\overleftarrow{2}\,\overleftarrow{1}\,\overrightarrow{4}\,\overleftarrow{5}$ with respect to the identity. Figure 7.10 shows the diagram arranged in a circle. Reality edges go along the circumference. Desire edges are chords. Notice that sometimes there are two parallel edges, one of type reality and one of type desire, between the same pair of vertices. This occurs exactly in the points that are *not* breakpoints. Reality equals desire there. On the other hand, there are no self-loops.

The reality and desire diagram is sometimes called *graph of reality and desire*. We prefer to use the term *diagram* to stress the fact that we draw it in a particular way. A graph can be drawn in any way as long as the vertices are correctly labeled and the edges make the correct connections. But our diagram $RD(\alpha)$ must be drawn in such a way as to follow the order of labels in the permutation α. This will be particularly important when we talk about crossing desire edges in some of the later sections. In spite of that, we will use the standard graph terminology when referring to the diagram, with the obvious meanings.

An important property for us is that every node has degree two in $RD(\alpha)$. This is because both A, the set of reality edges, and B, the set of desire edges, are *matchings* in $RD(\alpha)$. This means that each node is incident to exactly one edge from A and exactly one from B. The union $A \cup B$ forms a number of connected components, each one being a cycle with edges alternating between A and B. Furthermore, each such cycle has an even number of edges, half of them reality edges and half of them desire edges. The number of cycles of $RD(\alpha)$ is denoted by $c_\beta(\alpha)$, or just $c(\alpha)$ if β is implicit. Then note that $c_\beta(\beta) = n + 1$, because β has no breakpoints, so all the cycles are composed by

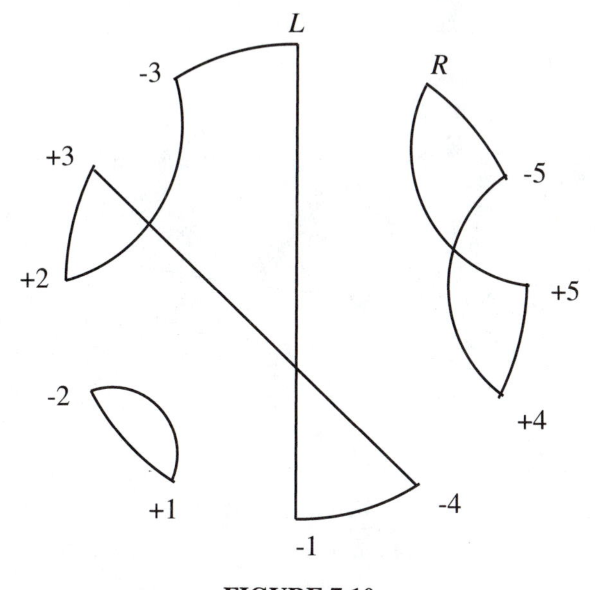

FIGURE 7.10

Diagram of reality and desire.

two parallel edges between the same pair of nodes. Because there are $2n + 2$ nodes, we have $n + 1$ cycles. Furthermore this is the only permutation for which $c_\beta(\alpha) = n + 1$.

So, in a way, we can see the process of sorting a permutation as a process of transforming $RD(\alpha)$ into a graph with as many cycles as possible. That does not necessarily mean that every reversal that increases the number of cycles is good — see Exercise 6. However, a very natural question in this context is, How does a reversal affect the cycles in $RD(\alpha)$? This is what the following result tells us. Observe first that a reversal is characterized by the two points where it "cuts" the current permutation, which correspond each to a reality edge.

THEOREM 7.1 Let (s, t) and (u, v) be two reality edges characterizing a reversal ρ, with (s, t) preceding (u, v) in the permutation α. Then $RD(\alpha\rho)$ differs from $RD(\alpha)$ as follows.

1. Reality edges (s, t) and (u, v) are replaced by (s, u) and (t, v).
2. Desire edges remain unchanged.
3. The section of the circle going from node t to node u, including these extremities, in counterclockwise direction, is reversed.

Proof. The transformation on the reality edges is the same as the one discussed in page 222, and the only difference is that here we are dealing with terminals instead of labels. In any case, no other reality edges are affected. Desire edges are not changed because they depend only on the target permutation. Finally, the reversal of terminals is a direct consequence of the action of a reversal on labels. Figures 7.11 and 7.12 show a diagram before and after a reversal. ■

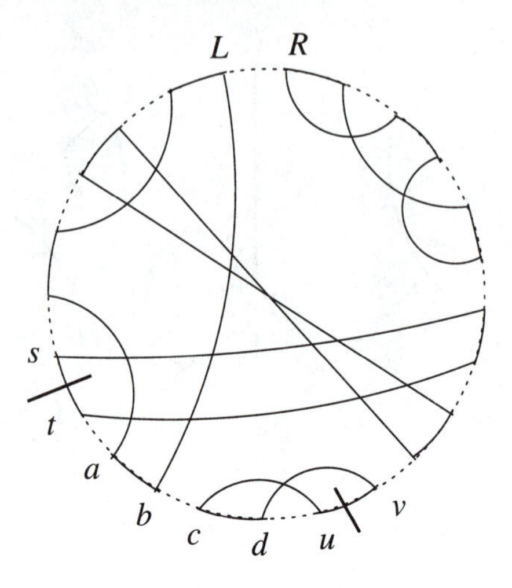

FIGURE 7.11

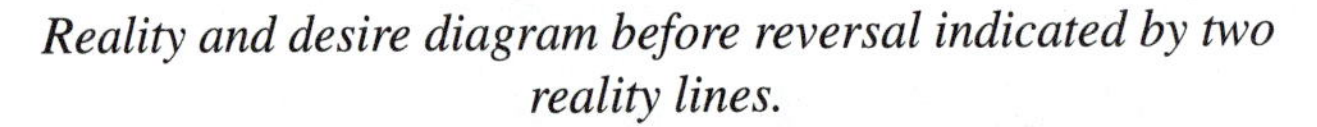
Reality and desire diagram before reversal indicated by two reality lines.

The following result tells us how the number of cycles is affected by a reversal. We need another important definition before stating and proving it. Let e and f be two reality edges belonging to the same cycle in $RD(\alpha)$. Think for a moment of both e and f as being oriented counterclockwise in the diagram. Each one of them induces an orientation of the common cycle. If the orientations induced by e and f coincide, we say that they are *convergent* or that they *converge*; otherwise we say they are *divergent,* or that they *diverge*.

Every time we need to think of reality edges as directed, we write them as an ordered pair. The first element is the tail and the second is the head of the edge. The orientation of reality edges is important to determine good cycles, as we will see below.

THEOREM 7.2 Let ρ be a reversal acting on two reality edges e and f of $RD(\alpha)$. Then

1. If e and f belong to different cycles, $c(\alpha\rho) = c(\alpha) - 1$.

2. If e and f belong to the same cycle and converge, then $c(\alpha\rho) = c(\alpha)$.

3. If e and f belong to the same cycle and diverge, then $c(\alpha\rho) = c(\alpha) + 1$.

Proof. Figure 7.13 will help to clarify the argument. The new edges in $RD(\alpha\rho)$ are indicated by dashed arrows there. Edges e and f are oriented as described above. Notice that we did not draw the cycles as part of the diagram, because here all that we need from $RD(\alpha)$ are the orientations on e and f.

As mentioned earlier, only the cycles containing e and f can be affected, as far as the number of cycles is concerned. The other cycles will not change as connected components. Let us then discuss each case separately.

If e and f belong to different cycles, these cycles are joined into a single one (see Figure 7.13a). Thus, the number of cycles decreases by one. If e and f converge, the new edges are such that the cycle remains connected (see Figure 7.13b). Thus the number

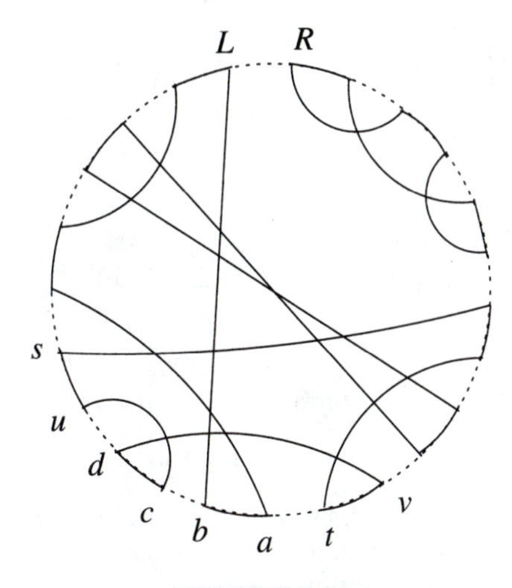

FIGURE 7.12

Reality and desire diagram of permutation obtained from Figure 7.11 after the indicated reversal.

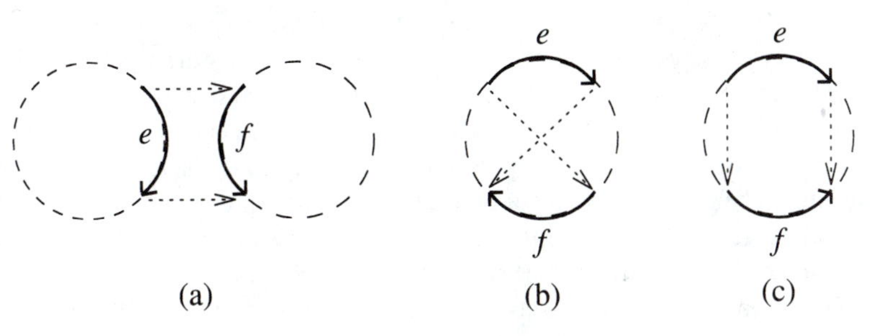

FIGURE 7.13

The effect of a reversal in the cycles of a reality–desire graph: (a) edges from different cycles, (b) convergent edges, and (c) divergent edges.

of cycles remains the same. If e and f diverge, the cycle gets broken into two pieces, causing an increase by one in the number of cycles (see Figure 7.13c). ■

Theorem 7.2 has an important consequence: The number of cycles changes by at most one with each reversal. This leads to another lower bound for the reversal distance.

Suppose that a series of reversals $\rho_1, \rho_2, \ldots, \rho_t$ takes α to β, that is,

$$\alpha\rho_1\rho_2 \ldots \rho_t = \beta.$$

Computing the number of cycles, we get

$$c(\alpha\rho_1\rho_2 \ldots \rho_t) = c(\beta) = n + 1. \tag{7.5}$$

On the other hand, by Theorem 7.2, we have

$$\begin{array}{rcll} c(\alpha\rho_1) & - & c(\alpha) & \leq 1, \\ c(\alpha\rho_1\rho_2) & - & c(\alpha\rho_1) & \leq 1, \\ & & & \vdots \\ c(\alpha\rho_1\ldots\rho_t) & - & c(\alpha\rho_1\ldots\rho_{t-1}) & \leq 1. \end{array}$$

Adding all this up and canceling terms, we get

$$n + 1 - c(\alpha) \leq t.$$

If $\rho_1, \rho_2, \ldots, \rho_t$ is an optimal sorting, then $t = d(\alpha)$ and

$$n + 1 - c(\alpha) \leq d(\alpha). \tag{7.6}$$

This lower bound is very good, unlike the one derived in Section 7.2.2. For most oriented permutations, it comes very close to the actual distance. The reason that it does not always work is explained in the following sections.

7.2.4 INTERLEAVING GRAPH

Based on Theorem 7.2, we can classify the cycles of $RD(\alpha)$ as *good* or *bad* in the following sense. A cycle is good if it has two divergent reality edges; otherwise it is called bad. Good cycles can be spotted in the reality and desire diagram because they have at least two desire edges that cross. We must be careful, however, because not all cycles with crossing desire edges are good. The presence of crossing edges is just an indication that the cycle might be good. When in doubt, we should check directly the definition; that is, we should look for divergent reality edges. The characterization of crossing edges relies strongly on the arrangement of the diagram around a circle. If we were allowed to draw the graph any way we wished, we could make edges cross or not at will. But in our diagrams desire edges must be chords, and although we sometimes draw them a bit curved when they would be too close to the circumference, their crossing properties relative to other chords are conserved. This classification does not apply to cycles with one reality edge. Such cycles do not represent breakpoints and never need to be touched in sorting a permutation. Let us call *proper* cycles the cycles with at least four edges.

It turns out that if we have only good cycles in a permutation α, then the lower bound (7.6) is actually an equality, and we can sort α increasing the number of cycles at the rate of one cycle per reversal. Of course, we must be careful in choosing the reversals. We will deal with this question later.

Even when there are bad cycles, that is, cycles in which all reality edges induce the same orientation, it is sometimes possible to sort at the rate of one cycle per reversal. This happens because a reversal can twist a bad cycle, transforming it into a good one. Thus, while we are breaking one cycle, we may be twisting another one, so that we always have a pair of divergent edges to do the next step, until we reach the target permutation. However, twisting one cycle while breaking another is possible only if the two cycles are such that some desire edge from one of the cycles crosses some desire edge from the other one. We say that two cycles in this situation *interleave*. Notice that we are relying again on the fact that desire edges are chords on a circle.

Therefore, it is important to verify which cycles interleave with which other cycles in RD(α). We construct a new graph called the *interleaving graph* of α with respect to β taking as nodes the proper cycles of RD(α) and making two nodes adjacent if and only if the corresponding cycles interleave. This graph is denoted by $I_\beta(\alpha)$. Apart from the graph structure, we also label each cycle as good or bad. The connected components of $I_\beta(\alpha)$ are classified based on the cycles they contain, as follows. If a component consists entirely of bad cycles, it is called a bad component. Otherwise, it is a good component. Thus, a component with at least one good cycle is good.

Examples of good and bad cycles and components are given next. The diagram in Figure 7.14 shows a permutation with six cycles, each one containing two reality edges and two desire edges. Cycles C and F are good and all the rest are bad. Good cycles are represented in gray. The same figure also shows the interleaving graph for this permutation. There are two good components and one bad component in this case. The only bad component is the one containing the bad cycles A and E.

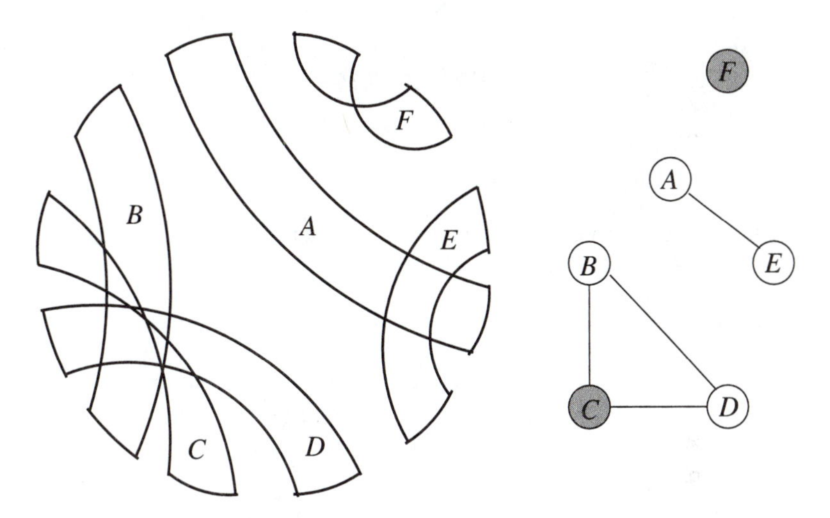

FIGURE 7.14

Reality and desire diagram of a permutation. Cycles are labeled with the letters A through F. The interleaving graph of this permutation appears on the right, with good cycles in gray.

Figure 7.15 shows a diagram with two interleaving bad cycles. There are no good cycles, and this means that no reversal will increase the number of cycles in this case. It follows that the reversal distance here is strictly greater than the lower bound $n+1-c(\alpha)$. Actually, this argument applies to any diagram that contains a bad component. Because reversals that act on a given cycle A can only twist cycles that interleave with A, a bad component will remain bad as long as reversals act on cycles of other components. The only way to sort a bad component is by using a reversal that does not increase the number of cycles. Thus the lower bound $n+1-c(\alpha)$ will not be tight for permutations with bad components.

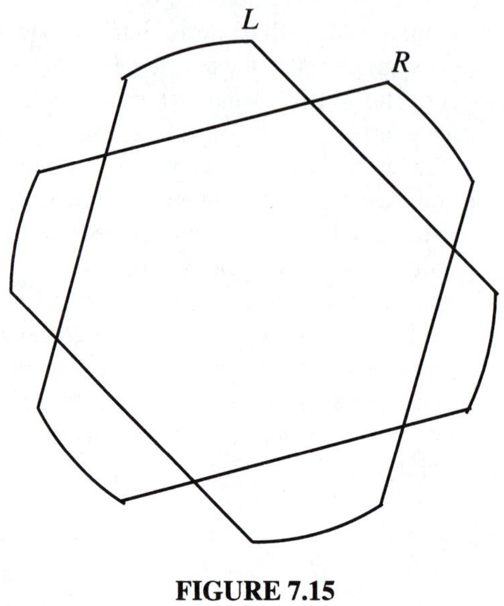

FIGURE 7.15

A diagram with no good cycles.

Another example is presented in Figure 7.16. There we have two proper cycles, B and C, with B a bad cycle and C a good one.

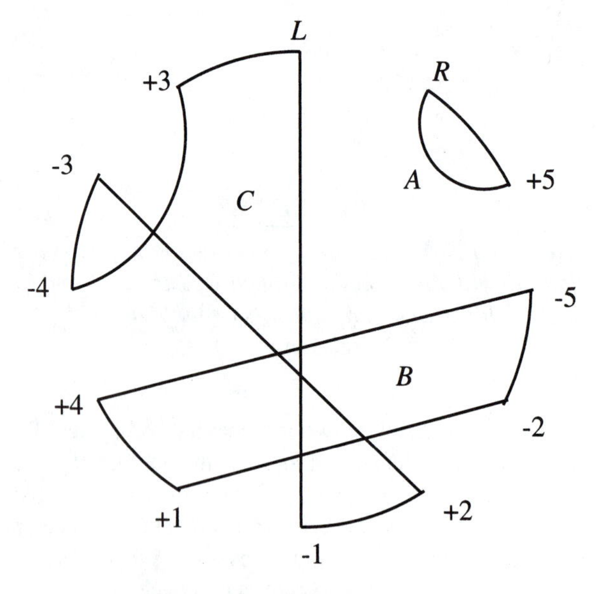

FIGURE 7.16

A diagram with three cycles. Notice that only proper cycles take part in the interleaving graph.

We now use this last example to illustrate the sorting of a good component. Recall that we need to choose two divergent edges in the same cycle to define a reversal that increases the number of cycles. In Figure 7.16 the only good cycle has three reality edges: $e = (L, +3)$, $f = (-3, -4)$, and $g = (-1, +2)$. Edges f and g converge, so this pair is not a good choice. Edges e and g diverge, and the reversal defined by them produces two good components, each containing a single cycle. Edges e and f diverge as well, and produce a single good component with two cycles. In both cases, we do not create bad components. Thus, there are two sorting reversals in this example.

We end this section with the following result, which summarizes our conclusions so far. The proof is simply a repetition of the arguments we used in the illustrative examples.

THEOREM 7.3 A reversal characterized by two divergent edges of the same cycle is a sorting reversal if and only if its application does not lead to the creation of bad components.

This theorem characterizes sorting reversals of a particular kind, but it does not tell us anything about their existence. In fact, it is true that as long as we have a good cycle, there will always be a sorting reversal of the kind that increases the number of cycles, but the proof of this will be omitted.

7.2.5 BAD COMPONENTS

Having sorted all the good components with the methods of Section 7.2.4, we now turn our attention to the bad ones. Our goal in this section is to classify bad components in a hierarchy and, based on that, to provide an exact formula for the reversal distance of a permutation with respect to a target.

Bad components are classified according to Figure 7.17. We first distinguish hurdles from nonhurdles. Among hurdles, we distinguish super hurdles from single hurdles. We define all these concepts below.

We say that a component B *separates* components A and C if all chords in $RD(\alpha)$ that link a terminal in A to a terminal in C cross a desire edge of B. Desire edges can be viewed as chords as well, so the usual definition of chord crossing applies here. Figure 7.18 illustrates this concept. Letters A to F indicate components, all of which are bad. In this example, B separates A and C. Other examples of separation include: E separates F and C, B separates F and D. Exercise 10 asks for all the separations in this diagram.

Separation is an important concept, in that a reversal through reality edges in different components A and C will result in every component B that separates A and C being twisted. A bad component becomes good when twisted, but a good component can remain good or become bad upon twisting. To avoid complications, we resort to twisting only when there are no good components around.

A *hurdle* is a bad component that does not separate any other two bad components. If a bad component separates others, then it is a *nonhurdle*. We denote by $h(\alpha)$ the number of hurdles of α.

We say that a hurdle A protects a nonhurdle B when removal of A would cause B to become a hurdle. In other words, B is protected by A when every time B separates

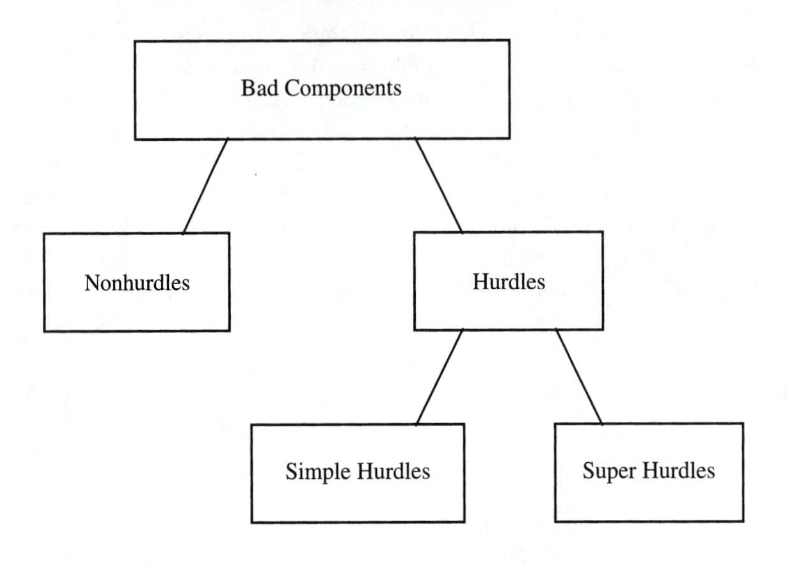

FIGURE 7.17

Classification of bad components.

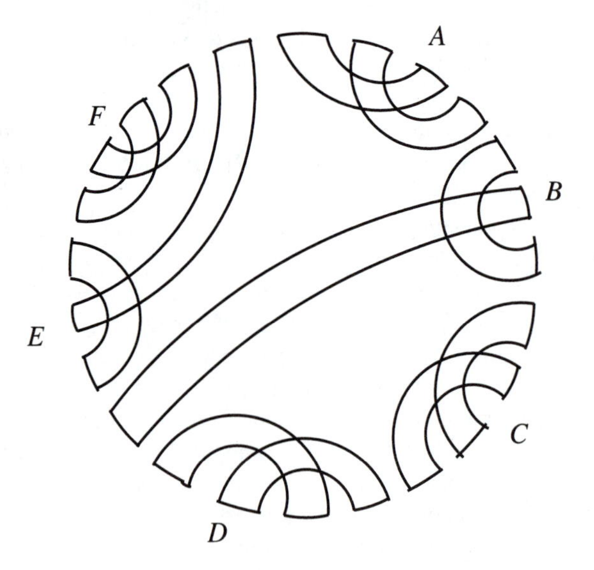

FIGURE 7.18

A diagram with no good cycles. Bad components that separate other bad components.

two bad components, A is one of them. In Figure 7.18, F protects E, and this is the only case of protection in this diagram.

A hurdle A is called a *super hurdle* if it protects some nonhurdle B. Otherwise it is called a *simple hurdle*. Notice that a hurdle can protect at most one nonhurdle. In Figure 7.18 the only super hurdle is F. All other hurdles are simple.

A permutation α is called a *fortress* when its reality and desire diagram contains an odd number of hurdles and all of them are super hurdles. Fortresses are permutations that require one extra reversal to sort, due to their special structure, as we will see below. A fortress must have at least 24 labels, given that we need at least two labels to make a proper cycle, four labels to make a bad component, eight labels to make a super hurdle and its corresponding protected nonhurdle, and at last three such super hurdles and accompanying hurdle pairs to make a fortress. Figure 7.19 shows the diagram of such a smallest possible fortress.

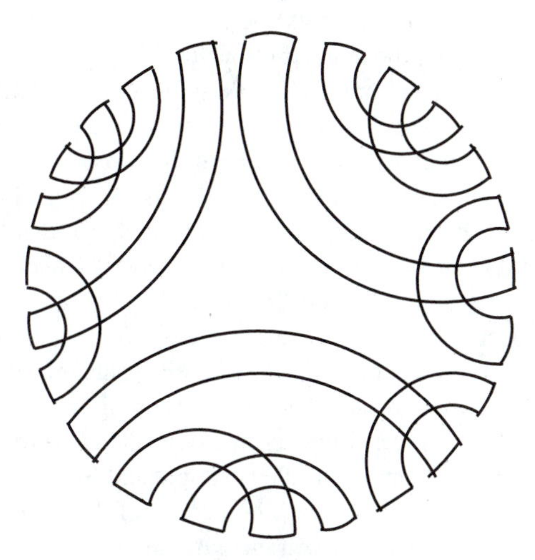

FIGURE 7.19

A smallest possible fortress.

We are now ready to write the exact formula for reversal distance of oriented permutations. We have

$$d(\alpha) = n + 1 - c(\alpha) + h(\alpha) + f(\alpha), \tag{7.7}$$

where $c(\alpha)$ is the number of cycles (including proper and nonproper), $h(\alpha)$ is the number of hurdles, and $f(\alpha)$ is 1 or 0 according to α being a fortress or not, respectively.

The term $n + 1 - c(\alpha)$ accounts for good components and any components that are initially bad but become good in the process of sorting. The term $h(\alpha)$ is needed because bad components sometimes need an extra reversal to make them good. We do not need to include nonhurdles as these will be twisted. Finally, an extra reversal is needed for fortresses, because a nonhurdle becomes a hurdle at some point in the process.

7.2.6 ALGORITHM

In this section we describe an algorithm for sorting an arbitrary oriented permutation, using the formula for reversal distance presented previously. The algorithm chooses at each step a sorting reversal, thereby guaranteeing that we reach the target permutation with the minimum number of reversals.

We will later analyze in detail the complexity of the algorithm. Let us now verify that it is indeed a correct procedure. Before explaining the algorithm, we define special kinds of reversals that are used in the sorting procedure.

As we saw, when there are no good cycles, we are forced to use either a reversal on two convergent edges or a reversal on edges of different cycles. In the first case, the number of cycles remains constant, whereas in the second case the number of cycles decreases by one.

When choosing a reversal on convergent edges, it is better to do this in a hurdle than in a nonhurdle. The reason is that this operation transforms the bad component into a good one, without modifying the number of cycles. If we get rid of a nonhurdle, we do not modify the number of hurdles or the fortress status of the current permutation, so the reversal distance will remain the same according to formula (7.7). It does not help to do it in a super hurdle either, because then the nonhurdle it protects will become a hurdle, keeping $h(\alpha)$ constant. Thus, every time we have to use a reversal on convergent edges we will do so on a simple hurdle. It does not matter which cycle in the simple hurdle we use — they are all equivalent for this purpose. We call a reversal characterized by reality edges of the same cycle in a hurdle a *hurdle cutting*.

Hurdle cutting does not change $c(\alpha)$ and decreases $h(\alpha)$ when the hurdle is simple, but we have to be careful or $f(\alpha)$ may increase. To eliminate this possibility, we will use hurdle cutting only when $h(\alpha)$ is odd. Then the remaining hurdles are even in number and we can rest assured that the resulting permutation is not a fortress. Therefore, hurdle cutting is a sorting reversal when the hurdle is simple and $h(\alpha)$ is odd.

The other possibility is to use a reversal on edges of different cycles. This will decrease the number of cycles, that is, it will increase the term $n + 1 - c(\alpha)$ in the reversal distance, but we can still make progress if we manage to reduce $h(\alpha)$ by two with this reversal. The only way to do it is by choosing edges in different hurdles. A reversal characterized by reality edges of different hurdles is called a *hurdle merging*. Its effect is that these two hurdles become good components, as well as any nonhurdle that separates them.

There is the danger of a hurdle merging transforming a nonhurdle into a hurdle (see Figure 7.20). This happens when the two hurdles being merged "protect" the nonhurdle in a similar way as a super hurdle protects a nonhurdle. To solve this problem we need an additional definition. We call two hurdles A and B *opposite* when we find the same number of hurdles when walking around the circle counterclockwise from A to B as we find walking from B to A counterclockwise. Notice that opposite hurdles exist only if $h(\alpha)$ is even. Choosing opposite hurdles A and B we are safe, because this guarantees that we are not creating new hurdles when merging A and B, and hence $n + 1 - c(\alpha) + h(\alpha)$ will decrease by one.

We still have to be careful with $f(\alpha)$. If we have even $h(\alpha)$, there is nothing to worry about, for hurdles merging on opposite hurdles will then decrease $h(\alpha)$ by exactly two,

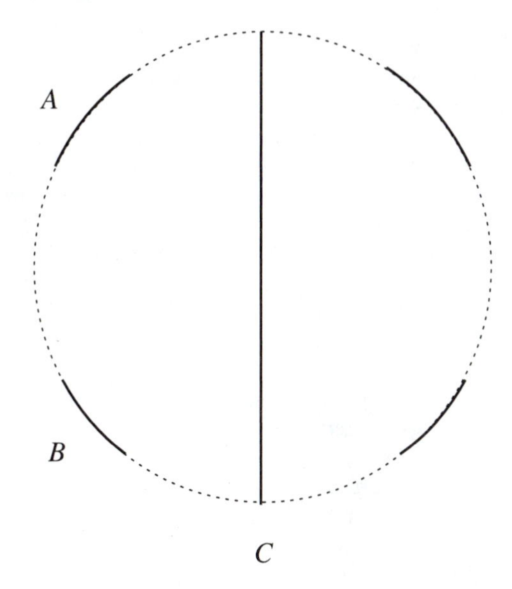

FIGURE 7.20

Merging hurdles A and B will transform nonhurdle C into a hurdle. A sorting reversal must merge either A or B with some hurdle separated from it by C.

keeping it even. If $h(\alpha)$ is odd and there is a simple hurdle, there are no opposite hurdles, but we apply a sorting hurdle cutting and then we have even $h(\alpha)$. If $h(\alpha)$ is odd and there are no simple hurdles, we have a fortress, so $f(\alpha)$ is already equal to 1 and cannot increase. We conclude that hurdles merging is a sorting reversal as long as we pick two opposite hurdles to merge.

We summarize all this in the routine *Sorting Reversal* presented in Figure 7.21. This function finds a sorting reversal given an oriented permutation α. The complete algorithm to sort by reversals consists merely of a loop in which we call the function and apply the returned reversal until we reach the target.

Let us now give a rough estimate of the running time in terms of the permutation size n. We start by analyzing the construction of all structures defined in this section that are relevant for the algorithm. Constructing the diagram $RD_\beta(\alpha)$ takes linear time, since all we have to do is to determine the reality and desire edges. Finding the cycles in this diagram also takes $O(n)$ time. For each cycle we must determine whether it is good or bad, and this takes $O(n)$ per cycle, for a total of $O(n^2)$. Next, we must determine the interleaving relationships between cycles. This can be done by examining each pair of desire edges and by storing the interleaving graph as an adjacency matrix in $O(n^2)$ time. Counting the number of good and bad components, nonhurdles, simple hurdles, and super hurdles can all be done in linear time once we have the information on good and bad cycles (see Exercise 11). Therefore, an overall time bound to construct the needed structures would be $O(n^2)$.

In algorithm *Sorting Reversal* the most time-consuming part is the one involving good components. We need to try several reversals and see whether the structures of the

Algorithm *Sorting Reversal*
 input: distinct permutations α and β
 output: a sorting reversal for α with target β
 if there is a good component in $RD_\beta(\alpha)$ **then**
  pick two divergent edges e, f in this component,
   making sure the corresponding reversal does not
   create any bad components
  return the reversal characterized by e and f
 else
  if $h(\alpha)$ is even **then**
   return merging of two opposite hurdles
  else
   if $h(\alpha)$ is odd and there is a simple hurdle
    return a reversal cutting this hurdle
   else
   // fortress
    return merging of any two hurdles

FIGURE 7.21

An algorithm that returns a sorting reversal for α with target β, provided $\alpha \neq \beta$.

resulting permutation — after the reversal is applied — have some property. Since a reversal is identified by a pair of edges, we have to try at most $O(n^2)$ reversals. For each one, $O(n^2)$ time has to be spent checking the resulting permutation. Hence, the worst case running time estimate for *Sorting Reversal* is $O(n^4)$. We need to call this function $d_\beta(\alpha)$ times in order to sort a permutation α with respect to β, so the final time bound is $O(n^5)$. Faster implementations have been proposed (see the bibliographic notes).

UNORIENTED BLOCKS

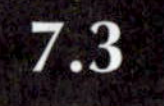

7.3

In this version of the problem the orientation of blocks in the genomes is unknown. All we know is their ordering. A similar theory can be developed for this case, with some results analogous to what we had in the oriented case, but with many differences as well. One very important difference is the fact that the unoriented case results in an NP-hard problem, whereas the oriented case, as we saw, can be solved in polynomial time.

We start by giving some definitions, most of which are analogous to the ones given for the oriented case. An *unoriented permutation* α is a mapping from $\{1, 2, \ldots, n\}$ to a set $\mathcal{L}$ of n labels. Usually $\mathcal{L}$ is the same as the domain of α. A *reversal* is defined as in the case of oriented permutations, except that it does not flip arrows, as there are no arrows. It just reverses the order of a segment of consecutive labels. Notice that here it does not make sense to have a unitary reversal, because it does not change α at all.

The goal is again to obtain a shortest possible series of reversals that transform α into β. If $\rho_1, \rho_2, \ldots, \rho_t$ is a shortest series of reversals such that

$$\alpha \rho_1 \rho_2 \ldots \rho_t = \beta,$$

then t is the *reversal distance* of α with respect to β, denoted by $d_\beta(\alpha)$. Sorting reversals are defined accordingly. The same argument used in the oriented case holds to show that $d_\beta(\alpha) = d_\alpha(\beta)$. If the target is clear from the context, we drop explicit mention to it. The identity in this context is the permutation defined by $I(i) = i$ for all i in $1..n$. The following example should clarify our discussion.

Example 7.1 Figure 7.22 shows two chromosomes with homologies between blocks indicated by lines. We assigned labels 1 through 8 to the blocks in the lower chromosome, and then transferred the labels to the upper one giving equal labels to homologous blocks. Thus we obtain a starting permutation in the upper chromosome and our goal is to sort it into the lower one, the identity. A series of four reversals that sort this permutation is given in Figure 7.23.

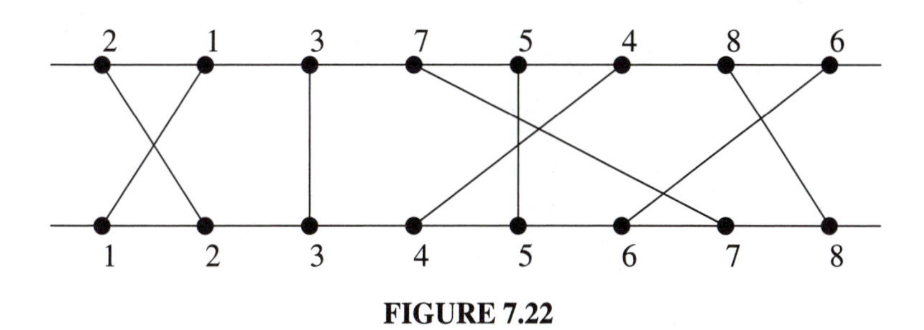

FIGURE 7.22

Two chromosomes with homologous blocks.

$\underline{2\ \ 1}\ \ 3\ \ 7\ \ 5\ \ 4\ \ 8\ \ 6$

$1\ \ 2\ \ 3\ \ \underline{7\ \ 5\ \ 4}\ \ 8\ \ 6$

$1\ \ 2\ \ 3\ \ 4\ \ 5\ \ 7\ \ \underline{8\ \ 6}$

$1\ \ 2\ \ 3\ \ 4\ \ 5\ \ \underline{7\ \ 6}\ \ 8$

$1\ \ 2\ \ 3\ \ 4\ \ 5\ \ 6\ \ 7\ \ 8$

FIGURE 7.23

Sorting permutation 21375486.

Can we obtain a shorter series of reversals? Not in this case. To make a convincing argument that four reversals are necessary, we look at the breakpoints of this permutation, just as we did in the oriented case. For simplicity, we assume that the target is the identity permutation for the rest of this chapter.

A breakpoint of an unoriented permutation α is a pair of labels adjacent in α but not in the target, or, equivalently, adjacent labels that are not consecutive. Breakpoints are to be considered in the extended version of α, which includes extra left (L) and right (R) elements. The expression $b(\alpha)$ denotes the number of breakpoints of α. The permutation in question has seven breakpoints, as shown in Figure 7.24. The only pairs that are not breakpoints are 2, 1 and 5, 4.

$$L \bullet 2 \quad 1 \bullet 3 \bullet 7 \bullet 5 \quad 4 \bullet 8 \bullet 6 \bullet R$$

FIGURE 7.24

Breakpoints of an unoriented permutation indicated by the symbol •.

A reversal can remove at most two breakpoints, just as with oriented permutations. Hence, the same bound we had previously for oriented permutations applies here, which is

$$d(\alpha) \geq \frac{b(\alpha)}{2}. \tag{7.8}$$

With this in mind, it now becomes obvious why one cannot sort the permutation α of Figure 7.24 with less than four reversals. From Equation (7.8) we see that $d(\alpha) \geq 3.5$, but, because $d(\alpha)$ is an integer, this implies $d(\alpha) \geq 4$. So, our solution with four reversals is optimal.

7.3.1 STRIPS

Consider the permutation α displayed below in its extended form, and with breakpoints marked by bullets:

$$L \bullet 4 \quad 5 \bullet 3 \quad 2 \quad 1 \bullet R$$

Observe that between two consecutive breakpoints we have a sequence of consecutive labels, either increasing or decreasing. A bit of thought persuades us that this is actually a general phenomenon. Indeed, if we have two adjacent labels that do not make a breakpoint, they must be of the form

$$\cdots x(x+1) \cdots$$

or

$$\cdots x(x-1) \cdots$$

In the first case, if $x+1$ is not part of a breakpoint, then the label to its right must be $x+2$, because its other consecutive label, x, has been used already. Thus, the increasing sequence extends to the right as far as the next breakpoint or until the end of the permutation is reached. A similar argument can be applied to the left of x and also to both sides of the second case above.

A sequence of consecutive labels surrounded by breakpoints but with no internal breakpoints is called a *strip*. We have two kinds of strips, increasing and decreasing ones. A single label surrounded by breakpoints is itself a strip, and in this case it is both an increasing *and* a decreasing strip. Exceptions to this rule are the labels L and R, which are always considered part of an increasing strip, even if they are all by themselves in a strip. In fact, we identify these extreme elements into a single one for the purpose of defining strips. So, for instance, if 0 is involved in a longer increasing strip $0, 1, \ldots$ and $n+1$ is also part of a longer strip $\ldots, n, n+1$, we consider these two sequences as a single strip. They are linked by the common element $L = R$.

Example 7.2 The following example has five strips: two increasing ones ($L = R$, 1, 2 and 5, 6), one decreasing strip (8, 7), and two strips that are both increasing and decreasing (3 and 4).

$$L \quad 1 \quad 2 \bullet 8 \quad 7 \bullet 3 \bullet 5 \quad 6 \bullet 4 \bullet R$$

Looking at strips can help us remove breakpoints, as the following result shows.

THEOREM 7.4 If label k belongs to a decreasing strip and $k-1$ belongs to an increasing strip, then there is a reversal that removes at least one breakpoint.

Proof. Labels $k-1$ and k cannot belong to the same strip, because the only strips at the same time increasing and decreasing are composed of a single element. Therefore, they belong to different strips. This in turn implies that each one is the last element in its strip, that is, each one is immediately followed by a breakpoint, as in Figure 7.25.

$$\cdots \; (k-1) \bullet \qquad \cdots \quad k \quad \bullet \qquad \cdots$$

$$\cdots \quad k \quad \bullet \qquad \cdots \; (k-1) \bullet \qquad \cdots$$

FIGURE 7.25

Possible relative positions of strips of k and $k-1$.

This figure presents the two possible schemes. The strip of $k-1$ may either follow or precede the strip of k in the permutation. In both cases, the reversal that cuts the two breakpoints immediately following $k-1$ and k brings these two labels side by side, thus reducing the number of breakpoints by at least one. ∎

We know that all permutations have at least one increasing strip, the one containing L (or R). In contrast, a permutation may not have any decreasing strip. If, however, a permutation does have a decreasing strip, then we can use the preceding result and conclude that there is a breakpoint-removing reversal. This happens because when there are both increasing and decreasing strips, they must "meet" somewhere, in the sense of their ordering in the identity permutation. Another way of seeing that is by taking the smallest label belonging to a decreasing strip, and calling it k. Certainly $k - 1$ belongs to an increasing strip, otherwise k would not be the minimum. A similar result holds with $k+1$ substituted for $k - 1$ in the statement, as the following theorem shows.

THEOREM 7.5 If label k belongs to a decreasing strip and $k+1$ belongs to an increasing strip, then there is a reversal that removes at least one breakpoint.

Proof. As in the previous theorem, k and $k+1$ cannot be in the same strip. Thus, each one is the first element of its strip, and is immediately preceded by a breakpoint. The reversal cutting through these two breakpoints reduces the total number of breakpoints, since k and $k+1$ will end up side by side. ■

These two results mean that, as long as we have decreasing strips, we can always reduce the number of breakpoints. Notice that this applies also to single-element strips. However, sometimes we do not have a decreasing strip. The following result provides a compensation when we have to create a permutation with no decreasing strips in the process of sorting.

THEOREM 7.6 Let α be a permutation with a decreasing strip. If all reversals that remove breakpoints from α leave no decreasing strips, then there is a reversal that removes two breakpoints from α.

Proof. Let k be the smallest label involved in a decreasing strip. As we saw in the proof of Theorem 7.4, there is a reversal ρ that unites k and $k-1$. If $k-1$ is to the right of k, ρ leaves a decreasing strip, contradicting our hypothesis. So, $k-1$ must be to the left of k (see Figure 7.26a).

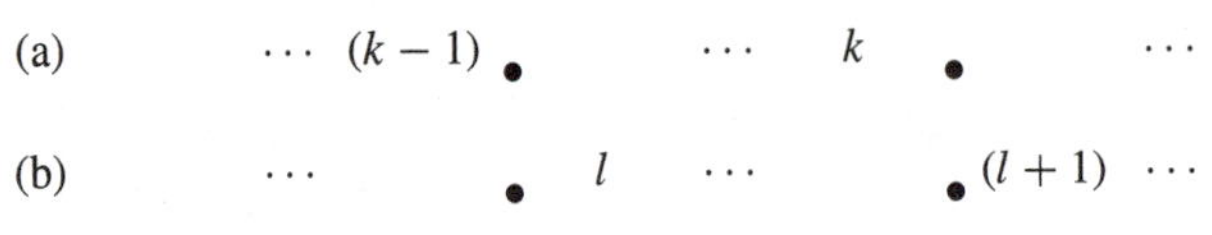

(a) $\cdots\ (k-1) \bullet \quad \cdots \quad k \quad \bullet \quad \cdots$

(b) $\cdots \quad \bullet\ l \quad \cdots \quad \bullet\ (l+1)\ \cdots$

FIGURE 7.26

Permutations without reversals that leave a decreasing strip.

Now let l be the largest label involved in a decreasing strip. From Theorem 7.5 there is a reversal σ that unites l and $l+1$. If $l+1$ is to the left of l, σ leaves a decreasing strip, so $l+1$ must be to the right of l (see Figure 7.26b). Observe that k must be inside the interval reversed by σ, otherwise σ would leave k's decreasing strip intact. Likewise, l must belong to the interval of ρ.

We claim that these two intervals are actually the same, that is, $\rho = \sigma$. Indeed, if this is not true, there is a strip that belongs to the interval of only one of these reversals. Suppose for a moment that this strip belongs to the interval of ρ. If the strip is increasing, ρ would make it into a decreasing strip. If it is decreasing, then it will survive in $\alpha\sigma$, because σ can only augment it. So, neither case is possible. A similar argument can be drawn if there is a strip exclusive to σ. Thus, the claim holds. But in this case this reversal removes two breakpoints, because it joins $k - 1$ with k and l with $l + 1$. ■

7.3.2 ALGORITHM

We are now ready to present an algorithm that sorts a permutation using at most twice as many reversals as the minimum possible, $d(\alpha)$. This is therefore an approximation algorithm. It relies on the theorems we just saw. The idea behind it is simple. There is a main loop, where we look at the current permutation and select the best possible reversal to apply to it, following certain criteria. We then update the current permutation and report the reversal applied to it. The loop stops when the current permutation is the identity.

The criteria we use to select reversals are based on the previous results. If the current permutation has a decreasing strip, we look for a reversal that reduces the number of breakpoints *and* leaves a decreasing strip. Good candidates for this are the reversals that appeared in the proofs of Theorems 7.4, 7.5, and 7.6. If no such reversal exists, we know there is a reversal that encompasses all the decreasing strips and removes two breakpoints. Finally, if the current permutation has no decreasing strips, we select a reversal that cuts two breakpoints. We know that there are at least two breakpoints, otherwise the current permutation would be the identity. No permutation has exactly one breakpoint, so all permutations different from the identity have at least two breakpoints. Furthermore a reversal that cuts through two breakpoints is guaranteed not to increase the total number of breakpoints. The resulting algorithm is shown in Figure 7.27.

We know that this algorithm terminates because the number of breakpoints is reduced at every iteration, except when we encounter permutations with no decreasing strips. But this situation cannot happen twice in a row. So we can certainly say that the number of breakpoints decreases at least one unit every two iterations. In fact, we have a larger average rate of breakpoints removed per iteration, as the following result shows.

THEOREM 7.7 The number of iterations in algorithm *Sorting Unoriented Permutation* (Figure 7.27) is less than or equal to the number of breakpoints in the initial permutation.

Proof. Our goal is to prove that, on average, each iteration removes at least one breakpoint. The only difficulty is caused by the reversals chosen when there are no decreasing strips. But each time we reach this situation, the *previous* iteration just removed two breakpoints, sustaining the average of at least one breakpoint removed per iteration. This argument does not work for the first iteration, if we happen to get an initial permutation with no decreasing strip. But then the very last iteration, that produces the identity, removes two breakpoints and compensates for this special case. ■

```
Algorithm Sorting Unoriented Permutation
    input: permutation α
    output: series of reversals that sort α
    list ← empty
    while α ≠ I do
        if α has a decreasing strip then
            k ← the smallest label in a decreasing strip
            ρ ← the reversal that cuts after k and after k − 1
            if αρ has no decreasing strip then
                l ← the largest label in a decreasing strip
                ρ ← the reversal that cuts before l and before l + 1
        else
            ρ ← the reversal that cuts the first two breakpoints
        α ← αρ
        list ← list + ρ
    return list
```

FIGURE 7.27

An algorithm for sorting unoriented permutation with a guarantee: The number of reversals used is not greater than twice the minimum.

SUMMARY

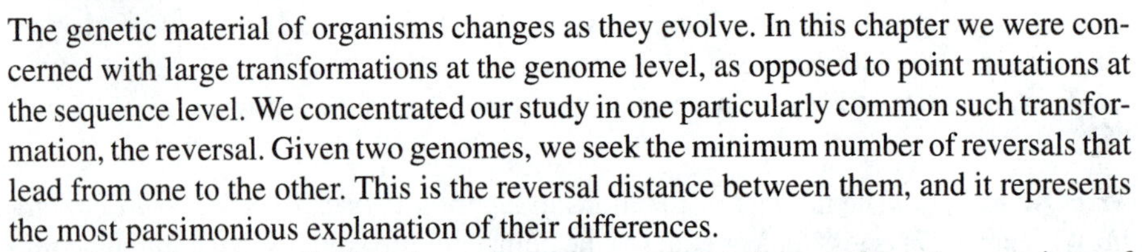

The genetic material of organisms changes as they evolve. In this chapter we were concerned with large transformations at the genome level, as opposed to point mutations at the sequence level. We concentrated our study in one particularly common such transformation, the reversal. Given two genomes, we seek the minimum number of reversals that lead from one to the other. This is the reversal distance between them, and it represents the most parsimonious explanation of their differences.

Depending on available information on gene orientation, we have two versions of this problem: one involving oriented permutations and another for the unoriented case. A remarkable fact is that, while the reversal distance can be computed in polynomial time for the former, the problem is NP-hard for the latter.

We presented an elegant theory that leads to a polynomial-time algorithm for the oriented case. This theory relies on simple concepts such as breakpoints and the reality and desire diagram, as well as on more complex ones such as hurdles and fortresses.

For the unoriented case, we developed a number of tools that provide important information on the actual distance, then presented an approximation algorithm that generates a series of reversals at most twice as long as the optimal one.

EXERCISES

1. Consider the following piece of DNA

$5'$ ... CCTCACTAGCGACCAGACTATAAATTACGTGACCTGT ... $3'$
$3'$ ... GGAGTGATCGCTGGTCTGATATTTAATGCACTGGACA ... $5'$

and the "genes" a = GGTCA, b = AATTAC, c = AGTCTGG, and d = CACTAG. Find out where these genes occur in the DNA piece and write a permutation over $\{a, b, c, d\}$ describing their position and orientation.

2. Find the breakpoints of $\overleftarrow{a}\ \overrightarrow{b}\ \overrightarrow{c}\ \overrightarrow{d}\ \overleftarrow{e}$ with respect to target $\overrightarrow{a}\ \overrightarrow{b}\ \overrightarrow{c}\ \overrightarrow{d}\ \overrightarrow{e}$.

3. We have two permutations α = □□□□□ and β = □□□□□. Someone erased the labels, leaving only the breakpoints of α with respect to β and of β with respect to α. Can you write oriented labels in the empty boxes so that the breakpoints are as assigned?

4. Reconstruct α and β knowing that $RD_\beta(\alpha)$ has the following edges.
Reality: $(L, -3)$, $(-1, -2)$, $(R, +1)$, $(+2, +3)$.
Desire: $(L, -1)$, $(-2, +1)$, $(R, +3)$, $(-3, +2)$.

5. Consider oriented permutations. Show a reversal that removes two breakpoints but is not a sorting reversal.

6. Consider oriented permutations. Show a reversal that increases the number of cycles in $RD(\alpha)$ but is not a sorting reversal.

7. Prove the following statement: An oriented permutation admits a reversal that increases the number of cycles if and only if it has at least one label with a right-to-left arrow. Assume that the target is the identity.

8. Show that every desire edge which does not coincide with a reality edge must cross some other desire edge in a reality and desire diagram.

9. Prove that $b_\beta(\alpha)/2 \le n + 1 - c_\beta(\alpha)$ for any oriented permutation α and target β.

10. Find all triples XYZ of bad component such that Y separates X and Z in the diagram of Figure 7.18.

11. Show how to compute the number of nonhurdles, simple hurdles, and super hurdles in $O(n)$ time given information on good and bad cycles and the interleaving graph.

12. Given an oriented permutation and a target, can you find *all* sorting reversals for them?

13. Prove that any unoriented permutation over n elements can be sorted with at most $n-1$ reversals.

14. Sort by reversals all the 24 unoriented permutations over $\{1, 2, 3, 4\}$. How many need three reversals?

15. Find an unoriented permutation that cannot be sorted optimally solely with reversals that cut breakpoints.

16. A *knot* of an unoriented permutation α over $[1..n]$ is an interval $[i..j]$ such that $i \le j$ and $i \le \alpha(k) \le j$ for every $k \in [i..j]$. Note that $[1..n]$ is a knot of every permutation. Prove that if $[1..i]$ is a knot of α for some $i < n$, then $d(\alpha) \le n - 2$.

17. A *Golan permutation* over $[1..n]$ is a permutation α with $[1..n]$ as its only knot and such that $|\alpha(i) - i| \le 2$ for all $i \in [1..n]$.

a. Show that the inverse of a Golan permutation is a Golan permutation.

b. Show that for every $n \geq 3$ there are only two Golan permutations over $[1..n]$, one being the inverse of the other.

c. Show that $b(\alpha) = n + 1$ if α is a Golan permutation and $n \geq 4$.

⋆⋆ **18.** Show that a Golan permutation α over $[1..n]$ satisfies $d(\alpha) = n - 1$.

⋆⋆ **19.** Show that the Golan permutations are the only permutations over $[1..n]$ with reversal distance equal to $n - 1$.

BIBLIOGRAPHIC NOTES

Most of this chapter is based on the work by Bafna and Pevzner [18], Hannenhalli and Pevzner [87], and Kececioglu and Sankoff [116]. With respect to these papers you will notice a number of differences in our notation, particularly the use of arrows instead of plus and minus signs and the absence of padding in our treatment. We also use the terms *unoriented* and *oriented* instead of their *unsigned* and *signed* to refer to permutations.

Berman and Hannenhalli [23] gave a more efficient implementation for algorithm *Sorting Reversal* of Figure 7.21. Hannenhalli and Pevzner [88] presented a polynomial time solution for the problem of genome rearrangements involving sets of chromosomes. Ferreti, Nadeau, and Sankoff [62] studied the problem of inferring evolution based on information about which genes occur in which chromosomes.

An analysis of genomic evolution is given by Sankoff [167]. Watterson, Ewens, and Hall [205] examined the problem of sorting circular permutations and posed a number of conjectures.

Biology papers in which genome rearrangements appear implicitly or explicitly include Koonin and Dolja's study of RNA viruses [117]; Palmer's [148] and Palmer, Osorio, and Thompson's [150] studies of chloroplast DNA; Fauron and Havlik's comparison of maize mitochondrial genomes [58]; Palmer and Herbon's analysis of several mitochondrial genomes in the *Brassica* family, which includes turnip and cabbage [149]; and Hoffmann, Boore, and Brown's study of a mollusk's mitochondrial DNA [95].

Caprara, Lancia, and Ng [31] propose an algorithm to sort unoriented permutations that achieves good performance in practice. Caprara [30] showed recently that this problem is NP-hard. Gates (of Microsoft fame) and Papadimitriou [69] worked on a related problem in which sorting must be done using prefix reversals only.

CHAPTER

8

MOLECULAR STRUCTURE PREDICTION

Up to now we have studied biomolecules considering only their basic, linear sequence. But as we saw in Chapter 1 a biomolecule such as RNA or a protein is actually a three-dimensional object. Knowing a biomolecule's precise spatial structure is one of the foremost goals of molecular biology, for it is this structure that determines the molecule's function. However, determining the three-dimensional structures of proteins and RNA using techniques such as x-ray crystallography and nuclear magnetic resonance has proved to be difficult, costly, and not always feasible. This presents a contrast to the ease with which it is possible to determine the primary sequence of these molecules (the bases in the case of the RNA and amino acids in the case of proteins). As a result, there currently exists a gap between the number of proteins for which the sequence is known (in the thousands) and the number for which the three-dimensional structure is known also (in the hundreds); this gap has been widening every year. Such a situation has created an intense search for *structure prediction methods:* methods that can predict the three-dimensional structure of a molecule based on its primary sequence. properties of its ranging from molecular dynamics simulation to neural net programs. In this book, however, we are primarily interested in algorithmic techniques. Such techniques apparently are not yet applicable in general to molecule structure prediction, since they have to rely on reasonably accurate models of the molecule folding process; no such general models as yet exist. Nevertheless, for special cases an algorithmic formulation can be given, and its results have been useful in practice for a number of years. Such is the case of RNA structure prediction, which we address in the next section. We then discuss the protein folding problem, the solution of which is a sort of "holy grail" in molecular biology. We describe the problem and present one algorithmic approach that has shown great promise.

RNA SECONDARY STRUCTURE PREDICTION

8.1

We recall from Chapter 1 that an RNA molecule is a single-stranded chain of the nucleotides A, C, G, U. The fact that RNA is single-stranded makes all the difference with respect to DNA: A nucleotide in one part of the molecule can base-pair with a complementary nucleotide in another part, and thus the molecule folds. Furthermore, the nucleotide sequence uniquely determines the way the molecule folds, and this is why we can attempt to predict RNA's structure by analyzing its sequence. However, the real three-dimensional structure is still too complicated for current prediction methods; instead, we concentrate on *secondary structure* prediction, which simply tries to determine which bases pair to which.

We consider an RNA molecule as a string of n characters $R = r_1 r_2 \ldots r_n$ such that $r_i \in \{A, C, G, U\}$. Typically n is in the hundreds, but could also be in the thousands. The secondary structure of the molecule is a collection S of pairs (r_i, r_j) of bases such that $1 \leq i < j \leq n$. If $(r_i, r_j) \in S$, in principle we should require that r_i be a complement to r_j and that $j - i > t$, for t a certain threshold (because it is known that an RNA molecule does not fold too sharply on itself). However, the prediction method to be used is based on the computation of minimum free-energy configurations; thus we can avoid the situations above simply by imposing large free-energies on the corresponding configurations.

There is one natural configuration that we shall exclude: the so-called *knots*. A knot exists when $(r_i, r_j) \in S$ and $(r_k, r_l) \in S$, and $i < k < j < l$. Without knots, S becomes a planar graph. Knots do occur in RNA molecules but their exclusion simplifies the problem. A better reason for excluding knots is that secondary structure predictions are used to infer RNA three-dimensional structures, and knots can be inferred at that stage as well.

Having laid the foundations, we now ask: What kind of algorithm could possibly be proposed to predict the secondary structure of an RNA molecule? Before we answer this question, let us examine the problem more closely. We have mentioned above that we will compute minimum free-energy structures. For this we need some way to attribute energy values to a given structure. We will simply assume that there exists one or more empirically derived functions that accomplish this task for us. Thus the simplest of all algorithms would be this: Enumerate all possible structures, and choose the one with lowest free energy. Unfortunately, the number of possible structures is exponential in the number of bases, and such an algorithm would not be practical for large molecules (but it is worthwhile to note that some researchers have actually developed and used such enumerative algorithms). Therefore we need to look at other, more efficient ways of determining the lowest free-energy structure. We present in what follows two formulations that yield simple dynamic programming algorithms for attaining this objective.

Independent Base Pairs

The first formulation we describe for finding a structure of lowest free-energy makes the assumption that in any RNA structure the energy of a base pair is independent of all the others. This is a gross approximation, but it is useful and will help us to understand the foundations of more sophisticated algorithms. With such an assumption, the total free-energy E of a structure S is given by

$$E(S) = \sum_{(r_i, r_j) \in S} \alpha(r_i, r_j), \tag{8.1}$$

where $\alpha(r_i, r_j)$ gives the free energy of base pair (r_i, r_j). We adopt the convention that $\alpha(r_i, r_j) < 0$, if $i \neq j$, and $\alpha(r_i, r_j) = 0$, if $i = j$.

The assumption that base-pair energies are independent provides the key to an efficient algorithm, since we can use solutions for smaller strings in determining the solutions for larger strings. Suppose for example that we want to determine a secondary structure $S_{i,j}$ of minimum energy for string $R_{i,j} = r_i r_{i+1} \ldots r_j$. Consider now what can happen to base r_j. Either it forms a pair with some base between i and j or it does not. If it does not, then $E(S_{i,j}) = E(S_{i,j-1})$. If r_j is base-paired with r_i, then we can say that $E(S_{i,j}) = \alpha(r_i, r_j) + E(S_{i+1,j-1})$. Finally, it could be that r_j is base-paired with some r_k, for $i < k < j$. In this case we can split the string in two: $R_{i,k-1}$ and $R_{k,j}$, and say that $E(S_{i,j}) = E(S_{i,k-1}) + E(S_{k,j})$, for some k. (We can do this because we have assumed there are no knots.) The problem here is that we do not know which k gives the lowest overall energy, so we should actually have $E(S_{i,j}) = \min\{E(S_{i,k-1}) + E(S_{k,j})\}$, where k should vary between i and j. What this all means is that we can compute the value of $E(S_{i,j})$ based on values for shorter strings. This situation is familiar to us: It means we can use a dynamic programming technique to solve the problem. In particular, we can write the above equations in the following way, which is very similar to the recurrence relations seen in Chapter 3:

$$E(S_{i,j}) = \min \begin{cases} E(S_{i+1,j-1}) + \alpha(r_i, r_j) \\ \min\{E(S_{i,k-1}) + E(S_{k,j})\} \text{ for } i < k \leq j. \end{cases} \tag{8.2}$$

(The case in which j is not base-paired is subsumed by the second case above, because when $k = j$ we have $E(S_{j,j}) = 0$.) The algorithm to obtain $E(S_{1,n})$ is very similar to those seen in Section 3.2 (just fill in values in an $n \times n$ matrix) and will not be repeated here. But notice that, as in Section 3.2, solving the recurrence provides us with a *value* for the lowest free-energy; the actual structure must be found by some postprocessing on the dynamic programming matrix, just as we did to get an alignment from the sequence comparison matrix.

Let us compute the running time of the above algorithm. In the basic sequence comparison algorithm we saw in Section 3.2 we spend constant time per element of the matrix; hence the running time is $O(n^2)$, for an $n \times n$ matrix. Here things are a little different because we have to compute sums using a variable number of previous entries of the matrix for every pair i, j, as indicated in Expression (8.2). The exact overall number of sums is given by

$$\sum_{j=1}^{n} \sum_{i=1}^{j-1} (j - i - 1), \tag{8.3}$$

which is $O(n^3)$, and this is the algorithm's running time.

Structures with Loops

Now let us relax the independence assumption to include the fact that the free energy of a base pair depends on adjacent base pairs. This relaxation leads us to consider RNA structures called *loops*. Suppose (r_i, r_j) is a base pair in S, and consider positions u, v, and w such that $i < u < v < w < j$. We say that base r_v is *accessible* from (r_i, r_j) if (r_u, r_w) is not a base pair in S for any u and w. A loop is the set of all bases accessible from a base pair (r_i, r_j). Figure 8.1 shows various kinds of loops. The assumption that no knots occur implies that any structure S decomposes string R into disjoint loops.

The *size* of a loop is the number of unpaired bases it contains. We generalize the concept of loop to include base pairs as loops of size zero. When several base pairs occur next to each other, we get the so-called "stacked pairs" or "helical regions." Thus the number of loops in a structure is given by the number of base pairs it contains.

An example of the dependence of free energy on the presence or absence of certain structures comes from the fact that helical regions have negative (stabilizing) free energy, whereas loops of size greater than zero have positive (destabilizing) energy. This has the following consequence: When (r_i, r_j) is the base pair that closes a loop, structure $S_{i+1,j-1}$ may not be the minimum free energy for substring $R_{i+1,j-1}$. In other words, it could be that there exists structure $S'_{i+1,j-1}$ that has lower energy than $S_{i+1,j-1}$ but when we add the energy of pair (r_i, r_j) to the energy $S_{i+1,j-1}$, the stabilizing energy of the pair more than makes up for the difference. This means that we have to formulate new recurrence relations in addition to those in (8.2).

These recurrence relations still contain some simplifications with respect to a fully general loop model. These simplifications are necessary to make the dynamic programming algorithm more efficient. It is possible to show that a general model leads to an algorithm that runs in exponential time.

The first simplification is that we will disregard the free energies of bases that do not belong to any loops. A second simplification is that in order to obtain the overall free energy of a structure we can simply add the energy of its components. Finally, we will assume we have to deal only with the kinds of loops depicted in Figure 8.1. Let us see the details of this.

We want to determine $S_{i,j}$ for string $R_{i,j}$. First suppose r_i is not base-paired. Then $E(S_{i,j}) = E(S_{i+1,j})$. Suppose now r_j is not base-paired. Similarly, $E(S_{i,j}) = E(S_{i,j-1})$. Now suppose r_i and r_j are base-paired, but not with each other. Then we can say (as we did for the situation without loops above) that $E(S_{i,j}) = \min\{E(S_{i,k}) + E(S_{k+1,j})\}$, for $i < k < j$, where r_k is any base between the pair of i and the pair of j. The final case happens when (r_i, r_j) is a base pair. This means that there may be one or more loops between i and j. Denote by $L_{i,j}$ the minimum free-energy structure in this case (to be detailed below). Writing up the full recurrence relation we get:

$$E(S_{i,j}) = \min \begin{cases} E(S_{i+1,j}) \\ E(S_{i,j-1}) \\ \min\{E(S_{i,k}) + E(S_{k+1,j})\} \text{ for } i < k < j \\ E(L_{i,j}). \end{cases} \tag{8.4}$$

Now let us detail the case of $L_{i,j}$. We have to account for the various kinds of loops shown in Figure 8.1, and for that we need to express free-energy values. As before, we will assume that experimental procedures have determined what these values are and that

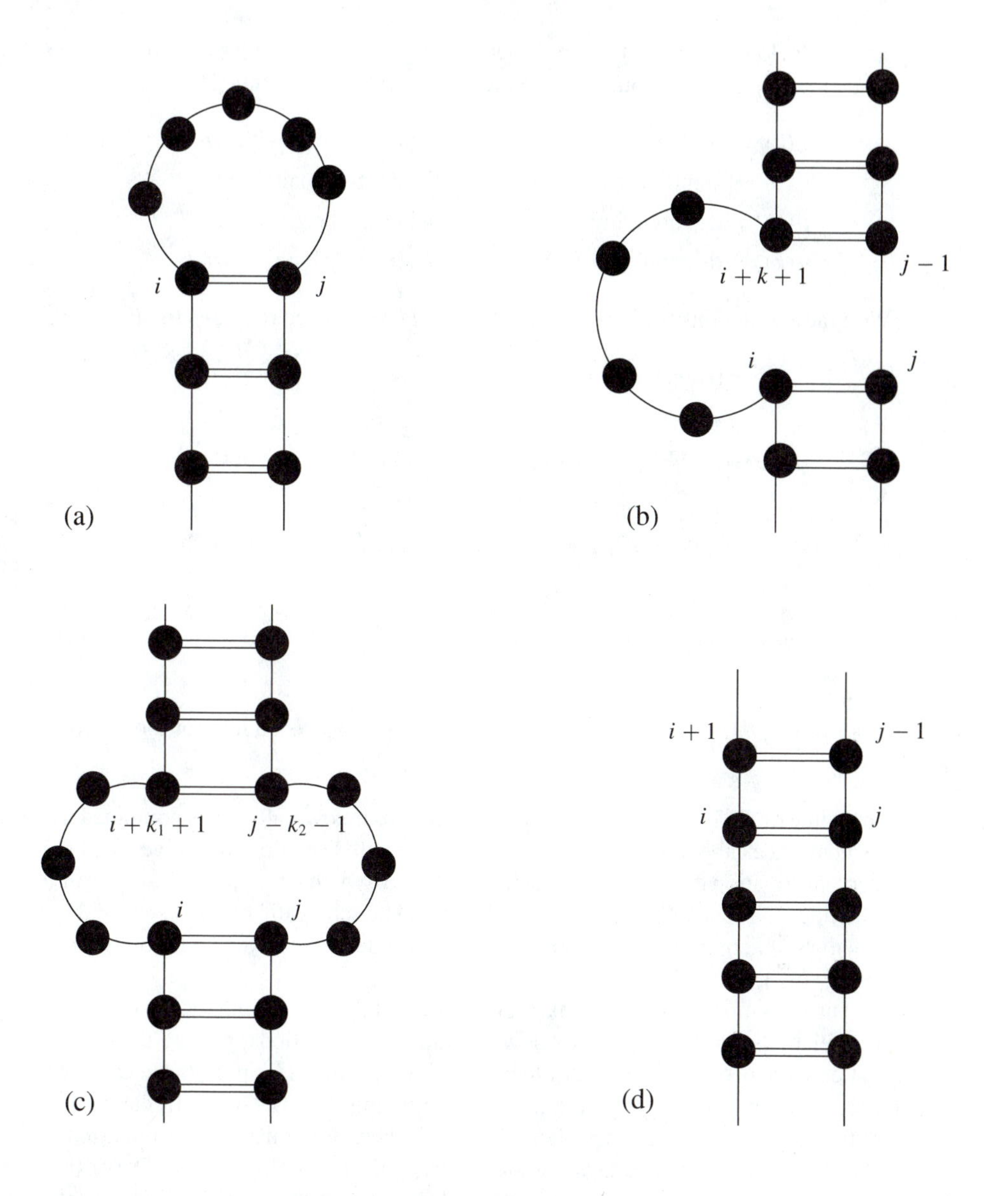

FIGURE 8.1

RNA secondary structures without knots. The bullets are ribonucleotides and the horizontal double lines show the base pairs. (a) hairpin loop; (b) bulge on i; (c) interior loop; (d) helical region. (Adapted from [105].)

they are available to us in the form of functions. Again, $\alpha(r_i, r_j)$ gives the free energy of base pair (r_i, r_j). In addition, we have these other free-energy functions:

$$\begin{aligned}
\xi(k) &= \text{destabilizing free energy of a hairpin loop with size } k \\
\eta &= \text{stabilizing free energy of adjacent base pairs} \\
\beta(k) &= \text{destabilizing free energy of a bulge of size } k \\
\gamma(k) &= \text{destabilizing free energy of an interior loop of size } k
\end{aligned}$$

With these functions we can write the following recurrence relations for $E(L_{i,j})$:

$$\alpha(r_i, r_j) + \xi(j - i - 1), \text{ if } L_{i,j} \text{ is a hairpin loop} \tag{8.5}$$

$$\alpha(r_i, r_j) + \eta + E(S_{i+1,j-1}), \text{ if } L_{i,j} \text{ is a helical region} \tag{8.6}$$

$$\min_{k \geq 1}\{\alpha(r_i, r_j) + \beta(k) + E(S_{i+k+1,j-1})\}, \text{ if } L_{i,j} \text{ is a bulge on } i \tag{8.7}$$

$$\min_{k \geq 1}\{\alpha(r_i, r_j) + \beta(k) + E(S_{i+1,j-k-1})\}, \text{ if } L_{i,j} \text{ is a bulge on } j \tag{8.8}$$

$$\min_{k_1, k_2 \geq 1}\{\alpha(r_i, r_j) + \gamma(k_1 + k_2) + E(S_{i+1+k_1,j-1-k_2})\}, \text{ if } L_{i,j} \text{ is an interior loop} \tag{8.9}$$

The value of $E(L_{i,j})$ is found by taking the minimum over all values computed using the recurrences above. These equations, together with Equation (8.4) give us all the necessary information to set up a familiar matrix-based dynamic programming algorithm for predicting RNA secondary structure with loops. The schematic matrix layout in Figure 8.2 shows which regions of the matrix must have their values known before we can compute $E(L_{i,j})$, for each of the recurrences involved.

Because the most complicated part of the algorithm is the computation of the recurrence relations for $E(L_{i,j})$, let us now examine these relations in more detail. The simplest cases happen when $L_{i,j}$ is a helical region or hairpin loop. In these cases we spend constant time per (i, j) pair, and thus the running time is $O(n^2)$. In the case of bulges, the parameter k ranges between 1 and $j - i$. Therefore this case is similar to the computation of minimum energies without loops, and the running time is $O(n^3)$ (see Equation 8.3). The final case is that of interior loops. Because we are dealing with the two parameters k_1 and k_2, in this case we have to look at previous values of E which appear in the submatrix delimited by i and j. Intuitively, this means that for every pair i, j we have to look at a submatrix having as many as $O(n^2)$ elements, giving a total of $O(n^4)$. More formally, we have:

$$\sum_{1 \leq i \leq j \leq n} \left(\sum_{i \leq i' \leq j' \leq j} c \right) = O(n^4),$$

for some constant c. The value of this constant is given by the time it takes to compute $\alpha(r_i, r_j) + \gamma(k_1 + k_2) + E(S_{i+1+k_1,j-1-k_2})$ for specific values of i, j, k_1, and k_2.

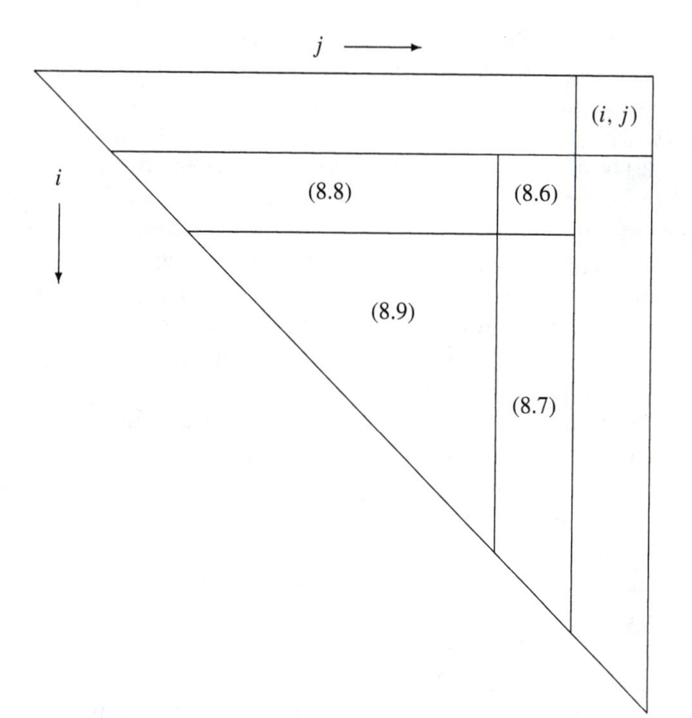

FIGURE 8.2

Matrix layout for computation of recurrence relations 8.6–8.9. The equation numbers label the matrix regions where previous values of $E(L_{i,j})$ must be known.

Improving the Efficiency of Interior Loop Computation

It is thus clear that in the framework presented, the computation of interior loops is the most expensive. We now show how its running time can be lowered to $O(n^3)$. This can be done using the fact that along diagonals the size of a loop is constant. Let us give an example.

The parameter for the γ function is the size of the interior loop, or k_1+k_2, as above. The constraints on k_1 and k_2 are that $2 \leq k_1 + k_2 \leq j - i - 2 - l_{\min}$, where $l_{\min}$ is the minimum loop size. Suppose we want to compute $E(L_{4,18})$ and that $l_{\min} = 1$. Then $2 \leq k_1 + k_2 \leq 11$. We could have then at a certain point $k_1 = 2$ and $k_2 = 6$, meaning that we need to evaluate $\gamma(8)$ and $E(S_{10,14})$, and at another point $k_1 = 5$ and $k_2 = 3$, meaning that we have to evaluate $\gamma(8)$ again and $E(S_{7,11})$. What is happening is that (10, 14) and (7, 11) lie along the same diagonal, and there loop sizes are constant.

This suggests the following trick. We store for each pair (i, j) a vector $V_{i,j}$ indexed by all possible loop sizes. There are at most $2n - 1$ diagonals, so this vector has at most that many elements. In each element we store the minimum value found for each diagonal. Then, when we have to evaluate $E(L_{i,j})$ for an interior loop and given pair (i, j), we need only compute the minimum of $\{\alpha(r_i, r_j) + \gamma(l) + V_{i,j}[l]\}$, for all possible values of l, the loop size. This will take $O(n)$ per pair (i, j), giving a total of $O(n^3)$. For this trick to work we must have an auxiliary matrix to store all $V_{i,j}$ vectors, and we have to update the vectors as we scan the original matrix M. Fortunately we can do this incrementally. For example, when we move from row i to row $i+1$, to each element in row $i+1$ there corresponds an element (a diagonal in M) from $V_{i,j}$. The elements of $V_{i+1,j}$ can be obtained by comparing the elements in row $i+1$ of M to elements in $V_{i,j}$, which also takes $O(n)$. Hence we also spend $O(n^3)$ overall updating the $V_{i,j}$ vectors. What this all means is that we can shave a factor of n from the running time for interior loop computation, but we double the storage space.

This running time can be further improved if we look carefully at the kinds of free-energy functions we have, similarly to the improvements obtained in Chapter 3 for gap-penalty functions. We will not describe these improvements, given that the resulting algorithms are much more complicated than those seen so far; some pointers are given in the bibliographic notes. We remark that with *linear* destabilizing functions the time above can be improved to $O(n^2)$ (for interior loops and bulges). Experiments, however, have shown that destabilizing functions grow logarithmically.

The algorithms that we have described for RNA structure prediction have one important flaw: They predict only one structure. In practice we would like to have a set of possible solutions, some of them possibly suboptimal in terms of free energy. The reason is the same as in other problems considered in this book: Given the abstractions in the model used to set up the algorithm, the optimal solution (or any optimal solution) is not necessarily the true structure. For this reason it would be important to extend the above algorithms to find optimal as well as suboptimal solutions within a certain energy threshold. A reference is given in the bibliographic notes regarding this extension.

8.2 THE PROTEIN FOLDING PROBLEM

We recall that in the protein folding problem the main goal is to determine the three-dimensional structure of a protein based on its amino acid sequence. This problem has been intensely researched since the early 1950s, and as yet no completely satisfactory solution has been found. There is a vast body of work on this topic, and this section is intended to be a very brief introduction. In the next section we present one particular approach to the problem, among the many that exist. We believe that these two sections contain enough material to give the reader a good start on the subject. References for further reading are given at the end of the chapter.

An important assumption of all protein folding prediction methods is that the amino acid sequence completely and uniquely determines the folding. This assumption is supported by the following experimental evidence. If we unfold a protein *in vitro,* such that no other substances are present, and then release it, the protein immediately folds back to the same three-dimensional structure it had before. The folding process takes less than a second. Thus it appears that all information necessary for the protein to achieve its "native" structure is contained in its amino acid sequence. (This is not true of all proteins because there are some that need "auxiliary molecules" to fold.)

Before discussing the difficulties of protein folding prediction, we briefly describe the main facts about protein three-dimensional structure. Recall from Chapter 1 that proteins have several levels of three-dimensional structure. Above the primary linear structure are the *secondary structures,* of which there are basically three kinds: the *α-helices,* the *β-sheets,* and structures that are neither helices nor sheets, called *loops.*

An α-helix is a simple helix having on average 10 residues, or three turns of the helix; some helices can have as many as 40 residues. Research has shown that some amino acids appear more frequently on helices than other amino acids, a fact that helps in secondary structure prediction but is not strong enough to allow accurate results.

A β-sheet consists of the binding between several sections of the amino acid sequence. Each participating section is called a β-strand and has generally between 5 and 10 residues. The strands become adjacent to each other, forming a kind of twisted sheet. Certain amino acids show a preference for being in β-strands, but again in this case these preferences are not so positive as to allow accurate prediction.

Loops are the sections of the sequence that connect the other two kinds of secondary structure. In contrast to helices and sheets, loops are not regular structures, both in shape and in size. In general, loops are outside a folded protein, whereas the other structures form the protein **core**. Proteins that are evolutionary related usually have the same helices and sheets, but may have very different loop structures. In this case we say that core segments are *conserved.* The concept of a core is very important for the next section.

Proteins have also higher-level structures. These are called *motifs* and *domains.* A motif is a simple combination of a few secondary structures that appears in several different proteins. An example of a motif is the helix–loop–helix. Research has shown that this motif is used as a binding site for calcium atoms, so this structure has a clear function. Other motifs appear to play no role at all in protein function. A domain is a more complex combination of secondary structures that by definition has a very specific function, and therefore contains an *active site.* An active site is the section of the protein where some binding to an external molecule can take place. A protein may have only one domain, or may contain several. All of them taken together form the protein's *tertiary structure.*

We can now state the folding problem in more detail. Given the amino acid sequence of a protein, we would like to process it and determine where exactly all of its α-helices, β-sheets, and loops are, and how they arrange themselves in motifs and domains. Given enough and reliable chemical and physical information about the each component amino acid, in principle it should be possible to compute the free energy of foldings. A prediction method could simply enumerate all possible foldings, compute the free energy of each, and choose the one with minimum free energy (assuming that such a folding *is* the protein's native structure, which is not an established fact). Now let us have an idea of how many possible foldings a typical protein can achieve. Recall from Section 1.2 that proteins can fold thanks in large part to the angles ϕ and ψ between the alpha carbon and neighboring atoms. Experiments have shown that these angles can assume only a few values independently of each other; for each pair of values in each residue we would have one *configuration.* Let us suppose that each angle can assume only three values. Therefore each residue can be in only one of nine configurations. For a protein with 100 residues, which can be considered small, there are thus 9^{100} possible configurations, an astronomical number that rules out enumerative algorithms.

To make matters worse, even if researchers agree that the native structure is the one with minimum free energy, there is no agreement on how to compute the energy of a configuration. There are far too many factors to take into account, such as shape, size, and polarity of the various component molecules, the relative strength of interactions at the molecular level, and so on. From current knowledge it is possible only to formulate some general rules, such as the requirement that hydrophobic amino acids stay "inside" the protein and hydrophilic amino acids stay "outside" the protein, in the case of proteins in solution (water is the universal solvent found both inside and outside cells).

Such a state of affairs has led researchers to concentrate on prediction methods for protein secondary structures. The observation above about the difficulties in assigning free-energy values to tertiary structures applies to secondary structures as well. Hence, no general and reliable dynamic programming approach is known for the prediction of protein secondary structure. What we find in practice is the use of an assortment of methods, ranging from programs based on neural nets to pattern-recognition methods based on statistical properties of residues in proteins with known structure. However, these methods are still not as reliable as scientist want them to be. For this reason still other techniques have been developed to help cope with the problem, and it is one such technique that is the topic of the next section.

PROTEIN THREADING

An early method used for secondary structure prediction was based on the idea that similar sequences should have similar structures. If we know the structure of protein A (from x-ray crystallography, say) and protein B is very similar to it at the sequence level, it seems reasonable to assume that B's structures will be the same or nearly the same as A's. Unfortunately this is not true in general. Similar proteins at the sequence level may have very different secondary structures. On the other hand it has been observed that certain proteins that are very different at the sequence level are structurally related in the following sense: Although they have different kinds of loops, they have very similar cores. These observations have led researchers to propose the *inverse protein folding problem*. Instead of trying to predict the structure from the sequence, we try to fit a known structure to a sequence.

To fit a structure to a sequence, we have to be more precise about what we mean by "structure." In this section, structure actually means *structural model,* and such a model can be obtained in the following way. Given a protein whose structure is known, we obtain a structural model by replacing amino acids by place-holders, but keeping associated with each place-holder some basic properties of the original amino acid. These properties can vary depending on the model adopted, but they should retain the fact that the original amino acid was in an α-helix or β-strand, or in a loop, and reflect spatial constraints of the structure such as distances to other amino acids, how much inside or outside the whole structure the place-holder is, and so on. Here we will not give details of such models but will assume instead that such a characterization of local structure is possible.

With a given structural model we can now get a protein with unknown structure and try to align it to the model, taking into consideration the properties recorded for that particular model. This special kind of alignment is known as *protein threading*. As we will see this technique results in a precise combinatorial optimization problem, since there can be many ways of threading one protein sequence into a model. As was the case with many other problems in this book, this one turns out to be NP-hard, too. However, the problem can be cast in such a way that a well-known technique for handling NP-hard problems can be used. The success obtained with this approach is our motivation for this section. We now proceed to describe the problem in more detail.

We asssume that the input is given in the following way:

1. A protein sequence A with n amino acids a_i.
2. A core structural model C, with m core segments C_i. For this model we have the following information:
   a. The length c_i of each core segment.
   b. Core segments C_i and C_{i+1} are connected by loop region λ_i, and for each λ_i we know its maximum ($l_i^{\max}$) and minimum ($l_i^{\min}$) lengths.
   c. The local structural environment for each amino acid position in the model (chemical properties, spatial constraints, etc).
3. A score function to evaluate a given threading.

The desired output is a set $T = \{t_1, t_2, \ldots, t_m\}$ of integers, such that the value of t_i indicates what amino acid from A occupies the first position from core segment i. It is thus clear that a threading is an alignment between the sequence and the core structural model.

In this formulation the score function obviously plays a key role. Given a threading and local structure environments, the score function should be able to evaluate sequence amino acids individually in their alloted positions in the model, as well as the resulting interactions between the various amino acids. Interactions can potentially be between pairs of amino acids, triples, quadruples, and so on.

If we ignore interactions between amino acids, while allowing variable-length loop regions, the problem can be solved by standard dynamic programming techniques. If we allow pairwise interactions, the problem becomes NP-hard (see the bibliographic notes). Instead of oversimplifying in order to fall back on the (by now) familiar dynamic programming algorithms, here we will allow pairwise interactions and show how to solve the problem exactly with a standard technique for handling NP-hard problems. This technique is known as *branch-and-bound*. We now digress in order to explain this technique and then we will see how it can be applied to solve the threading problem with pairwise interactions.

Branch-and-Bound

In a combinatorial optimization problem we want to find an optimum solution among many possibilities in the *solution space*. The optimum refers to the maximum (or minimum) value of some function f that can be computed for each candidate solution. For most problems the solution space is exponentially large, although for problems in class P the optimum can be found in polynomial time. In the case of NP-hard problems, however, no algorithm is known that can search the solution space and find the optimum in polynomial time. Therefore the only alternative to obtain the optimum solution *for sure* is to enumerate and evaluate each possible solution one by one and choose the best. However, we can try to speed up this process a bit by applying the following technique.

First, it should be possible to separate the solution space into subspaces according to some constraints. For example, given solution space S, we could perhaps partition it into two subsets S_1 and S_2. S_1 could consist of all the solutions that have a certain property, and S_2 all of those that do not. This is the branching part of branch-and-bound.

Second, it should be possible to obtain a bound on the value of the optimization function f for all candidate solutions in a set. This bound can be used to eliminate many solutions from consideration in the following way. Suppose that we have a *minimization* problem and that we already know the value $f(s)$ of some candidate solution s. Now consider a subset X of the solution space, and suppose that we can obtain a lower bound l on the value of *any* solution from X. Finally suppose that $l > f(s)$. This means that any solution from X cannot hope to improve on the solution we already have, and should therefore be discarded. If on the other hand $l < f(s)$, we should explore set X, because the optimum may lie there.

Thus we see that the key ingredients in a branch-and-bound minimization algorithm are a rule to split the solution space in some efficient way and a lower bound on the value of the optimization function for candidate solutions. The lower bound should be as close to the actual function value as possible, so that we can find the optimum faster than with a weaker lower bound; it should also be efficient to compute. However, we should keep in mind that the worst-case running time of branch-and-bound algorithms applied to NP-hard problems is always exponential.

We will now apply these concepts to the protein threading problem. We first must define the possible solutions, and for this we must list the constraints faced by each possible threading. The spacing constraints can be given by the following inequalities:

$$1 + \sum_{j<i} (c_j + l_j^{\min}) \leq t_i \leq n + 1 - \sum_{j \geq i} (c_j + l_j^{\min}).$$

Another constraint comes from the fact that any threading should obey the order of core segments in the model. (This could be relaxed in a more general approach.) This results in:

$$t_i + c_i + l_i^{\min} \leq t_{i+1} \leq t_i + c_i + l_i^{\max}.$$

These constraints imply that each t_i must lie in an interval given by values b_i and e_i. Thus a set of possible threadings can be expressed by a collection of intervals, one for each of the m t_i's. Figure 8.3 illustrates this.

The branching part of the algorithm can now be explained. Choose one core segment, say i, and a position u_i inside the interval $[b_i, e_i]$. Split the solution set into three sets: one in which the new interval for t_i will be $[b_i, u_i - 1]$, another in which the new interval for t_i will be $[u_i + 1, e_i]$, and another in which t_i is forced to be equal to u_i. We have not specified which core segment to choose nor which position inside the segment's interval. These choices can be made based on considerations such as interval size.

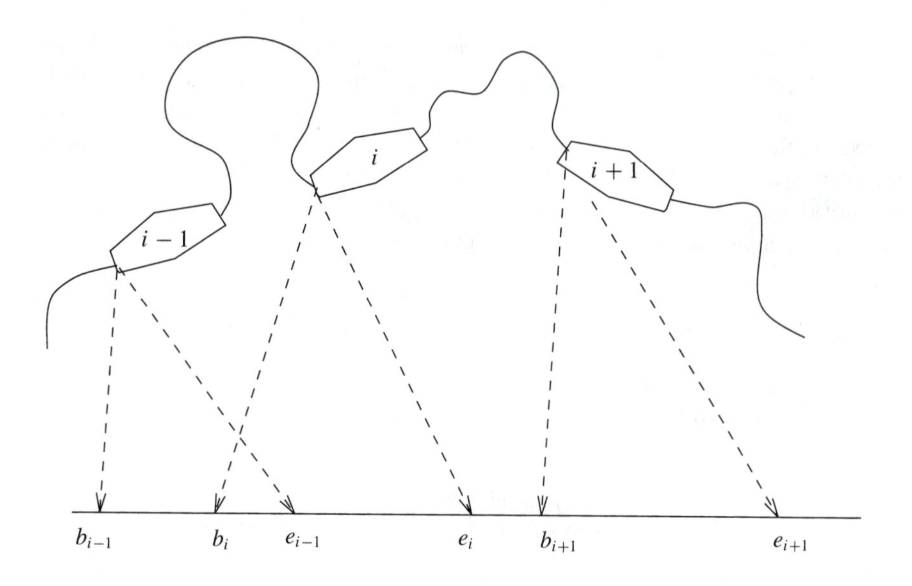

FIGURE 8.3

Sets of threadings. Core segments are the polygons labeled by i − 1, i, and i + 1. Between them are loop regions. The line at the bottom is the sequence being threaded, and the dashed arrows show the intervals where the first position of each core segment can assume values in the sequence, hence depicting sets of possible threadings.

We now turn to the lower bound. The score function can be given by the following equation, assuming T is some threading:

$$f(T) = \sum_i g_1(i, t_i) + \sum_i \sum_{j>i} g_2(i, j, t_i, t_j). \tag{8.10}$$

Note that the score function deals with core segments rather than individual amino acids. The reason is that the value of t_i determines the position of all amino acids from sequence A in core segment i, hence the pair of arguments (i, t_i) encapsulates all that is happening with segment i. The first term in the right-hand side above gives the contribution of each individual core segment to the total, whereas the second term gives the contribution of pairs of core segments. Functions g_1 and g_2 are part of the input.

Given a set of threadings $\mathcal{T}$, a simple lower bound is given by:

$$\min_{T \in \mathcal{T}} f(T) = \min_{T \in \mathcal{T}} \sum_i \left[g_1(i, t_i) + \sum_{j>i} g_2(i, j, t_i, t_j) \right] \tag{8.11}$$

$$\geq \sum_i \left[\min_{b_i \leq x \leq e_i} g_1(i, x) + \sum_{j>i} \min_{\substack{b_i \leq y \leq e_i \\ b_j \leq z \leq e_j}} g_2(i, j, y, z) \right]. \tag{8.12}$$

This lower bound is obtained by summing up the lower bounds on each term separately. A stronger lower bound is possible (see Exercise 7). It is desirable to obtain a lower bound that corresponds to one particular threading, because this threading can then be used in the splitting process. Suppose threading T currently has the lowest value for the lower bound. We choose one core segment C_i from it and split it at point u_i given by the threading. The core segment chosen can be the one with the largest interval size. A description of the algorithm is given in Figure 8.4.

```
Algorithm Protein Threading
    input: protein sequence, core structural model, score functions,
        lower bound function f_lb
    output: a threading T
    𝒯 ← all possible threadings
    lb ← f_lb(𝒯)
    // We use a priority queue Q to keep threadings
    // In Q each element is a set of threadings and its lower bound
    Insert(Q, 𝒯, lb)
    while TRUE do
        // We remove a candidate set of threadings from Q
        𝒯_c ← RemoveMin(Q)
        if |𝒯_c| = 1 then
            return 𝒯_c
            // The only remaining threading is the solution
        else
            Split(𝒯_c)
            for each new subset 𝒯_i from 𝒯_c do
                lb_i ← f_lb(𝒯_i)
                Insert(Q, 𝒯_i, lb_i)
```

FIGURE 8.4

The branch-and-bound protein threading algorithm.

This finishes the description of branch-and-bound applied to protein threading. We close this section with a few comments regarding the usage of such a technique. First, this is a recent development and must therefore undergo many tests. These tests can be roughly divided in three categories. In the first, after a structural model is obtained from a structure, we perform a *self-threading;* that is, we thread the very protein sequence that gave rise to the structural model. Another test involves proteins that are genetically close (and hence have similar sequences) and share a similar structure (called *protein homologs*). A final test involves proteins that have different sequences but are structurally similar (called *structural analogs*). In these tests all structures are known beforehand, so we know exactly what the result should be and can detect anomalies. Encouraging results have been obtained, in that the structure obtained is usually very close to correct. When there are errors, it is relatively easy to insert additional constraints to correct them. This is another advantage of this technique, and it can thus use the extensive knowledge on the quirks of protein structure that has been accumulating over the years.

An application of this technique to the real world should work like this. A database of structural models is maintained. A researcher comes up with a new protein sequence and submits it to the database server. The server then threads the sequence on all models from the database, returning one or a few models where a good fit was found. Note that this would mean solving one protein threading problem for each structural model in the database, a task that requires a fast solution to each such problem.

SUMMARY

In this chapter we presented a brief sampling of algorithmic results for molecular structure prediction. We studied two dynamic programming algorithms for RNA secondary structure prediction, and showed how to improve the running time of the more complex algorithm from $O(n^4)$ to $O(n^3)$. We then briefly explained the protein folding problem, pointing out that extensive research has led to many approaches to the problem, of which we presented one, called protein threading. This approach allows an algorithmic treatment based on a familiar technique for handling NP-hard problems.

EXERCISES

1. Given RNA sequence AUGGCAUCCGUA, find a structure with the greatest number of base pairs.
2. Give an example of an RNA sequence where a knot may appear.
3. Write out in detail the dynamic programming algorithm for RNA structure prediction where the energy of each base pair is independent of all the others.
4. Give the algorithm that will retrieve the RNA secondary structure computed using recurrence relation (8.2).
5. We mentioned that the kind of destabilizing function used in RNA secondary structure prediction may influence the running time of the dynamic programming algorithm. Explain why this is so using as examples functions that are linear, logarithmic, quadratic, or simple table look-up.
6. Prove that the protein threading problem is NP-complete.
7. Propose a stronger lower bound for the threading score function.
8. List other NP-hard problems in this book and try to design branch-and-bound algorithms for them.

★9. The computation of the lower bound (8.10) requires the computation of all points that fall within some interval (in the case of g_1), or all points that fall in some area (in the case of g_2). How can we preprocess all points to answer these queries efficiently?

BIBLIOGRAPHIC NOTES

The overall organization of Section 8.1 was based primarily on the paper by Zuker [207], which contains much additional material. The improvement in running time for interior loop computation is due to Waterman and Smith [201]. Further improvements in running time for special classes of functions were achieved by Eppstein, Galil, and Giancarlo [55] and by Larmore and Schieber [121]. Work on computing optimal and suboptimal structures within a certain energy threshold has been done by Zucker [206]. A good survey on RNA structure prediction in general is given by Turner and Sugimoto [186].

The protein folding problem has an immense body of literature. Here we point out a few important recent surveys. Richards [161] presents an overview for a general audience. Creighton [38] presents a technical survey. The different molecular interactions that play a role in protein folding are discussed by Dill [48]. Our discussion of protein folding was based in part on the paper by Kanehisa and DeLisi [105]. Two of our figures were based on figures from this paper. The computational difficulty of protein folding prediction has been studied by Unger and Moult [189] and Fraenkel [65], who prove that obtaining minimum-energy foldings of proteins based on simple lattice models is an NP-hard problem. Berger [21] and Berger and Wilson [22] present a successful approach for protein motif recognition.

The section on protein threading was based on the work by Lathrop and Smith [123]. Figure 8.3 was based on a similar figure from their paper. This paper contains much additional material on the tests performed as well as detailed description of the various techniques used to speed up computation. Lathrop [122] proved that the protein threading problem with variable-length loop regions and pairwise amino acid interactions is NP-complete.

CHAPTER

9

EPILOGUE: COMPUTING WITH DNA

Throughout this book we have seen many algorithms that help solve problems in molecular biology. In this chapter we present surprising results that go in the opposite direction: how to use molecular biology to solve hard algorithmic problems. These results have created an exciting new field, called *DNA computing,* whose ultimate aim is the creation of very efficient biomolecular computers. It is still far from being a practical field, but intense research is addressing each of its shortcomings, and new developments are certain to appear in the near future.

9.1 THE HAMILTONIAN PATH PROBLEM

In this section we describe how to solve the Hamiltonian path problem in directed graphs with DNA molecules in a laboratory using a method whose number of laboratory procedures grows linearly with the number of vertices in the graph. Recall from Section 2.3 that the Hamiltonian path problem is NP-complete. This method is *not* a proof that P = NP, because at its heart the method is a brute-force algorithm executing an exponential number of operations. What enables the method to have a linear number of laboratory procedures is the fact that an exponential number of operations is done *in parallel* in the first such procedure.

Formally, the problem to be solved is this. Given a directed graph $G = (V, E)$, such that $|V| = n$ and $|E| = m$, and two distinguished vertices s and t, verify whether the graph has a path $(s, v_1, v_2, \ldots, t)$ whose length (in number of edges) is $n-1$ and whose vertices are all distinct. We denote this problem by HDPP, the Hamiltonian directed path problem. See an example in Figure 9.1. A brute-force algorithm is very simple. Generate all possible paths with exactly $n - 1$ edges, and then verify whether one of them obeys the problem constraints. There are at most $(n-2)!$ such paths, because for the first vertex on the path we can choose $(n - 2)$ vertices, for the second vertex on the path we have $(n - 3)$ choices, and so on.

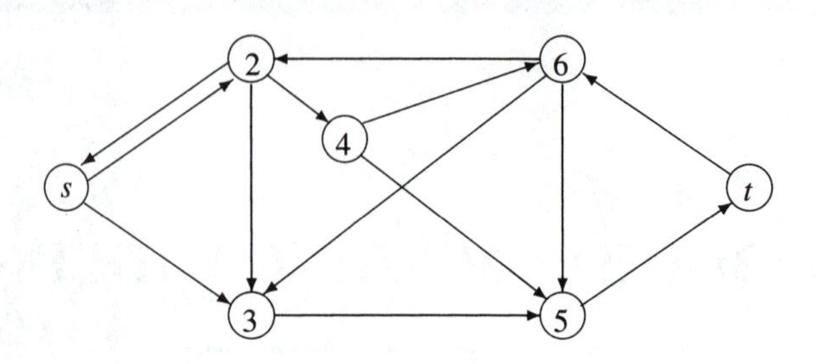

FIGURE 9.1

A directed graph. An st-Hamiltonian path is $(s, 2, 4, 6, 3, 5, t)$.

In the method to be presented below paths are not generated as above. Rather, in the first step an enormous number of *random* paths are created. The fact that random paths are being generated makes possible the generation of each path independent of all the others, and hence all of them can be created *simultaneously*. On the other hand, this makes the algorithm probabilistic. The complete algorithm is as follows:

1. Generate random paths.
2. From all paths created in the previous step, keep only those that start at s and end at t.
3. From all remaining paths, keep only those that visit exactly n vertices.
4. From all remaining paths, keep only those that visit each vertex at least once.
5. If any path remains, return yes; otherwise return no.

We now describe how each of these steps can be implemented in the laboratory.

Random Path Generation

We assume that random single-stranded DNA sequences with 20 nucleotides are available, and that the generation of astronomical numbers (more than 10^{10}) of copies of short DNA strands is easy to do. Recall that $\overline{S}$ represents the reverse complement of DNA string S.

- *Vertex representation:* Choose at random n distinct single-stranded 20-character DNA sequences, and to each vertex v associate sequence S_v. For each such sequence obtain its reverse complement $\overline{S}_v$. Generate many copies of each $\overline{S}_v$ sequence in test tube T_1.
- *Edge representation:* If $(u, v) \in E$, then build sequence S_{uv} by concatenating the 10-base suffix of S_u to the 10-base prefix of S_v. Suffixes and prefixes are distinguished by the 5′-3′ direction of DNA strands. When $u = s$ the whole of S_s is used, and hence S_{sv} has 30 bases. Similarly for the case when $v = t$. Note that with this construction, $S_{uv} \neq S_{vu}$. Generate many copies of each S_{uv} sequence in test tube T_2.

- *Path construction:* Pour T_1 and T_2 into T_3. We assume that in T_3 any binding between sequences can freely take place (that is, in T_3 many *ligase reactions* will take place).

By executing the three steps above we get many random paths. Why? We will explain by an example. Consider the sequences $\overline{S}_u$, $\overline{S}_v$, $\overline{S}_w$, S_{uv} and S_{vw}, for u, v and w distinct vertices. Because many copies of these sequences are floating about in tube T_3, with high probability the 10-base suffix of one $\overline{S}_u$ sequence will bind to the 10-base prefix of one S_{uv} sequence (since one is complementary to the other); at the same time the 10-base suffix of the same sequence S_{uv} binds to the 10-base prefix of one $\overline{S}_v$ sequence, whose 10-base suffix binds to the 10-base prefix of one S_{vw} sequence. And still at the same time the 10-base suffix of this S_{vw} sequence binds to the 10-base prefix of one $\overline{S}_w$ sequence. The final double strand thus obtained encodes path (u, v, w) in G. Let us make this clear by seeing what happens in the example of Figure 9.1. That graph has $n = 7$ vertices. Suppose that the sequences chosen to represent vertices 2, 4, and 5 are the following:

$S_2 =$ GTCACACTTCGGACTGACCT
$S_4 =$ TGTGCTATGGGAACTCAGCG
$S_5 =$ CACGTAAGACGGAGGAAAAA

The reverse complement of these sequences are:

$\overline{S}_2 =$ AGGTCAGTCCGAAGTGTGAC
$\overline{S}_4 =$ CGCTGAGTTCCCATAGCACA
$\overline{S}_5 =$ TTTTTCCTCCGTCTTACGTG

We build edges (2, 4) and (4, 5) obtaining the following sequences:

$(2, 4) =$ GGACTGACCTTGTGCTATGG
$(4, 5) =$ GAACTCAGCGCACGTAAGAC

Finally, the path (2, 4, 5) will be encoded by the following double strand:

```
                       (2,4)
5'
GTCACACTTCGGACTGACCTTGTGCTATGG...
CAGTGTGAAGCCTGACTGGA ACACGATACCCTTGAGTCGC...
      ← S̄2                       S̄4

                                  (4,5)
                                                       3'
                       ...GAACTCAGCGCACGTAAGAC GGAGGAAAAA
                                ...GTGCATTCTG CCTCCTTTTT
                                              S̄5  ←
```

Further steps of the algorithm depend on the assumption that with high probability *all* possible paths in the graph with n vertices will be represented by at least one double strand. Note also that the number 20 was chosen so as to make it very unlikely that two sequences assigned to two different vertices would share long subsequences. Should this happen, we might have double-stranded sequences that would not encode any paths in the graph.

Remaining Steps in the Algorithm

In the remaining steps we place the contents of T_3 through successive biochemical sieves that select paths with desired properties. Step (2) is implemented by a polymerase chain reaction (see Section 1.5.2) using primers S_s and $\overline{S}_t$. Step (3) is implemented by separating double-stranded DNA having exactly $20n$ bases, which therefore represent paths with exactly n vertices. Step (4) is implemented by retaining strands where $\overline{S}_u$ is present for each $u \in V$ different from s and t. In the final step we simply see what has remained through a conventional gel experiment.

We can analyze this algorithm as follows. We need a number of operations proportional to the size of the graph in order to build it. After that we need to perform a number of operations that are proportional to the number of vertices in the graph, mainly because of step (4) above. In that step we have to check sequentially the presence of each vertex-representing sequence in the test tube.

SATISFIABILITY

9.2

In this section we describe how we can use an approach similar to the one described in the previous section to solve the *satisfiability problem* (SAT). This is a very important problem in the theory of computing, since it was the very first one to be shown NP-complete. Moreover, a generalization of the algorithm described here leads to a general approach for solving any NP-complete problem directly (see Exercise 4).

To explain the SAT problem, we need a few definitions. A *boolean variable* is a variable that can take either 0 or 1 as a value. Boolean variables may be put together in a formula using the operators $\vee$ (logical *or*), $\wedge$ (logical *and*), and parentheses. Furthermore we can also use a variable x in its negated form, $\overline{x}$. This means that if x's value is 1, then $\overline{x}$'s value is 0, and vice versa. The actual way in which we use a variable is called a *literal*. It can be shown that any boolean formula can be expressed by a collection of *clauses,* where each clause is a collection of literals connected only by the operator $\vee$, each clause is delimited by parentheses, and clauses are connected to each other by the operator $\wedge$ (we say that such formulas are in *conjunctive normal form*). An example of such a formula is $F = (x_1 \vee x_2) \wedge (\overline{x}_2 \vee x_3) \wedge (\overline{x}_3)$. We will consider formulas constructed using n variables, m clauses, and a fixed number of literals per clause. The SAT problem consists of determining whether there is an assignment of values to the variables that *satisfies* a given formula, in other words, an assignment that makes the formula evaluate to 1. In the above example, $x_1 = 1$, $x_2 = 0$, and $x_3 = 0$ satisfies F, whereas any other assignment does not. In general, a formula may have more than one satisfying assignment or it may have none.

An algorithm to test whether a formula is satisfiable is this: Try out all possible values for the variables. If there are n variables, there are 2^n such assignments. A more systematic way of trying out all possible assignments, and the one actually used to solve SAT with DNA, is as follows:

1. Generate all possible 2^n assignments and place them in set S_0.

2. From all assignments in set S_0 keep only those that satisfy clause 1, and place them in set S_1.

3. From all assignments in set S_1 keep only those that satisfy clause 2, and place them in set S_2.

...

m. From all assignments in set S_{m-1} keep only those that satisfy clause m, and place them in set S_m.

After step m, perform one more operation: If set S_m is nonempty return yes; otherwise return no.

To implement this using DNA, we must have a way of encoding boolean variable assignments. We do this by using a graph. We will have one graph for each distinct number of variables; thus G_n will be the graph to be used in a problem with n variables. In Figure 9.2 we show G_n for $n = 3$. Notice that it is a directed graph and that it has three kinds of vertices: the v_i vertices, $1 \leq i \leq n + 1$, the x_j vertices, and the $\overline{x}_j$ vertices, $1 \leq j \leq n$. Each possible assignment will correspond to one particular $(v_1\text{-}v_{n+1})$-path in this graph in the following way. Starting from vertex v_1, we can go either to x_1 or to $\overline{x}_1$. Choosing x_1 means assigning 1 to variable x_1; choosing $\overline{x}_1$ means assigning 0 to x_1. Then the path must go through v_2, and again we have a choice, which will correspond to the value assigned to variable x_2, and so on. Thus a $(v_1\text{-}v_{n+1})$-path in the graph encodes an n-bit number, which corresponds to one particular assignment of boolean values to n variables.

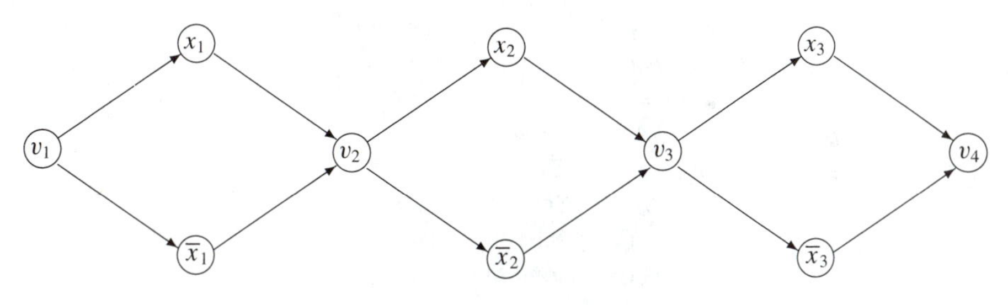

FIGURE 9.2

An example of a graph whose $(v_1 - v_{n+1})$-paths encode boolean variable assignments.

Having determined this encoding, the next step is the construction of all paths in graph G_n, which we do in the same manner as was done for the HDDP (see previous section). The starting point in our DNA SAT algorithm is thus a test tube T_0 containing an enormous number of sequences representing paths in G_n, and with high probability sequences representing *all* $(v_1\text{-}v_{n+1})$-paths will be present in T_0.

For the implementation of the next steps we need two kinds of operations on test tubes: *Detect* and *Extract*. *Detect* is an operation already employed in the HDPP problem (step 5 in that algorithm); it simply reports whether a test tube has any DNA sequence at all. The *Extract* operation is similar to the operations employed in the HDDP algorithm, in the sense that we use it to extract sequences having a certain property. Here we are interested in only one particular property: We apply *Extract* to test tube T in order to extract the DNA sequences in T that contain a given precise DNA substring of fixed length. It is not required that all such sequences be extracted from T; thus it is not a problem if among the remaining sequences in T there are still some that contain the given substring. But it is a problem if we extract sequences that do *not* have the desired property, or if we fail to extract any relevant sequences even if T contains some. Even though these failures can happen in practice, in this section we assume *Extract* works as described.

We will now describe how we deal with the first clause in our formula F. Let us assume that this clause is $(x_1 \vee \overline{x}_2)$. Consider the first literal in this clause, x_1. Because it is not negated, an assignment that makes $x_1 = 1$ is a candidate for the desired solution. Thus the corresponding n-bit number has a 1 in its first position (first from left to right). Similarly, an assignment that makes $x_2 = 0$ is another candidate. By combining these two, we can say that only the n-bit numbers with a 1 in their first position, or with a 0 in the second position are the numbers that can possibly satisfy F; all others are out. This means that we should extract from T_0 all sequences encoding n-bit numbers that satisfy this requirement. Those sequences will then be the input for the treatment of the next clause.

From the above description it is clear that we need the *Extract* operation in the following format: *Extract*(T, i, b), where T is a test tube, i is the index of a variable, and b is a boolean value. We can now sketch more formally the steps that take care of the first clause, assuming that it has 2 literals:

// T_0, T_1, A_1, A'_1, A_2 *are test tubes*
// v_1 *is the first literal*
if $v_1 = x_1$ **then**
    $A_1 \leftarrow$*Extract*$(T_0, 1, 1)$
else // $v_1 = \overline{x}_1$
    $A_1 \leftarrow$ *Extract*$(T_0, 1, 0)$
$A'_1 \leftarrow T_0$ // *the remainder*
// v_2 *is the second literal*
if $v_2 = x_2$ **then**
    $A_2 \leftarrow$*Extract*$(A'_1, 1, 1)$
else $A_2 \leftarrow$*Extract*$(A'_1, 1, 0)$
$T_1 \leftarrow A_1 + A_2$ // *We mix the tubes*

Notice that when dealing with the second literal we extracted sequences from the remainder of the first extraction. This is essential to make the algorithm work. If the clause were $(x_1 \vee \overline{x}_2)$ it should be clear that in tube T_1 we end up with all sequences encoding n-bit numbers that have a 1 in the first position (which were in tube A_1) and all sequences encoding n-bit numbers that have a 0 in the second position (which were in tube A_2). Using tube T_1 we are ready to proceed to the next clause, where the steps are similar to the ones described. The whole algorithm is given in Figure 9.3.

This algorithm clearly executes $O(m)$ *Extract* operations and only one *Detect* operation.

```
Algorithm DNA Satisfiability
    input: test tube T_0 containing sequences that encode all n-bit numbers,
           and boolean formula F in conjunctive normal form
           (n variables and m clauses)
    output: YES if F is satisfiable, NO if not.
    // We use a fixed number of auxiliary test tubes A_i
    for i ← 1 to m do
        A'_0 ← T_{i-1}
        for j ← 1 to l_i do    // l_i is the number of literals in clause i
            if v_j = x_j then
                A_j ← Extract(A'_{j-1}, j, 1)
            else A_j ←Extract(A'_{j-1}, j, 0)
            A'_j ← A'_{j-1}    // remainder
        T_i ← A_1 + A_2 + ... + A_{l_i}
    if Detect(T_m) then return YES else return NO
```

FIGURE 9.3

The DNA algorithm to solve the SAT problem.

9.3 PROBLEMS AND PROMISES

When the algorithm from Section 9.1 was first used (on a graph of seven vertices) it was successful and demonstrated the feasibility of the approach. But there are several problems that need to be solved before the promise of DNA computing can be fulfilled. We list here some of the difficulties.

Is the DNA approach competitive with state-of-the-art algorithms on electronic computers for solving the HDPP? Not at the moment, by a *very* long shot, for two reasons. First, only small HDPP instances can currently be solved by this approach. For a graph with n vertices, let us assume conservatively that we need at least 2^n molecules, each one encoding one distinct path. So, for a graph with 100 vertices, we would need about 10^{30} molecules. In one gram of water there are no more than 10^{23} molecules. Thus to solve an instance with 100 vertices, we would need 10^7 grams, or 10 tons of water, at the very least. The second reason is that the laboratory procedures to be used, although few in number, are time-consuming. However, it is expected that much of it can be automated. On the other hand, there are computer programs today that can solve instances of the traveling salesman problem (which is, of course, at least as hard as the HDPP) with 100 vertices in a matter of minutes; in addition, much larger instances (with thousands of vertices) have also been solved, although it does not follow that these programs can quickly solve *any* large instance.

What are the errors that can occur? As we have seen throughout this book, biochemical experiments always contain errors; and that certainly applies to the procedures described here. Note that there is a randomizing component to the approach right at the start, but the biochemical errors are much more of a concern. These can happen as we try to isolate the strands that interest us, using the *Extract* operation, as mentioned above.

We may miss a "good" strand, or a "bad" strand may be extracted where none should. There are various techniques that can be used to decrease the possibility of such events happening, and this is a current research topic. Another source of errors comes from the fact that DNA in vitro decays through time, so laboratory procedures cannot take too long.

Can this approach be used for other problems? The descriptions above are suited to solve just the HDPP and SAT. Even though every NP-complete problem is transformable into any other in polynomial time, it would be nice if we could use these ideas to solve any NP-complete problem directly. The SAT algorithm is actually a step in this direction; see Exercise 4. The ultimate goal, however, is a general-purpose computer based on DNA. The results of this chapter suggest that such a computer is in principle possible, and that it could be much faster and use much less energy than conventional computers. If and when this dream comes true, maybe we will be able to use such a machine to decipher the remaining mysteries of molecular biology.

EXERCISES

1. In the HDPP DNA algorithm, suppose the input graph has 1,000 vertices. After the algorithm is finished, and assuming a path was found, how can we determine the path itself?
2. Show how we can use DNA to check whether a directed graph has a Hamiltonian cycle.
3. Determine the probability that a nonvalid path may be created in the HDDP algorithm for a graph with seven vertices. That is, determine the probability that the 10-base suffix of a random 20-base sequence is the reverse complement of the 10-base suffix of another random 20-base sequence. How does this probability change with the number of bases in the sequence? How does this probability change with the number of vertices in the graph?

★ 4. Generalize the DNA algorithm for the SAT problem so that any boolean formula can be solved. A general boolean formula is defined by the following recursive definition. (a) Any variable x is a formula; (b) if F us a formula so is $\overline{F}$; (c) if F_1 and F_2 are formulas, then so are $F_1 \vee F_2$ and $F_1 \wedge F_2$.

BIBLIOGRAPHIC NOTES AND FURTHER SOURCES

The breakthrough that created the field of DNA computing is due to Adleman [2], who showed how to solve the HDPP using DNA molecules. Lipton [128] then extended the result to the SAT problem and used that to prove that any NP-complete search problem can be solved directly by a DNA approach. Karp, Kenyon, and Waarts [112] studied methods that deal with errors inherent in DNA computations, and they pointed to many other papers dealing with this topic. Although the field of DNA computing wasn't born until the end of 1994, the related literature is already growing fast.

In this final chapter we take the opportunity to provide pointers to sources for further information on computational molecular biology as covered in this book. One very important book on the subject, Waterman [199], contains topics that we do not cover, in particular with respect to statistical issues. The one journal dedicated to the area, *The Journal of Computational Biology,* started publication in 1994 and can be reached at:

http://brut.gdb.org/compbio/jcb/

Theoretical computer science and mathematics journals, such as *The Journal of Algorithms, The SIAM Journal on Computing, The SIAM Journal on Discrete Mathematics,* and *Advances in Applied Mathematics,* run occasional papers on computational biology. Other relevant journals are *The Bulletin of Mathematical Biology* and *Computer Applications in the Biosciences.*

More up-to-date research results can be found in proceedings from research conferences. There is one annual conference dedicated to the area, the *International Conference on Computational Molecular Biology (RECOMB);* the first one is scheduled for January 1997. Another important meeting is the annual conference *Intelligent Systems for Molecular Biology,* where one finds papers using artificial intelligence techniques, among others. The annual conference *Combinatorial Pattern Matching* is also an important forum for research results in algorithmic techniques for computational biology.

As with journals, important conferences on the theory of computing present occasional papers on computational biology. Among them are the *ACM-SIAM Symposium on Discrete Algorithms,* the *ACM Symposium on the Theory of Computing,* the *IEEE Symposium on Foundations of Computer Science,* and the *International Colloquium on Automata, Languages and Programming.*

The very latest is, of course, the World Wide Web. Some WWW sites were given earlier in the text (look under "WWW site" in the index). Here we mention course material that was available in 1996. At least three courses on computational biology have material available through the Web. One of them is coordinated by Dalit Naor and Ron Shamir from Tel Aviv University. More information on this course can be found at:

http://www.math.tau.ac.il/~shamir/algmb.html

Another set of lecture notes, by Richard Karp, Larry Ruzzo, and Martin Tompa from the University of Washington, is available at:

http://www.cs.washington.edu/education/courses/590bi/

Finally, the "Virtual School of Natural Sciences BioComputing Division," a group of research faculty and students coordinated by Robert Giegerich and Georg Fuellen from the University of Bielefeld, offers course material at:

http://www.techfak.uni-bielefeld.de/bcd/Curric/welcome.html

ANSWERS TO SELECTED EXERCISES

Chapter 1

1. Sequence as is:

rf1:	TAA	TCG	AAT	GGG
	stop	Ser	Asn	Gly
rf2:	AAT	CGA	ATG	GGC
	Asn	Arg	Met	Gly
rf3:	ATC	GAA	TGG	
	Ile	Glu	Val	

Reverse Complement: GCCCATTCGATTA

rf1:	GCC	CAT	TCG	ATT
	Ala	His	Ser	Ile
rf2:	CCC	ATT	CGA	TTA
	Pro	Ile	Arg	Leu
rf3:	CCA	TTC	GAT	
	Pro	Phe	Asp	

2. The strand that starts with AUG is the coding strand. The top strand goes $5'$ to $3'$, and the bottom one goes $3'$ to $5'$.

   a. AUGAUACCGACGUACGGCAUUUAA

   b. M I P T Y G I

3. There are twelve possibilities: Eight of them can be represented by CTNATGAAR, where N stands for any base and R stands for either A or G. The remaining four are given by TTRATGAAR.

5. A 4-cutter cuts once every $4^4 = 256$ bases. So, we expect $40000/256 = 156.25$ cuts, that is, about 157 pieces.

6. Hungarian grapevine chrome mosaic nepovirus.

Chapter 2

1. It has three distinct substrings with one character, three with two, three with three, two with four, and one with five, for a total of 12 distinct substrings.

4. A complete directed graph has $n(n-1)$ edges, and half this much if it is undirected. An n-vertex tree has $n-1$ edges.

7. If one algorithm runs in $n^2/2$ time and the other in $10n$ time, then for $n = 4$ the quadratic algorithm takes 8 units versus 40 for the linear one.

8. The $O(\sqrt{n})$ algorithm is faster.

12. Section 4.3.5 presents a method that can be generalized for any acyclic graph as follows. Repeat the following step: Find a source in the graph and remove it. The sequence of removed sources is a topological ordering. Sources are vertices with indegree one.

13. Start with one set for each node. For each edge, if its extremes are not in the same sets, join their sets. After all edges are processed, the resulting sets are the connected components.

15. Given a graph G and an integer K, is there a vertex cover of size at most K in G?

17. Need to show that HC can be reduced to HP. Given a procedure for HP, test HC as follows. For each edge in the input graph, remove it from the graph and call the procedure for HP. HC is true if and only if at least one of these answers from HP is true.

20. *Hint:* Use an $n \times K$ array where position (i, j) means that there is a solution involving objects up to i and weight exactly j.

Chapter 3

1. Problem (1) : *KBand*
Problem (2) : semiglobal comparison
Problem (3) : filter and then semiglobal
Problem (4) : local comparison
Problem (5) : database search

2. Score is 8.

3.
```
AAAG    AAAG    AAAG
AC-G    A-CG    -ACG
```

4. Try sequences CGCG and GCGC.

7.
```
ACGGAGG
ACGTAGG
```

8. We have always *global* $\leq$ *semiglobal* $\leq$ *local*, because every global alignment is a semiglobal alignment, and every semiglobal alignment is a local alignment.

11. The alignment

```
GAGTTATCCGCCATC
AAGAGTTATCCGCCA
```

has score -7. No alignment with $k \leq 1$ — at most one pair of gaps — has a better score. But the following alignment, with two pairs of gaps, has a score of $+9$:

```
--GAGTTATCCGCCATC
AAGAGTTATCCGCCA--
```

14. It needs $O(\sum_{i<j} n_i n_j)$ space to hold the total score arrays c_{xy}.

15. Let D be a diagonal matrix with $D_{aa} = 1/p_a$. Scores are symmetric if $M^k D$ is symmetric, that is, if $(M^k D)^t = M^k D$, where the superscript t indicates transpose. The text shows the result for $k = 1$, that is,

$$(MD)^t = DM^t = MD.$$

Assuming the result for $k-1$ is true, we have

$$(M^k D)^t = (MM^{k-1}D)^t = (M^{k-1}D)^t M^t = M^{k-1}DM^t = M^{k-1}MD = M^k D.$$

16. No. The actual figure is lower than 2% because there is a small chance of an amino acid changing back to its initial kind after two mutations.

17. *Coix lacrima-jobi* α-coixin genes.

18. Use double induction. First prove that $w(k+1) \leq w(k)+w(1)$ by induction on k, and then $w(k+m) \leq w(k)+w(m)$ by induction on m.

20. The gap penalty function must be additive, that is, $w(x+y) = w(x)+w(y)$.

22. One possible such system would be $p(a,b) = 2$ if $a = b$, $p(a,b) = 1$ if $a \neq b$, and $g = 0$.

Chapter 4

1. One possible assembly follows.

```
CCTCGAGTTAA-----GCCCGCGGCTTCAACGGAT---------------
-------TTAAGTACTGCCCG------------ATCTGTGTCGGG-----
---------AAGTACTGCCCGCG-------------TGTGTCGGGAGTCG
-CTCGAGTTAAGTA---CCCGCGGCTTCAACGGATCTGTG----------
--------------------------------------------------
CCTCGAGTTAAGTACTGCCCGCGGCTTCAACGGATCTGTGTCGGGAGTCG
```

2. Minimum is 1, maximum is 3, average is 2.22.

3. The smallest possible is $\epsilon = 1/6$.

4. The values are $\epsilon = 1/6$ for Figure 4.3 and $\epsilon = 1/5$ for Figure 4.4.

5. It would resolve the ambiguity because the only layout compatible with this fragment is the bottom one in Figure 4.8.

6. The SCS is TTAAATA.

7. Sequence at source vertex: AC. Sequence at sink vertex: GT. Sequences in the middle: CG, CAG, and CTG.

8. *Hint:* Use suffix trees.

10. AAAAAAAAAA.

12. For $\epsilon = 0.1$ we have

```
--CGTT
ATC---
-TCG--
------
ATCGTT,
```

and, for $\epsilon = 0.25$,

```
AACG
ATC-
-TCG
----
ATCG.
```

13. The minimum number of contigs is one, as follows.

```
TCCCTACTT------
------AATCCGGTT
----GACATC-GGT-.
```

14. *Hint:* Do not take a substring-free collection.

15. Here is one collection with three strings: $\mathtt{AC}^k\mathtt{G}^k$, $\mathtt{C}^k\mathtt{T}$, $\mathtt{G}^k\mathtt{AC}^k$. Call them f, g, h, respectively. Then $w(h, f) = k+1$, $w(f, h) = k$, and $w(h, g) = k$ are the only nonzero edges. The heaviest one invalidates the other two, which together produce a path of weight $2k$, against $k + 1$ of the single maximum weight edge.

16. Fill in the blanks with Y and Y, or with Z and X, in this order.

17. Use the vector sequence as a query sequence and the collection of fragments as your database.

18. Given an instance $\mathcal{F}$ for SCS, replace each character x in every fragment $f \in \mathcal{F}$ by $\mathtt{AA}x\mathtt{CC}$. Give the resulting collection to a RECONSTRUCTION solver, together with $\epsilon = 0$. The resulting reconstruction will have to use all the fragments in the same orientation, as $\overline{\mathtt{AA}x\mathtt{CC}} = \mathtt{GG}\overline{x}\mathtt{TT}$ will not match anything in direct orientation.

Chapter 5

1. Enzyme A: 3, 10, 8, 6. Enzyme B: 5, 7, 4, 11.

2. Enzyme A: 4, 7, 5, 12, 8. Enzyme B: 6, 4, 9, 3, 10, 4.

3. 3, 7, 2, 9, 8.

9. A permutation for the columns: 3, 5, 1, 8, 10, 4, 7, 6, 9, 2.

11. See [67] or [74].

21. Denote end probes by the number 2. The problem now is to obtain a column permutation such that all 1s in each row are consecutive and possibly flanked by one or two 2s. It makes the problem a little easier but not much. See [9].

20. This is the cycle basis algorithm from [79].

23. Choose k to have a small A_k and a large B_k, and choose another splitter i so that $|A_i| \approx |B_i|$.

Chapter 6

3. Use parenthesis notation. For example, $(a((b)(c(d)(e))))$ could be a representation for a tree with a as root, b and c as children, and d and e as children of c.

4. Sort columns interpreting each column as a binary number, and remove columns that are equal. The new matrix admits a perfect phylogeny if and only if the old does too.

8. The input has a perfect phylogeny. Characters c_1 and c_2 are compatible if we take 1 as the ancestral state.

6. For a simple algorithm see [82].

10. See [107].

11. See [106].

13. Character pairwise compatibility (to define the edges of the graph) should be determined using the algorithm from Section 6.3.

15. Run the algorithm and then compare pairwise distances on the tree with distances on the matrix. If they are equal, the matrix is additive; otherwise it is not.

Chapter 7

1. $\overrightarrow{d}\ \overleftarrow{c}\ \overrightarrow{b}\ \overleftarrow{a}$.

2. $L\overleftarrow{a}$, $\overleftarrow{a}\,\overrightarrow{b}$, $\overrightarrow{d}\,\overleftarrow{e}$, $\overleftarrow{e}R$.

4. $\alpha = (\overrightarrow{3}, \overleftarrow{2}, \overrightarrow{1}), \beta = (\overrightarrow{1}, \overrightarrow{2}, \overrightarrow{3})$.

5. $\overrightarrow{4}, \overleftarrow{1}, \overleftarrow{3}, \overleftarrow{5}, \overrightarrow{2}$.

6. $\overrightarrow{4}, \overleftarrow{1}, \overleftarrow{3}, \overleftarrow{5}, \overrightarrow{2}$.

10. ABC, ABD, AEF, BEF, CBE, CBF, CEF, DBE, DBF, DEF, and their reverses.

11. Walk around the reality and desire diagram and write down a circular list recording the components that each reality edge belongs to. Record bad components only; if a reality edge belongs to a good component, just ignore it. Also, avoid recording the same component name twice in a row. For instance, for the diagram of Figure 7.18 we would get the circular list $EFEBDCBA$ and then back to E. Given this list, it is easy to compute the information sought: Hurdles correspond to elements that appear just once; a hurdle X protects nonhurdle Y when the list contains YXY and these are the only occurrences of Y in the list.

Chapter 8

5. See Section 3.3.3.

6. Reduce from the SAT problem (defined in Section 9.2) restricted as follows: Every clause has exactly three literals, and we look for a satisfying assignment where in each clause exactly one literal makes the clause true. Details can be found in [122].

9. Use techniques for *range-searching*. For one dimension a simple binary search tree suffices. For two dimensions, use a 2D tree. See [170].

Chapter 9

1. Such a path will be encoded by a sequence having 20,000 base pairs. To determine the path, we would need to apply the techniques of Chapter 4.

4. See [128].

REFERENCES

1. M. A. Adams, C. Fields, and J. C. Venter, editors. *Automated DNA Sequencing and Analysis.* New York: Academic Press, 1994.

2. L. M. Adleman. Molecular computation of solutions to combinatorial problems. *Science*, 266:1021–1024, 1994.

3. R. Agarwala, V. Bafna, M. Farach, B. Narayanan, M. Paterson, and M. Thorup. On the approximability of numerical taxonomy. In *Proceedings of the Seventh Annual ACM-SIAM Symposium on Discrete Algorithms*, pages 365–372, 1996.

4. R. Agarwala and D. Fernández-Baca. A polynomial-time algorithm for the phylogeny problem when the number of character states is fixed. *SIAM Journal on Computing*, 23(6):1216–1224, 1994.

5. R. Agarwala, D. Fernandez-Baca, and G. Slutzki. Fast algorithms for inferring evolutionary trees. In *Proceedings of the 30th Allerton Conference on Communication, Control, and Computation*, pages 594–603, 1992.

6. A. V. Aho. Algorithms for finding patterns in strings. In J. van Leeuwen, editor, *Handbook of Theoretical Computer Science*, volume A, pages 255–300. Amsterdam/Cambridge, MA: Elsevier/MIT Press, 1990.

7. B. Alberts, D. Bray, J. Lewis, M. Raff, K. Roberts, and J. D. Watson. *Molecular Biology of the Cell.* New York & London: Garland Publishing, 1994.

8. F. Alizadeh, R. M. Karp, L. A. Newberg, and D. K. Weisser. Physical mapping of chromosomes: A combinatorial problem in molecular biology. *Algorithmica*, 13(1/2):52–76, 1995.

9. F. Alizadeh, R. M. Karp, D. K. Weisser, and G. Zweig. Physical mapping of chromossomes using unique probes. *Journal of Computational Biology*, 2(2):159–184, 1995.

10. S. F. Altschul. Amino acid substitution matrices from an information theoretical perspective. *Journal of Molecular Biology*, 219:555–565, 1991.

11. S. F. Altschul, M. S. Boguski, W. Gish, and J. C. Wootton. Issues in searching molecular sequence databases. *Nature Genetics*, 6:119–129, Feb. 1994.

12. S. F. Altschul, W. Gish, W. Miller, E. W. Myers, and D. J. Lipman. A basic local alignment search tool. *Journal of Molecular Biology*, 215:403–410, 1990.

13. S. F. Altschul and D. J. Lipman. Trees, stars, and multiple biological sequence alignment. *SIAM Journal on Applied Mathematics*, 49(1):197–209, 1989.

14. A. Amir and D. Keselman. Maximum agreement subtrees in multiple evolutionary trees. In

Proceedings of the IEEE Thirty-Fifth Annual Symposium on Foundations of Computer Science, pages 758–769, 1994.

15. C. Armen and C. Stein. Short superstrings and the structure of overlapping strings. *Journal of Computational Biology*, 2(2):307–332, 1995.

16. C. Armen and C. Stein. A 2 2/3-approximation algorithm for the shortest superstring problem. In *Proceedings of the Seventh Symposium on Combinatorial Pattern Matching*, volume 1075 of *Lecture Notes in Computer Science*, pages 87–103. Berlin: Springer-Verlag, 1996.

17. H. Atlan and M. Koppel. The cellular computer DNA: program or data? *Bulletin of Mathematical Biology*, 52(3):335–348, 1990.

18. V. Bafna and P. A. Pevzner. Genome rearrangements and sorting by reversals. *SIAM Journal on Computing*, 25(2):272–289, 1996.

19. G. I. Bell. The human genome: an introduction. In Bell and Marr [20], pages 3–12.

20. G. I. Bell and T. M. Marr, editors. *Computers and DNA*. Reading, MA: Addison-Wesley, 1990.

21. B. Berger. Algorithms for protein structural motif recognition. *Journal of Computational Biology*, 2(1):125–138, 1995.

22. B. Berger and D. B. Wilson. Improved algorithms for protein motif recognition. In *Proceedings of the Sixth Annual ACM-SIAM Symposium on Discrete Algorithms*, pages 58–67, 1995.

23. P. Berman and S. Hannenhalli. Fast sorting by reversal. In *Proceedings of the Seventh Symposium on Combinatorial Pattern Matching*, volume 1075 of *Lecture Notes in Computer Science*, pages 168–185. Berlin: Springer-Verlag, 1996.

24. A. Blum, T. Jiang, M. Li, J. Tromp, and M. Yannakakis. Linear approximation of shortest superstrings. In *Proceedings of the Twenty-Third Annual ACM Symposium on Theory of Computing*, pages 328–336, 1991.

25. H. L. Bodlaender, M. R. Fellows, and T. J. Warnow. Two strikes against perfect phylogeny. In *Proceedings of the International Colloquium on Automata, Languages and Programming*, volume 623 of *Lecture Notes in Computer Science*, pages 273–283. Berlin: Springer-Verlag, 1993.

26. H. L. Bodlaender and T. Kloks. A simple linear time algorithm for triangulating three-colored graphs. *Journal of Algorithms*, 15:160–172, 1993.

27. K. S. Booth and G. S. Lueker. Testing of the consecutive ones property, interval graphs, and graph planarity using PQ-tree algorithms. *Journal of Computer and System Sciences*, 13(3):335–379, 1976.

28. C. Branden and J. Tooze. *Introduction to protein structure*. New York & London: Garland Publishing, 1991.

29. P. Buneman. A characterisation of rigid circuit graphs. *Discrete Mathematics*, 9:205–212, 1974.

30. A. Caprara. Sorting by reversals is difficult. In *Proceedings of the First Annual International Conference on Computational Molecular Biology*, 1997.

31. A. Caprara, G. Lancia, and S.-K. Ng. A column-generation-based branch-and-bound algorithm for sorting by reversals. Presented at the 4th DIMACS Implementation Challenge Workshop, 1995.

32. H. Carrillo and D. Lipman. The multiple sequence alignment problem in biology. *SIAM Journal on Applied Mathematics*, 48(5):1073–1082, Oct. 1988.

33. S. C. Chan, A. K. C. Wong, and D. K. Y. Chiu. A survey of multiple sequence comparison methods. *Bulletin of Mathematical Biology*, 54(4):563–598, 1992.

34. W. I. Chang and E. L. Lawler. Approximate string matching in sublinear expected time. In *Proceedings of the IEEE Thirty-First Annual Symposium on Foundations of Computer Science*, pages 116–124, 1990.

35. K.-M. Chao and W. Miller. Linear-space algorithms that build local alignments from fragments. *Algorithmica*, 13:106–134, 1995.

36. K.-M. Chao, J. Zhang, J. Ostell, and W. Miller. A local alignment tool for very long DNA sequences. *Computer Applications in the Biosciences*, 11(2):147–153, 1995.

37. T. H. Cormen, C. E. Leiserson, and R. L. Rivest. *Introduction to Algorithms*. Cambridge, MA/New York: MIT Press/McGraw-Hill, 1990.

38. T. E. Creighton. Protein folding. *Biochemistry Journal*, 270:1–16, 1990.

39. M. Crochemore and W. Rytter. *Text Algorithms*. Oxford: Oxford University Press, 1994.

40. J. C. Culberson and P. Rudnicki. A fast algorithm for constructing trees from distance matrices. *Information Processing Letters*, 30:215–220, 1989.

41. A. Czumaj, L. Gasienec, M. Piotrow, and W. Rytter. Parallel and sequential approximations of shortest superstrings. In *Proceedings of the Fourth Scandinavian Workshop on Algorithm Theory*, pages 95–106, 1994.

42. W. E. Day. Computational complexity of inferring phylogenies from dissimilarity matrices. *Bulletin of Mathematical Biology*, 49(4):461–467, 1987.

43. W. E. Day, D. S. Johnson, and D. Sankoff. The computational complexity of inferring rooted phylogenies by parsimony. *Mathematical Biosciences*, 81:33–42, 1986.

44. W. E. Day and D. Sankoff. Computational complexity of inferring phylogenies by compatibility. *Systematic Zoology*, 35(2):224–229, 1986.

45. M. Dayhoff, R. M. Schwartz, and B. C. Orcutt. A model of evolutionary change in proteins. In M. Dayhoff, editor, *Atlas of Protein Sequence and Structure*, volume 5, pages 345–352. National Biomedical Research Foundation, Silver Spring, MD, 1978. Supplement 3.

46. S. Dean and R. Staden. A sequence assembly and editing program for efficient management of large projects. *Nucleic Acids Research*, 19(14):3907–3911, 1991.

47. J. Devereux, P. Haeberli, and D. Smithies. A comprehensive set of sequence analysis programs for the VAX. *Nucleic Acids Research*, 12:387–395, 1984.

48. K. A. Dill. Dominant forces in protein folding. *Biochemistry*, 29(31):7133–7155, 1990.

49. A. J. Dobson. Unrooted trees for numerical taxonomy. *Journal of Applied Probability*, 11:32–42, 1974.

50. R. F. Doolittle. Proteins. *Scientific American*, 253(4):74–83, Oct. 1985.

51. R. F. Doolittle, editor. *Molecular Evolution: Computer Analysis of Protein and Nucleic Acid Sequences*, volume 183 of *Methods in Enzymology*. New York: Academic Press, 1990.

52. A. Dress and M. Steel. Convex tree realizations of partitions. *Applied Mathematics Letters*, 5(3):3–6, 1992.

53. M. L. Engle and C. Burks. Artificially generated data sets for testing DNA fragment assembly algorithms. *Genomics*, 16:286–288, 1993.

54. M. L. Engle and C. Burks. Genfrag 2.1: New features for more robust fragment assembly benchmarks. *Computer Applications in the Biosciences*, 10:567–568, 1994.

55. D. Eppstein, Z. Galil, and R. Giancarlo. Speeding up dynamic programming. In *Proceedings of the IEEE Twenty-Ninth Annual Symposium on Foundations of Computer Science*, pages 488–495, 1988.

56. M. Farach, S. Kannan, and T. Warnow. A robust model for finding optimal evolutionary trees. *Algorithmica*, 13:155–179, 1995.

57. M. Farach, T. M. Przytycka, and M. Thorup. On the agreement of many trees. Unpublished manuscript, 1995.

58. C. Fauron and M. Havlik. The maize mitochondrial genome of the normal type and the cytoplasmic male sterile type have very different organization. *Current Genetics*, 15:149–154, 1989.

59. M. R. Fellows, M. T. Hallett, and H. T. Wareham. DNA physical mapping: Three ways difficult. In *Proceedings of the First Annual European Symposium on Algorithms*, volume 726 of *Lecture Notes in Computer Science*, pages 157–168. Berlin: Springer-Verlag, 1993.

60. J. Felsenstein. Numerical methods for inferring evolutionary trees. *The Quarterly Review of Biology*, 57(4):379–404, 1982.

61. J. Felsenstein. PHYLIP — Phylogeny Inference Package (Version 3.2). *Cladistics*, 5:164–166, 1989.

62. V. Ferreti, J. H. Nadeau, and D. Sankoff. Original synteny. In *Proceedings of the Seventh Symposium on Combinatorial Pattern Matching*, number 1075 in Lecture Notes on Computer Science, pages 159–167. Berlin: Springer-Verlag, 1996.

63. J. Fickett. Fast optimal alignment. *Nucleic Acids Research*, 12(1):175–179, 1984.

64. L. R. Foulds and R. L. Graham. The Steiner problem in phylogeny is NP-complete. *Advances in Applied Mathematics*, 3:43–49, 1982.

65. A. S. Fraenkel. Complexity of protein folding. *Bulletin of Mathematical Biology*, 55(6):1199–1210, 1993.

66. K. A. Frenkel. The human genome project and informatics. *Communications of the ACM*, 34(11), Nov. 1991.

67. D. R. Fulkerson and O. A. Gross. Incidence matrices and interval graphs. *Pacific Journal of Mathematics*, 15(3):835–855, 1965.

68. M. R. Garey and D. S. Johnson. *Computers and Intractability: A Guide to the Theory of NP-Completeness*. New York: Freeman, 1979.

69. W. H. Gates and C. H. Papadimitriou. Bounds for sorting by prefix reversal. *Discrete Mathematics*, 27:47–57, 1979.

70. D. G. George, W. C. Barker, and L. T. Hunt. Mutation data matrix and its uses. In Doolittle [51], pages 333–351.

71. T. Gingerias, J. Milazzo, D. Sciaky, and R. Roberts. Computer programs for assembly of DNA sequences. *Nucleic Acids Research*, 7:529–545, 1979.

72. P. W. Goldberg, M. C. Golumbic, H. Kaplan, and R. Shamir. Three strikes against physical mapping of DNA. Unpublished manuscript, 1993.

73. L. Goldstein and M. S. Waterman. Mapping DNA by stochastic relaxation. *Advances in Applied Mathematics*, 8:194–207, 1987.

74. M. C. Golumbic. *Algorithmic Graph Theory and Perfect Graphs*. New York: Academic Press, 1980.

75. M. C. Golumbic, H. Kaplan, and R. Shamir. On the complexity of physical mapping. *Advances in Applied Mathematics*, 15:251–261, 1994.

76. G. H. Gonnet and R. Baeza-Yates. *Handbook of Algorithms and Data Structures,* 2nd ed. Reading, MA: Addison-Wesley, 1991.

77. O. Gotoh. Optimal alignments between groups of sequences and its application to multiple sequence alignment. *Computer Applications in the Biosciences*, 9(3):361–370, 1993.

78. D. Greenberg and S. Istrail. The chimeric mapping problem: Algorithmic strategies and performance evaluation on synthetic genomic data. *Computers and Chemistry*, 18(3):207–220, 1994.

79. D. Greenberg and S. Istrail. Physical mapping by STS hybrdization: Algorithmic strategies and the challenge of software evaluation. *Journal of Computational Biology*, 2(2):219–274, 1995.

80. A. Grigoriev, R. Mott, and H. Lehrach. An algorithm to detect chimeric clones and random noise in genomic mapping. *Genomics*, 22:482–486, 1994.

81. S. K. Gupta, J. Kececioglu, and A. A. Schäffer. Improving the practical space and time efficiency of the shortest-paths approach to sum-of-pairs multiple sequence alignment. *Journal of Computational Biology*, 2(3):459–472, 1995.

82. D. Gusfield. Efficient algorithms for inferring evolutionary trees. *Networks*, 21:19–28, 1991.

83. D. Gusfield. Efficient methods for multiple sequence alignment with guaranteed error bounds. *Bulletin of Mathematical Biology*, 55(1):141–154, 1993.

84. D. Gusfield. Faster implementation of a shortest superstring approximation. *Information Processing Letters*, 51:271–274, 1994.

85. D. Gusfield. *Algorithms on Strings, Trees, and Sequences: Computer Science and Computational Biology*. Cambridge, UK: Cambridge University Press, 1997. Forthcoming.

86. D. Gusfield, G. M. Landau, and B. Schieber. An efficient algorithm for the all pairs suffix–prefix problem. *Information Processing Letters*, 41:181–185, 1992.

87. S. Hannenhalli and P. A. Pevzner. Transforming cabbage into turnip (polynomial algorithm for sorting signed permutations by reversals). In *Proceedings of the Twenty-Seventh Annual ACM Symposium on Theory of Computing*, pages 178–189, 1995.

88. S. Hannenhalli and P. A. Pevzner. Transforming men into mice (polynomial algorithm for genomic distance problem). In *Proceedings of the IEEE Thirty-Sixth Annual Symposium on Foundations of Computer Science*, pages 581–592, 1995.

89. J. Hein. A new method that simultaneously aligns and reconstructs ancestral sequences for any number of homologous sequences, when the phylogeny is given. *Molecular Biology and Evolution*, 6(6):649–668, 1989.

90. J. Hein. An optimal algorithm to reconstruct trees from additive distance data. *Bulletin of Mathematical Biology*, 51(5):597–603, 1989.

91. J. Hein. A tree reconstruction method that is economical in the number of pairwise comparisons used. *Molecular Biology and Evolution*, 6(6):669–684, 1989.

92. J. Hein. Unified approach to alignment and phylogenies. In Doolittle [51], pages 626–645.

93. S. Henikoff and J. G. Henikoff. Amino acid substitution matrices from protein blocks. *Proceedings of the National Academy of Sciences of the U.S.A.*, 89:10915–10919, 1992.

94. D. Hirschberg. A linear space algorithm for computing maximal common subsequences. *Communications of the ACM*, 18:341–343, 1975.

95. R. J. Hoffmann, J. L. Boore, and W. M. Brown. A novel mitochondrial genome organization for the blue mussel, *mytilus edulis*. *Genetics*, 131:397–412, 1992.

96. D. Hofstadter. *Gödel, Escher, Bach*. New York: Basic Books, 1979.

97. W.-L. Hsu. A simple test for the consecutive ones property. In *Proceedings of the International Symposium on Algorithms & Computation (*ISAAC*)*, 1992.

98. X. Huang. A contig assembly program based on sensitive detection of fragment overlaps. *Genomics*, 14:18–25, 1992.

99. X. Huang. An improved sequence assembly program. *Genomics*, 33:21–31, 1996.

100. X. Huang, R. C. Hardison, and W. Miller. A space-efficient algorithm for local similarities. *Computer Applications in the Biosciences*, 6(4):373–381, 1990.

101. R. Idury and A. Schäffer. Triangulating three-colored graphs in linear time and linear space. *SIAM Journal on Discrete Mathematics*, 6(2), 1993.

102. T. Jiang, E. Lawler, and L. Wang. Aligning sequences via an evolutionary tree: complexity and approximation. In *Proceedings of the Twenty-Sixth Annual ACM Symposium on Theory of Computing*, pages 760–769, 1994.

103. R. Jones, W. Taylor, IV, X. Zhang, J. P. Mesirov, and E. Lander. Protein sequence comparison on the connection machine CM-2. In Bell and Marr [20], pages 99–108.

104. D. Joseph, J. Meidanis, and P. Tiwari. Determining DNA sequence similarity using maximum independent set algorithms for interval graphs. In *Proceedings of the Third Scandinavian Workshop on Algorithm Theory*, volume 621 of *Lecture Notes in Computer Science*, pages 326–337. Berlin: Springer-Verlag, 1992.

105. M. Kanehisa and C. DeLisi. The prediction of a protein and nucleic acid structure: problems and prospects. In G. Koch and M. Hazewinkel, editors, *Mathematics of Biology*, pages 115–137. Dordrecht: D. Reidel, 1985.

106. S. K. Kannan and T. J. Warnow. Triangulating 3-colored graphs. *SIAM Journal on Discrete Mathematics*, 5(2):249–258, 1992.

107. S. K. Kannan and T. J. Warnow. Inferring evolutionary history from DNA sequences. *SIAM Journal on Computing*, 23(4):713–737, 1994.

108. H. Kaplan, R. Shamir, and R. E. Tarjan. Tractability of parameterized completion problems on chordal and interval graphs: Minimum fill-in and physical mapping. In *Proceedings of the IEEE Thirty-Fifth Annual Symposium on Foundations of Computer Science*, pages 780–791, 1994.

109. S. Karlin and S. F. Altschul. Methods for assessing the statistical significance of molecular sequence features by using general scoring schemes. *Proceedings of the National Academy of Sciences of the U.S.A.*, 87:2264–2268, 1990.

110. S. Karlin, A. Dembo, and T. Kawabata. Statistical composition of high-scoring segments from molecular sequences. *Annals of Statistics*, 18(2):571–581, 1990.

111. R. M. Karp. Mapping the genome: some combinatorial problems arising in molecular biology. In *Proceedings of the Twenty-Fifth Annual ACM Symposium on Theory of Computing*, pages 278–285, 1993.

112. R. M. Karp, C. Kenyon, and O. Waarts. Error-resilient DNA computation. In *Proceedings of the Seventh Annual ACM-SIAM Symposium on Discrete Algorithms*, pages 458–467, 1996.

113. J. Kececioglu. The maximum weight trace problem in multiple sequence alignment. In *Proceedings of the Fourth Symposium on Combinatorial Pattern Matching*, volume 684 of *Lecture Notes in Computer Science*, pages 106–119. Berlin: Springer-Verlag, 1993.

114. J. D. Kececioglu. *Exact and approximation algorithms for DNA sequence reconstruction*. Ph.D. thesis, University of Arizona, 1991.

115. J. D. Kececioglu and E. W. Myers. Combinatorial algorithms for DNA sequence assembly. *Algorithmica*, 13:7–51, 1995.

116. J. D. Kececioglu and D. Sankoff. Exact and approximate algorithms for sorting by reversals, with application to genome rearrangement. *Algorithmica*, 13:180–210, 1995.

117. E. V. Koonin and V. V. Dolja. Evolution and taxonomy of positive-strand RNA viruses: implications of comparative analysis of amino acid sequences. *Critical Reviews in Biochemistry and Molecular Biology*, 28(5):375–430, 1993.

118. R. Kosaraju, J. Park, and C. Stein. Long tours and short superstrings. In *Proceedings of the IEEE Thirty-Fifth Annual Symposium on Foundations of Computer Science*, pages 166–177, 1994.

119. E. S. Lander. Analysis with restriction enzymes. In Waterman [196], pages 35–51.

120. E. S. Lander and M. S. Waterman. Genomic mapping by fingerprinting random clones: a mathematical analysis. *Genomics*, 2:231–239, 1988.

121. L. Larmore and B. Schieber. On-line dynamic programming with applications to the prediction of RNA secondary structure. In *Proceedings of the First Annual ACM-SIAM Symposium on Discrete Algorithms*, pages 503–512, 1990.

122. R. H. Lathrop. The protein threading problem with sequence amino acid interaction preferences is NP-complete. *Protein Engineering*, 7(9):1059–1068, 1994.

123. R. H. Lathrop and T. F. Smith. Global optimum protein threading with gapped alignment and empirical pair score functions. *Journal of Molecular Biology*, 255(4):641–665, 1996.

124. B. Lewin. *Genes V*. Oxford: Oxford University Press, 1994.

125. R. Lewontin. *Biology as Ideology*. New York: HarperPerennial, 1993.

126. M. Li. Towards a DNA sequencing theory (learning a string). In *Proceedings of the IEEE Thirty-First Annual Symposium on Foundations of Computer Science*, pages 125–134, 1990.

127. D. J. Lipman and W. R. Pearson. Rapid and sensitive protein similarity search. *Science*, 227:1435–1441, 1985.

128. R. J. Lipton. Using DNA to solve NP-complete problems. *Science*, 268:542–545, 1995.

129. U. Manber. *Introduction to Algorithms*. Reading, MA: Addison-Wesley, 1989.

130. U. Manber and E. W. Myers. Suffix arrays: A new method for on-line string searches. In *Proceedings of the First Annual ACM-SIAM Symposium on Discrete Algorithms*, pages 319–327, 1990.

131. C. K. Mathews and K. E. van Holde. *Biochemistry*. Redwood City, CA: Benjamin/Cummings, 1990.

132. F. R. McMorris. On the compatibility of binary qualitative taxonomic characters. *Bulletin of Mathematical Biology*, 39:133–138, 1977.

133. F. R. McMorris, T. Warnow, and T. Wimer. Triangulating vertex-colored graphs. *SIAM Journal on Discrete Mathematics*, 7(2), May 1994.

134. J. Meidanis. Distance and similarity in the presence of nonincreasing gap-weighting functions. In *Proceedings of the Second South American Workshop on String Processing*, pages 27–37, Valparaíso, Chile, Apr. 1995.

135. J. Meidanis and E. G. Munuera. A simple linear time algorithm for binary phylogeny. In *Proceedings of the Fifteenth International Conference of the Chilean Computing Society*, pages 275–283, 1995.

136. J. Meidanis and E. G. Munuera. A theory for the consecutive ones property. In *Proceedings of the Third South American Workshop on String Processing*, volume 4 of *International Informatics Series*, pages 194–202. Carleton University Press, 1996.

137. J. Meidanis and J. C. Setubal. Multiple alignment of biological sequences with gap flexibility. In *Proceedings of Latin American Theoretical Informatics*, volume 911 of *Lecture Notes in Computer Science*, pages 411–426. Berlin: Springer-Verlag, 1995.

138. J. Messing, R. Crea, and P. H. Seeburg. A system for shotgun DNA sequencing. *Nucleic Acids Research*, 9:309–321, 1981.

139. W. Miller. Building multiple alignments from pairwise alignments. *Computer Applications in the Biosciences*, 9(2):169–176, 1993.

140. W. Miller and E. W. Myers. Sequence comparison with concave weighting functions. *Bulletin of Mathematical Biology*, 50(2):97–120, 1988.

141. K. B. Mullis. The unusual origin of the polymerase chain reaction. *Scientific American*, 262(4):56–65, Apr. 1990.

142. E. W. Myers. An $O(ND)$ difference algorithm and its variations. *Algorithmica*, 1:251–266, 1986.

143. E. W. Myers. Advances in sequence assembly. In Adams et al. [1], pages 231–238.

144. E. W. Myers. Toward simplifying and accurately formulating fragment assembly. *Journal of Computational Biology*, 2(2):275–290, 1995.

145. E. W. Myers and W. Miller. Optimal alignments in linear space. *Computer Applications in the Biosciences*, 4(1):11–17, 1988.

146. S. B. Needleman and C. D. Wunsch. A general method applicable to the search for similarities in the amino acid sequence of two proteins. *Journal of Molecular Biology*, 48:443–453, 1970.

147. M. Nei. *Molecular Evolutionary Genetics*. New York: Columbia University Press, 1987.

148. J. D. Palmer. Chloroplast DNA evolution and biosystematic uses of chloroplast DNA variation. *The American Naturalist*, 130:S6–S29, 1987. Supplement.

149. J. D. Palmer and L. A. Herbon. Unicircular structure of the *brassica hirta* mitochondrial genome. *Current Genetics*, 11:565–570, 1987.

150. J. D. Palmer, B. Osorio, and W. F. Thompson. Evolutionary significance of inversions in legume chloroplast DNAs. *Current Genetics*, 14:65–74, 1988.

151. C. H. Papadimitriou. *Computational Complexity*. Reading, MA: Addison-Wesley, 1994.

152. C. H. Papadimitriou and K. Steiglitz. *Combinatorial Optimization: Algorithms and Complexity*. Englewood Cliffs, NJ: Prentice-Hall, 1982.

153. W. R. Pearson. Rapid and sensitive sequence comparison with FASTP and FASTA. In Doolittle [51], pages 63–98.

154. W. R. Pearson. Searching protein sequence libraries: Comparison of the sensitivity and selectivity of the Smith-Waterman and FASTA algorithms. *Genomics*, 11:635–650, 1991.

155. W. R. Pearson and D. J. Lipman. Improved tools for biological sequence comparison. *Proceedings of the National Academy of Sciences of the U.S.A.*, 85:2444–2448, 1988.

156. W. R. Pearson and W. Miller. Dynamic programming algorithms for biological sequence comparison. In L. Brand and M. L. Johnson, editors, *Numerical Computer Methods*, volume 210 of *Methods in Enzymology*, pages 575–601. New York: Academic Press, 1992.

157. H. Peltola, H. Söderlund, J. Tarhio, and E. Ukkonen. Algorithms for some string matching problems arising in molecular genetics. In *Information Processing 83:Proceedings of the International Federation for Information Processing (IFIP) Ninth World Computer Congress*, pages 53–64. Amsterdam: North Holland, 1983.

158. H. Peltola, H. Söderlund, and E. Ukkonen. SEQAIDS: A DNA sequence assembling program based on a mathematical model. *Nucleic Acids Research*, 12:307–321, 1984.

159. D. Penny, M. D. Hendy, and M. A. Steel. Progress with methods for constructing evolutionary trees. *Trends in Ecology and Evolution*, 7(3):73–79, 1992.

160. P. A. Pevzner. DNA physical mapping and alternating Eulerian cycles in colored graphs. *Algorithmica*, 13(1/2):77–105, 1995.

161. F. M. Richards. The protein folding problem. *Scientific American*, 264(1):54–63, Jan. 1991.

162. R. J. Robbins. Challenges in the human genome project. *IEEE Engineering in Medicine and Biology*, 11(1):25–34, Mar. 1992.

163. K. H. Rosen. *Discrete Mathematics and Its Applications*, 2nd ed. New York: McGraw-Hill, 1991.

164. M. Rosenberg and D. Court. Regulatory sequences involved in the promotion and termination of RNA transcription. *Annual Review of Genetics*, 13:319–353, 1979.

165. I. Rosenfeld, E. Ziff, and V. van Loon. *DNA for beginners*. Writers and Readers, 1984.

166. D. Sankoff. Minimal mutation trees of sequences. *SIAM Journal on Applied Mathematics*, 28:35–42, 1975.

167. D. Sankoff. Analytical approaches to genomic evolution. *Biochimie*, 75(409–413), 1993.

168. D. Sankoff and J. B. Kruskal. *Time Warps, String Edits, and Macromolecules: the Theory and Practice of Sequence Comparison*. Reading, MA: Addison-Wesley, 1983.

169. W. Schmitt and M. S. Waterman. Multiple solutions of DNA restriction mapping problems. *Advances in Applied Mathematics*, 12:412–427, 1991.

170. R. Sedgewick. *Algorithms*, 2nd ed. Reading, MA: Addison-Wesley, 1988.

171. D. Seto, B. Koop, and L. Hood. An experimentally derived data set constructed for testing large-scale DNA sequence assembly algorithms. *Genomics*, 15:673–676, 1993.

172. I. Simon. Sequence comparison: some theory and some practice. In *Proceedings of the LITP Spring School on Theoretical Computer Science*, volume 377 of *Lecture Notes in Computer Science*, pages 79–92. Berlin: Springer-Verlag, 1987.

173. J. Sims, D. Capon, and D. Dressler. *dna*G (Primase)-dependent origins of DNA replication. *Journal of Biolical Chemistry*, 254:12615–12628, 1979.

174. S. S. Skiena and G. Sundaram. A partial digest approach to restriction site mapping. *Bulletin of Mathematical Biology*, 56(2):275–294, 1994.

175. T. F. Smith and M. S. Waterman. Identification of common molecular subsequences. *Journal of Molecular Biology*, 147:195–197, 1981.

176. T. F. Smith, M. S. Waterman, and W. M. Fitch. Comparative biosequence metrics. *Journal of Molecular Evolution*, 18:38–46, 1981.

177. C. Soderlund and C. Burks. GRAM and genfragII: solving and testing the single-digest, partially ordered restriction map problem. *Computer Applications in the Biosciences*, 10(3):349–358, 1994.

178. R. Staden. A strategy of DNA sequencing employing computer programs. *Nucleic Acids Research*, 6:2601–2610, 1979.

179. M. A. Steel. The complexity of reconstructing trees from qualitative characters and subtrees. *Journal of Classification*, 9:91–116, 1992.

180. G. A. Stephen. *String Searching Algorithms*. Singapore: World Scientific, 1994.

181. D. L. Swofford and W. P. Maddison. Reconstructing ancestral character states under Wagner parsimony. *Mathematical Biosciences*, 87:199–229, 1987.

182. D. L. Swofford and G. J. Olsen. Phylogeny reconstruction. In D. M. Hillis and C. Moritz, editors, *Molecular Systematics*, pages 411–501. Sunderland, MA: Sinauer Associates, 1990.

183. R. Tamarin. *Principles of Genetics*. Dubuque, IA: Wm. C. Brown, 1991.

184. R. E. Tarjan. *Data Structures and Network Algorithms*. CBMS-NSF Regional conference series in applied mathematics. Society for Industrial and Applied Mathematics, 1983.

185. S.-H. Teng and F. Yao. Approximating shortest superstrings. In *Proceedings of the IEEE Thirty-Fourth Annual Symposium on Foundations of Computer Science*, pages 158–165, 1993.

186. D. H. Turner and N. Sugimoto. RNA structure prediction. *Annual Review of Biophysics and Biophysical Chemistry*, 17:167–192, 1988.

187. J. S. Turner. Approximation algorithms for the shortest common superstring problem. *Information and Computation*, 83:1–20, 1989.

188. E. Ukkonen. Algorithms for approximate string matching. *Information and Control*, 64:100–118, 1985.

189. R. Unger and J. Moult. Finding the lowest free energy conformation of a protein is an NP-hard problem: proof and implications. *Bulletin of Mathematical Biology*, 55(6):1183–1198, 1993.

190. M. Vingron and A. von Haeseler. Towards integration of multiple alignment and phylogenetic tree construction. Unpublished manuscript, 1995.

191. G. von Heijne. *Sequence Analysis in Molecular Biology: Treasure Trove or Trivial Pursuit?* New York: Academic Press, 1987.

192. L. Wang and D. Gusfield. Improved approximation algorithms for tree alignment. In *Proceedings of the Seventh Symposium on Combinatorial Pattern Matching*, volume 1075 of *Lecture Notes in Computer Science*, pages 220–233. Berlin: Springer-Verlag, 1996.

193. L. Wang and T. Jiang. On the complexity of multiple sequence alignment. *Journal of Computational Biology*, 1(4):337–348, 1994.

194. T. J. Warnow. Constructing phylogenetic trees efficiently using compatibility criteria. Unpublished manuscript, 1993.

195. T. J. Warnow. Tree compatibility and inferring evolutionary history. *Journal of Algorithms*, 16:388–407, 1994.

196. M. S. Waterman, editor. *Mathematical Methods for DNA Sequences*. Boca Raton, FL: CRC Press, 1989.

197. M. S. Waterman. Sequence alignments. In Waterman [196], pages 53–92.

198. M. S. Waterman. Parametric and ensemble sequence alignment algorithms. *Bulletin of Mathematical Biology*, 56(4):743–767, 1994.

199. M. S. Waterman. *Introduction to Computational Biology*. London: Chapman & Hall, 1995.

200. M. S. Waterman and J. R. Griggs. Interval graphs and maps of DNA. *Bulletin of Mathematical Biology*, 48(2):189–195, 1986.

201. M. S. Waterman and T. F. Smith. Rapid dynamic programming algorithms for RNA secondary structure. *Advances in Applied Mathematics*, 7:455–464, 1986.

202. M. S. Waterman, T. F. Smith, M. Singh, and W. A. Beyer. Additive evolutionary trees. *Journal of Theoretical Biology*, 64:199–213, 1977.

203. J. D. Watson et al. *Molecular Biology of the Gene*, volume 1. Redwood City, CA: Benjamin/Cummings, 1987.

204. J. D. Watson et al. *Molecular Biology of the Gene*, volume 2. Redwood City, CA: Benjamin/Cummings, 1987.

205. G. A. Watterson, W. J. Ewens, and T. E. Hall. The chromosome inversion problem. *Journal of Theoretical Biology*, 99:1–7, 1982.

206. M. Zuker. On finding all suboptimal foldings of an RNA molecule. *Science*, 244:48–52, 1989.

207. M. Zuker. The use of dynamic programming algorithms in RNA secondary structure prediction. In Waterman [196], pages 159–185.

Index